Allelochemicals: Biological Control of Plant Pathogens and Diseases

Disease Management of Fruits and Vegetables

VOLUME 2

Series Editor:

K.G. Mukerji, *University of Delhi, Delhi, India*

Allelochemicals: Biological Control of Plant Pathogens and Diseases

Edited by

INDERJIT
University of Delhi, India

and

K.G. MUKERJI
University of Delhi, India

Springer

A C.I.P. Catalogue record for this book is available from the Library of Congress.

ISBN-10 1-4020-4445-3 (HB)
ISBN-13 978-1-4020-4445-8 (HB)
ISBN-10 1-4020-4447-X (e-book)
ISBN-13 978-1-4020-4447-2 (e-book)

Published by Springer,
P.O. Box 17, 3300 AA Dordrecht, The Netherlands.

www.springer.com

Printed on acid-free paper

Cover photo:
Bio Control of powdery mildew pathogen *Phyllactinia dalbergae* on *Dalbergia sisoo* by hyperparasite *Cladosporium spongiosum*.
(Microphotograph taken by Prof. K.G. Mukerji and Mr. S.K. Das)

Printed in the Netherlands.

CONTENT

Preface vii

1. **Discovery and Evaluation of Natural Product-based Fungicides for Disease Control of Small Fruits** 1
   David E. Wedge and Barbara J. Smith
2. **Allelochemicals as Biopesticides for Management of Plant-parasitic Nematodes** 15
   Nancy Kokalis-Burelle and Rodrigo Rodríguez-Kábana
3. **Allelopathic Organisms and Molecules: Promising Bioregulators for the Control of Plant Diseases, Weeds, and Other Pests** 31
   Ana Luisa Anaya
4. **The Impact of Pathogens on Plant Interference and Allelopathy** 79
   Scott W. Mattner
5. **Allelopathy for Weed Control in Aquatic and Wetland Systems** 103
   Ramanathan Kathiresan, Clifford H. Koger , Krishna N. Reddy
6. **Bacterial Root Zone Communities, Beneficial Allelopathies and Plant Disease Control** 123
   Antony V. Sturz
7. **The Role of Allelopathic Bacteria in Weed Management** 143
   Robert J. Kremer
8. **The Allelopathic Potential of Ginsenosides** 157
   Mark A. Bernards, Lina F. Yousef, Robert W. Nicol
9. **Antimicrobial and Nematicidal Substances from the Root of Chicory (*Cichorium intybus*)** 177
   Hiroyuki Nishimura and Atsushi Satoh
10. **Disease Resistance in Plants through Mycorrhizal Fungi Induced Allelochemicals** 181
    Ren-sen Zeng
11. **Allelochemicals from *Ageratum conyzoides* L. and *Oryza sativa* L. and their Effects on Related Pathogens** 193
    Chuihua Kong

Index 207

List of Contributors 213

Preface

Biological control of plant diseases and plant pathogens is of great significance in forestry and agriculture. There is great incentive to discover biologically active natural products from higher plants that are better than synthetic agrochemicals and are much safer, from health and environmental considerations. The development of natural products as herbicides, fungicides, and their role in biological control of plant disease promises to reduce environmental and health hazards. Allelopathic techniques offer real promise in solving several problems linked with biological control of plant pests. The allelopathic effect of plants on microorganisms, and microorganisms on microorganisms is of great environmental and economic significance. This book is organized around the premise that allelochemicals can be employed for biological control of plant pathogens and plant diseases. Specifically, this volume focuses on (i) discovery and development of natural product based fungicides for agriculture, (ii) direct use of allelochemicals as well as indirect effects through cover crops and organic amendments for plant parasitic pest control and (iii) application of allelopathy in the pest management.

In an effort to address above points, contributing authors provided up-to-date reviews and discussion on allelochemicals-related biological control of plant diseases and pathogens. Chapters 1 - 3 discuss discovery and development of allelochmicals and their role in the management of plant diseases. Chapter 4 discusses the effects of pathogens on the competitiveness and allelopathic ability of their hosts. Chapter 5 highlights the importance of allelopathy for weed control in aquatic ecosystems. Chapters 6-7 deal with bacterial potential in weed management and plant disease control. Chapter 8 describes the role of organic compound ginsenosides from rhizosphere soil and root exudates of american ginseng plant in control of fungal diseases. Antimicrobial and nematicidal substances from the rhizome of chicory has been discussed in Chapter 9. The role of allelochemicals induced in mycorrhizal plants in imparting disease resistance is given in Chapter 10. The last chapter discusses the biocontrol of plant pathogens and diseases by allelochmicals from *Ageratum conyzoides* a weed and rice plants has been highlighted in Chapter 11.

We are grateful to all authors for providing their valuable work to this volume. The articles are original and some have been written for the first time in any book. We are indebted to the following referees for their constructive comments and suggestions: Ana L. Anaya, Mark Bernards, Nancy Kokalis Burelle, Chester L. Foy, John M. Halbrendt, Robert Kremer, Azim Mallik, Susan Meyer, Reid J. Smeda, Tony Sturz, David Wedge and Jeff Weidenhamer. The editorial help of Ineke Ravesloot, Publishing Department, Springer is sincerely appreciated. It is our hope that this book will serve scientific community well, and equally hope that the book will stimulate young students to work on biological control of plant pathogens and diseases through natural allelochemicals.

Inderjit and K.G. Mukerji
October 2005

DAVID E. WEDGE[1] and BARBARA J. SMITH[2]

DISCOVERY AND EVALUATION OF NATURAL PRODUCT-BASED FUNGICIDES FOR DISEASE CONTROL OF SMALL FRUITS

[1]*United States Department of Agriculture, Agricultural Research Service, Natural Products Utilization Research Unit, The Thad Cochran National Center for Natural Products Research, University of Mississippi, University, MS 38677, USA; and* [2]*Small Fruit Research Station, 306 S. High St., Poplarville, MS 39470, USA*
Email : dwedge@olemiss.edu

Abstract. The continuing development of fungicide resistance in plant and human pathogens necessitates the discovery and development of new fungicides. Discovery and evaluation of natural product fungicides is largely dependent upon the availability of miniaturized antifungal bioassays. Essentials for natural product bioassays include sensitivity to microgram quantities, selectivity to determine optimum target pathogens, and adaptability to complex mixtures. Experimental accuracy and precision must be stable between assays over time. These assays should be relevant to potential pathogen target sites in the natural infection process of the host and applicable to the agrochemical industry. Bioassays should take advantage of current high-throughput technology available to evaluate dose-response relationships, commercial fungicides standards, modes of action, and structure activity studies. The focus of this chapter is the evaluation of natural product based fungicides for agriculture and we will provide a review of bioautography prescreens and microtiter assays (secondary assays). Also presented is more detailed information on newer techniques such as the detached leaf assays for evaluating fungicides against strawberry anthracnose (*Colletotrichum* spp.) and field plot trials for gray mold (*Botrytis*) and anthracnose control in strawberry.

1. INTRODUCTION

Since the early 1970s, agriculture worldwide has struggled with the evolution of pathogen resistance to disease control agents. Increased necessity for repeated chemical applications, development of pesticide cross-resistance, and disease resistance management strategies have characterized the use of agricultural chemicals to-date. As a consequence, producers are currently attempting to control agricultural pests with a decreasing arsenal of effective crop protection chemicals. In addition, the desire for safer pesticides with less environmental impact has become a major public concern. Particularly desirable is the discovery of novel pesticidal agents from new chemical classes that are able to operate using different modes of action and, consequently, against plant pathogens with resistance to currently used chemistries. In this regard, evaluating natural products and extracts as a source of new pesticides is one strategy for the discovery of new chemical moieties that have not previously been created by synthetic chemists.

Inderjit and K.G. Mukerji (eds.),
Allelochemicals: Biological Control of Plant Pathogens and Diseases, 1–14.

Antibiotics, antineoplastics, herbicides, and insecticides often originate from plant and microbial chemical defense mechanisms (Wedge and Camper, 2000). Secondary metabolites, once considered unimportant products, are now thought to mediate chemical defense mechanisms by providing chemical barriers against animal and microbial predators (Agrios, 1997; Wedge and Camper, 2000). Plants produce numerous chemicals for defense and communication, and can elicit their own form of offensive chemical warfare by targeting the proliferation of pathogens. These chemicals may have general or specific activity against key target sites in bacteria, fungi, and viruses. Exploiting the chemical warfare that occurs between plants and their pathogens shows promise in providing new natural products for new anti-infectives for human, plant and animal health. The successful development of strobilurin fungicides and spinosad insecticides has continued the interest in natural products as crop protectants. The importance and future of natural product agrochemistry is emphasized by the fact that 21 companies have filed 255 patent applications primarily for use of the strobilurin class of fungicides (Qo I MET complex 3 inhibitors).

1.1. Direct Acting Defense Chemicals

Since the discovery of the vinca alkaloids in 1963, many of the known antitubulin agents used in today's cancer chemotherapy arsenal are products of plant and fungal secondary metabolism. Since 1991, 16 of 43 new pharmaceuticals were derived from natural products. In certain therapeutic areas 78% of the antibacterials and 74% of the anticancer compounds are natural products or have been derived from natural products (Roughi, 2003). These "natural products" are probably defense chemicals that target and inhibit cell division in invading pathogens (Wedge and Camper, 2000). Therefore, it is reasonable to hypothesize that plants and certain fungi can produce chemicals, such as resveratrol and strobilurin, that act directly in their defense by inhibiting pathogen proliferation, or indirectly by disrupting chemical signal processes related to growth and development of pathogens or herbivores (Wedge and Camper 2000).

1.2. Indirect Acting Defense Chemicals

Plant resistance to pathogens is considered to be systemically induced by some endogenous signal molecule produced at the infection site that is then translocated to other parts of the plant (Oku, 1994). The search for, and identification of, the putative signal molecule(s) is of great interest to many plant scientists because such moieties have possible uses as "natural product" disease control agents. However, research indicates that no single compound is involved, but rather a complex signal transduction pathway, which, in plants, can be mediated by a number of compounds that appears to influence octadecanoid metabolism. In response to wounding or pathogen attack, fatty acids of the jasmonate cascade are formed from membrane-bound α-linolenic acid by lipoxygenase-mediated peroxidation (Vick and Zimmerman, 1984). Analogous to the prostaglandin cascade in mammals, α-linolenic acid is thought to participate in a lipid-based signaling system where jasmonates induce the synthesis of a family of wound-inducible defensive proteinase inhibitors (Farmer and Ryan, 1992) and low and high molecular weight phytoalexins such as flavonoids, alkaloids, and terpenoids (Gundlach et al., 1992; Mueller et al., 1993).

Several plant and bacterial natural products have novel applications as plant protectants through the induction of systemic acquired resistance (SAR) processes. Commercial products that appear to induce SAR include Messenger® (EDEN Biosciences, Inc., Bothell, WA) and the bioprotectant fungicides Serenade® (AgraQuest, Davis, CA), Sonata® (AgraQuest, Davis, CA), and Milsana® (KHH BioSci, Inc., Raleigh, NC). Messenger is a harpin protein which switches on natural plant defenses in response to bacterial leaf spot, wilt, and blight and fungal diseases such as *Botrytis* brunch rot, and powdery mildew. Serenade is a microbial-protectant derived from *Bacillus subtilis,* with SAR activity that controls *Botrytis*, powdery and downy mildews, early blight, and bacterial spot. Sonata is also a microbial-biopesticide with activity against *Botrytis*, downy and powdery mildews, rusts, *Sclerotinia* blight, and rots. Milsana® is an extract from *Reynoutria sachalinensis* (giant knotweed) that induces phytoalexins able to confer resistance to powdery mildew and other diseases such as by *Botrytis*. However, elicitors with no innate antifungal activity will not appear active in most current screening high throughput screening systems.

Many experimental approaches have been used to screen compounds and extracts from plants and microorganisms in order to discover new antifungal compounds. The focus of this paper is on laboratory methods and field procedures that we use to evaluate naturally occurring antifungal compounds produced by plants, pathogens, and other terrestrial and marine organisms. As part of a program to discover and develop naturally occurring fungicides, several new in vitro detection systems and a detached leaf assay were developed to evaluate small amounts of compound. Our fungicide field protocol used to test potential lead fungicides using commercial strawberry is also presented.

2. MATERIAL AND METHODS

2.1. Pathogen Production

Isolates of *Colletotrichum acutatum*, *C. fragariae* and *C. gloeosporioides, Phomopsis viticola* and *P. obscurans, Botrytis cinerea* and *Fusarium oxysporum* are maintained on silica gel at 4-6 °C. The three *Colletotrichum* species and *P. obscurans* strain were isolated from strawberry (*Fragaria x ananassa* Duchesne). *Phomopsis viticola* and *B. cinerea* were isolated from commercial grape (*Vitis vinifera* L.) and *F. oxysporum* from orchid (*Cynoches* sp.). Fungal cultures were grown on potato-dextrose agar (PDA, Difco, Detroit, MI) in 9-cm petri dishes and incubated in a growth chamber at 24 ± 2 °C under cool-white fluorescent lights (55 ± 5 μmols/m^2/sec) with a 12h photoperiod.

Conidia were harvested from 7-10 day-old cultures by flooding plates with 5 mL of sterile distilled water and softly brushing the colonies with an L-shaped glass rod. Aqueous conidial suspensions are filtered through sterile Miracloth (Calbiochem-Novabiochem Corp., La Jolla CA) to remove hyphae. Conidial concentrations were determined photometrically (Espinel-Ingrof and Kekering, 1991; Wedge and Kuhajek, 1988) from a standard curve based on the absorbance at 625 nm, and suspensions are then adjusted with sterile distilled water to a concentration of $1.0x10^6$ conidia/mL.

Standard conidial concentrations are determined from a standard curve for each fungal species. Standard turbidity curves were periodically validated using both a Bechman/Coulter Z1 (Fullerton, CA) particle counter and hemocytometer counts. Conidial and mycelial growth are evaluated using a Packard Spectra Count (PerkinElmer Life and Analytical Sciences, Inc., Boston, MA). Conidial growth and germ tube development were evaluated using an Olympus IX 70 (Olympus Industrial America, Inc., Melville, New York) inverted microscope and recorded with a Olympus DP12 digital camera as appropriate for compounds that affect spore germination and early germ tube development.

2.2. *Direct Bioautography*

A number of bioautography techniques were used as primary screening systems to detect antifungal compounds. Matrix, one-dimensional, and two-dimensional bioautography protocols on silica gel thin layer chromatography (TLC) plates and *Colletotrichum* spp. as the test organisms were used to identify the antifungal activity according to published methods (Homan and Fuchs, 1970; Moore, 1996; Wedge and Nagle, 2000). Matrix bioautography is used to screen large numbers of crude extract at 20mg/mL. One-dimensional TLC (1D-TLC) and two-dimensional TLC (2D-TLC) are subsequently used to separate and identify the number of antifungal agents in extract. Modification of these procedures can be used to visually evaluate natural chemical defenses in disease resistant and susceptible plant cultivars (Vincent *et al.* 1999).

A 2D-TLC direct bioautography method was used to evaluate active crude or partially purified extracts. This protocol utilizes two sequential TLC runs in which the TLC plates are developed once with a polar solvent, turned 90°, and then developed a second time with a non-polar solvent system (Wedge and Nagle, 2000). The method takes advantage of the resolving power of 2D-TLC to separate chemically diverse mixtures found in crude extracts. Two-dimensional TLC bioautography is well suited for resolving extracts containing lipophilic natural products that are difficult to separate by single elution TLC.

Each plate was subsequently sprayed with a spore suspension (10^5 spores/mL) of the test fungus and incubated in a moisture chamber for 3 days at 26 °C with a 12h photoperiod. Clear zones of fungal growth inhibition on the TLC plate indicate the

---

Mention of trade names or commercial products in this article is solely for the purpose of providing specific information and does not imply recommendation or endorsement by the U. S. Department of Agriculture.

presence of antifungal constituents in each extract. Inhibition of fungal growth was evaluated 3-4 days after treatment. Antifungal metabolites can be readily located on the plates by visually observing clear zones where the active compounds inhibit fungal growth (Tellez, 2000,Vincent et al., 1999). The 2-D TLC method eliminates the need for the development of large numbers of plates in multiple solvent systems, reduces the amount of waste solvents for disposal, and substantially reduces the time required to identify active compounds.

2.3. 96-Well Microbiassay

The quick discovery and evaluation of natural product fungicides is heavily dependent upon miniaturized antifungal bioassay techniques. A reference method [M27-A from the National Committee for Clinical Laboratory Standards (NCCLS)] for broth-dilution antifungal susceptibility testing of yeast was adapted for evaluation of antifungal compounds against sporulating filamentous fungi (Wedge and Kuhajek, 1998).

This standardized 96-well microtiter plate assay was developed for the detection of natural product fungicidal agents, and can also be used to evaluate purified antifungal agents. In our study a 96-well microtiter assay was used to determine sensitivity *of C. acutatum, C. fragariae, C. gloeosporioides, F. oxysporum, B. cinerea, P. obscurans,* and *P. viticola* to the various antifungal agents in comparison with the commercial fungicides. Fungicides such as benomyl, azoxystrobin and captan with different modes of action were used as standards in these assays. Each fungal species was challenged in a dose-response format so that in the final test, compound concentrations of 0.3, 3.0, and 30.0 μM were achieved (in duplicate) in the different columns of the 96-well plates.

Fungal growth was evaluated by measuring absorbance of each well at 620 nm at 0, 24, 48, and 72 hr, with the exception of tests with *P. obscurans* and *P. viticola,* where the data are recorded for up to 120 hr. Treatments are repeated so that mean absorbance values and standard errors can be calculated and are used to evaluate fungal growth. Differences in spore germination and mycelial growth in each of the wells in the 96-well plate demonstrate sensitivity to particular concentrations of compound and can indicate fungistatic or fungicidal effects. The microtiter assay can also be used to compare the sensitivity of fungal plant pathogens to natural and synthetic compounds with industry standards (Wedge et al., 2001).

A novel application of the microbioassay was also developed for the discovery of compounds that inhibit *Phytophthora* spp. This protocol used the 96-well format for high-throughput capability and a standardized method for quantification of initial zoospore concentrations for maximum reproducibility. Zoospore suspensions were quantifiable between 0.7 and 1.5×10^5 zoospores/mL at an absorbance value of 620 nm. Subsequent growth of mycelia was monitored by measuring absorbance (620 nm) at 24-hour intervals for 96 hr. Full- and half-strength preparations of each of three media (V8 broth, Roswell Park Memorial Institute mycological broth, and mineral salts medium), and four zoospore concentrations (10, 100, 1000, and 10,000 zoospores/mL) were evaluated. Both full- and half-strength Roswell Park Memorial Institute mycological broth were identified as suitable synthetic media for growing *P. nicotianae*, and 1000 zoospores/mL was established as the optimum initial concentration (Kuhajek et al., 2003).

2.4. Detached Leaf Assay for Fungicide Evaluations In Planta

Anthracnose susceptible 'Chandler' strawberry plants were grown in 10 x 10 cm plastic pots in a 1:1 (v/v) mixture of Jiffy-Mix (JPA, West Chicago, IL), and pasteurized sand in the greenhouse for a minimum of six weeks before inoculation. The plants were grown under standard conditions of a 16-hr day length and 24 °C temperature. Growth parameters are varied as needed to accommodate the needs of particular studies.

Whole leaves (petiole and leaflets) were cut from plants no more than four hr before treatment or inoculation. Only the second or third youngest leaf on a plant was used for the fungicide assay, and only leaves with no visible signs of injury or symptoms of disease were collected for testing. Immediately after collection, the leaves were placed in a tray lined with moist paper towels and the tray is closed to retained near 100% RH and maintained at a cool temperature (~ 12 °C). To test for protective fungicide activity, treatment compounds were applied to the upper surface of each of the three leaflets on a leaf using a chromatography sprayer. After treatment, the base of each leaf stem was inserted into sterile distilled water in a 100 x 10 mm tissue culture tube. Each upper surface of each treated leaflet was inoculated with conidia from the test fungal isolate within 24 hr of treatment. Inoculated leaves were subsequently incubated in a dew chamber for 48 hour at 100% RH, 30 °C and then maintained at 25 °C in a moist chamber at 100% RH for 10 days. Sterile distilled water was added to each tube as needed to maintain the surface of the water above the base of the petiole. If a compound was to be tested for curative activity, the leaflets were inoculated 24 hr before the fungicidal compound is applied.

Experimental compounds were evaluated in a dose-response format. Azoxystrobin or other fungicides that have both protective and curative activity were used as a standard, and a solvent control was used in every study. The number of disease lesions per leaf was used to determine the ability of the test compounds to prevent infections. The size of the lesions was used to determine the curative activity of compounds. Each fungicide concentration is replicated four times and the experiment is repeated at least once. This bioassay does not differentiate between direct effects on the fungus and indirect effects through induction of plant defenses. However, if a compound is much more active in this *in vivo* assay than than the *in vitro* microtiter assay, 'induction' activity is indicated.

2.5. 2002-03 Experimental Field Studies at Hammond, Louisiana

Strawberry plots were established and maintained following standard horticulture practices used by commercial strawberry farmers in Lousisana. The following fungicides were applied to strawberries: fenhexamid (Elevate®, Arvesta, San Francisco, CA) at 1.6 kg/H, fenhexamid + captan (CaptEvate®, Arvesta, San Francisco, CA) at a low rate (3.9 kg/H) and a high rate (5.82 kg/H), azoxystrobin (Quadris®, Syngenta Crop Protection, Greensboro, NC) at 0.56 kg/H, cyprodinil + fludioxonil (Switch®, Syngenta Crop Protection, Greensboro, NC) at 0.8 kg/H, captan (Micro Flo Company, Memphis, TN) at 3.4 kg/H, mylclobutanil (Nova®, Dow AgroSciences, Indianapolis, IN) at 0.28 kg/H, pyrimethanil (Scala®, Bayer CropScience, Research Triangle Park, NC) at 1.9 kg/H, fenhexamid at 1.68 kg/H + captan at 3.4 kg/H, pyraclostrobin (Cabrio®, BASF, Research Triangle Park, NC) at 0.20 kg/H, pyraclostrobin + boscalid (Pristine, BASF, Research Triangle Park, NC) at 0.62 kg/H, boscalid (Emerald®, BASF, Research Triangle Park, NC) at 0..59 kg/H, and the experimental fungicide, CAY-1 at 0.37 and 0.74 kg/H. Fungicide treatments were applied every 7-10 days (or as soon after rainfall as possible) starting at or near full-bloom stage and continuing until 1 week before harvesting ceased.

Berries were harvested twice a week throughout the entire season beginning on January 30 and continuing until April 24 in 2003. Total fruit count and weights were obtained for both marketable and cull fruits, along with average weight per berry and percentages of marketable, physical cull, and diseased cull fruit per acre. Fruit were separated into marketable and cull classes, with the culls being further divided into physical (deformed or small fruit) culls and diseased culls. Diseases were identified and counted on three occasions including berries with gray mold (*B. cinerea*), anthracnose (*C. acutatum)*, leather rot (*Phytophthora cactorum)*, stem end rot (*Gnomonia comari)*, and other rots that occurred during the picking season. Plants were rated on April 24, 2003 for incidence of foliar disease, number of dead plants due to anthracnose, crown rot and plant vigor. Diseases were rated on a scale of 0-5; 0 plants showed no leaf symptoms and 5 plants were completely defoliated. Plant vigor was rated on a scale of 0-5; 0 were dead plants and 5 were extremely vigorous plants.

3. RESULTS AND DISCUSSION

Results from the bioautography studies suggest that chemically different populations of microorganisms and plants can easily be distinguished by their characteristic "2D-TLC fingerprint" and antifungal zone patterns (Wedge and Nagle, 2000). Thus, this system represents a very useful technique for the identification and selection of unique chemotypes from microbial isolates (Nagle and Paul, 1999) and plant extract(?) collections.

Chromatographic properties such as relative polarity, UV absorbance, chemical reactivity associated with each active metabolite provide valuable information that allows for rapid dereplication of known or nuisance compounds. When strains of different phytopathogenic fungi with dissimilar fungicide resistance profiles are inoculated onto replicate bioautography plates prepared from any given extract containing active metabolites, it is possible to visually observe distinct differences in the sensitivity of each fungal pathogen to single metabolites. These differences in pathogen sensitivity (fungicide resistance) can be observed by direct comparison of inhibition zone dimensions produced by active metabolites and control standards against each pathogenic strain tested. Chemical profiles provide valuable information for the rapid selection of specific antifungal metabolites with unique activity against fungicide-resistant pathogens and identify new compounds with novel mechanisms of action.

Using modifications of the methods described previously, bioautography techniques were successful in allowing us to track the path of naturally occurring fungitoxic compounds present in strawberry leaves. Our studies indicated that concentrations of fungitoxic compounds vary between anthracnose resistant and susceptible cultivars and are present in different amounts in vegetative tissues of different ages. Using leaves of the anthracnose-susceptible cultivar Chandler and the anthracnose-resistant cultivar Sweet Charlie, we isolated and demonstrated the presence of three antifungal compounds. While the mechanism of strawberry anthracnose resistance is unknown, results from this study indicate that anthracnose resistance in strawberry may depend on the concentration of two constitutive antifungal compounds and the elicitation of a third compound in younger leaves.

These two constitutive antifungal compounds were exhibited in both 'Chandler' and 'Sweet Charlie' plants but 'Sweet Charlie' plants produced approximately 15 times more antifungal activity than 'Chandler' plants. Fungal growth inhibition associated with extracts from 'Chandler' plants appeared to be temporary. A third compound, detected exclusively in 'Sweet Charlie' plants, was produced only after young leaves were sprayed with a commercially available elicitor of antifungal compounds (Vincent et al., 1999).

The antifungal activity of 32 naturally occurring quinones of four major classes: 1,4-naphthoquinones, 1,2-naphthoquinones, 1,4-benzoquinones, anthraquinones, and other miscellaneous compounds from our natural products collection were tested for antifungal activity using bioautography. Bioautography allowed for the rapid evaluation of quinones which demonstrated good to moderate antifungal activity against *Colletotrichum* spp. *Colletotrichum fragariae* appeared to be the most sensitive species to quinone-based chemistry, *C. gloeosporioides* of intermediate sensitivity, and *C. acutatum* was the least sensitive species to these naturally occurring compounds (Meazza et al., 2003).

Bioassay-directed isolation of antifungal compounds from an ethyl acetate extract of *Ruta graveolens* leaves yielded two furanocoumarins, one quinoline alkaloid, and four quinolone alkaloids, including a novel compound, 1-methyl-2-[6'-(3",4"-methylenedioxyphenyl)hexyl]-4-quinolone. Antifungal activities of the isolated compounds, together with 7-hydroxycoumarin, 4-hydroxycoumarin, and 7-methoxycoumarin which are known to occur in Rutaceae species, were evaluated using bioautography and microbioassay procedures. Four of the alkaloids had moderate activity against *Colletotrichum* species, including a benomyl-resistant *C. acutatum*. These compounds and the furanocoumarins 5- and 8-methoxypsoralen had moderate activity against *Fusarium oxysporum*. The novel quinolone alkaloid was highly active against *Botrytis cinerea*. *Phomopsis* species were much more sensitive to most of the compounds tested, with *P. viticola* being highly sensitive to all of the compounds screened (Oliva et al., 2003).

Hexane and EtOAc phases of MeOH extract of *Macaranga monandra* demonstrated fungal growth inhibition in *C. acutatum, C. fragariae, C. gloeosporioides, F. oxysporum, B. cinerea, Phomopsis obscurans* and *P. viticola.* Bioassay-guided fractionation resulted in the isolation of two active clerodane-type diterpenes that were elucidated by spectral methods as kolavenic acid and 2-oxo-kolavenic acid. The 96-well microbioassay revealed that kolavenic acid and 2-oxo-kolavenic acid produced moderate growth inhibition in *P. viticola* and *B. cinerea* (Salah et al., 2003).

The application of the microbioassay to *Phytophthora nicotianae* was effectively used to determine EC_{50} values (i.e., effective concentration for 50% growth reduction) for eight commercial antifungal compounds (azoxystrobin, fosetyl-aluminum, etridiazole, metalaxyl, pentachloronitrobenzene, pimaricin, and propamocarb). These EC_{50} values were compared to those obtained using conventional plate methods by measuring linear growth of mycelia on fungicide-amended medium. The microbioassay proved to be a rapid, reproducible, and efficient method for testing the efficacy of compounds against *P. nicotianae* and should be effective for other species of *Phytophthora* as well. The assay requires relatively small amounts of a test compound and was suitable for the evaluation of natural product samples (Kuhajek et al., 2003).

CAY-1 is a fungicidal steroidal saponin (Mol wt. 1243) isolated and identified by DeLucca et al. (2002) from the ground fruit of cayenne pepper (*Capsicum frutescens*). CAY-1 was lethal to germinating conidia of *Aspergillus flavus, A. fumigatus, A. parasiticus* and *A. niger*. It was also active against agricultural and medicinally important fungi and yeast. *In vitro* dose-response assays with CAY-1 against plant pathogenic fungi showed that 3.0 μM inhibited growth of *C. gloeosporioides* and *C. acutatum* by 100% and *C. fragariae, P. obscurans*, and *P. viticola* by 90%. Sampangine and several alkaloid analogs were isolated from the root bark of *Cleistopholis patens*. These novel broad-spectrum compounds showed promising antifungal activity against several serious pathogenic fungi of plants including *B. cinerea, C. fragariae, C. acutatum, C. gloeosporioides*, and *F. oxysporium* (Wedge and Nagle, 2003).

Detached leaf assays provide us with the opportunity to evaluate new fungicides directly on the leaf surface in a dose-response format (Table 1). This assay allowed us to benchmark potential lead compounds such as CAY-1 and sampangine with a commercial standard (azoxystrobin) of known mode of action (Q_o I inhibitor). The number of diseased lesions was used to determine effective concentrations needed for disease control. Lesion size is used to determine the relative effectiveness of the systemic activity that produced curative activity 24 hrs after inoculation. The detached leaf assay was also used to establish experimental field rates for future studies. Study of 'protectant' activity indicated that 1250 ppm. CAY-1 or sampangine appeared to be an effective concentration for disease control of anthracnose on the leaf surface, or between 100-1000 times the concentration required for *in vitro* activity (Post, Table 1).

Table 1. Anthracnose disease severity and phytotoxicity (Phyto) scores of detached leaves following inoculation with Colletotrichum fragariae isolate CF-75, pre and post treatment with commercial and experimental fungicides.

	Azoxystrobin		CAY-1		Sampangine	
Level	Disease[1]	Phyto[1]	Disease	Phyto	Disease	Phyto
Fungicide applied to the upper leaf surface. Leaves not inoculated.						
None	0.29 a[2]	0.04 A	0.29 a	0.04 a	0.29 a	0.04 a
Solvent	0.29 a	0.15 A	0.29 a	0.15 a	0.29 a	0.15 a
625	0.17 a	0.00 A	0.10 a	0.00 a	0.08 a	0.08 a
1250	0.19 a	0.08 A	0.21 a	0.04 a	0.13 a	0.04 a
2500	0.08 a	0.04 A	0.21 a	0.00 a	0.17 a	0.08 a
lsd	0.34	0.20	0.32	0.16	0.30	0.19
Pr>F	0.70	0.67	0.75	0.52	0.50	0.80
Fungicide applied to the upper leaf surface 24 hr before inoculation (Post).						
None	0.75 a	0.33 A	0.75 ab	0.33 a	0.75 a	0.33 a
Solvent	0.83 a	0.00 B	0.83 a	0.00 b	0.83 a	0.00 b
625	0.42 ab	0.08 B	0.33 bc	0.00 b	0.25 b	0.00 b
1250	0.08 b	0.17 ab	0.21 c	0.00 b	0.17 b	0.00 b
2500	0.42 ab	0.00 b	0.21 c	0.08 b	0.29 b	0.00 b
lsd	0.54	0.23	0.42	0.17	0.41	0.17
Pr>F	0.06	0.04	0.02	0.00	0.01	0.00
Fungicide applied to the upper leaf surface 24 hr after inoculation (Pre).						
None	0.58 a	0.08 b	0.58 abc	0.08 b	0.58 a	0.08 b
Solvent	0.25 a	0.33 a	0.25 c	0.33 ab	0.25 a	0.33 ab
625	0.33 a	0.08 b	0.92 ab	0.58 a	0.75 a	0.08 b
1250	0.83 a	0.08 b	1.00 a	0.25 ab	0.75 a	0.08 b
2500	0.67 a	0.08 b	0.83 ab	0.08 b	0.58 a	0.50 a
lsd	0.60		0.23	0.47	0.57	0.36
Pr>F	0.27	0.14	0.04	0.28	0.36	0.08
Fungicide applied to the lower leaf surface (translaminar) 24 hr before inoculation.						
625	0.33 a	0.04 a	. .	. .	0.13 a	0.08 a
1250	0.08 a	0.04 a	. .	. .	0.25 a	0.08 a
2500	0.17 a	0.00 a	. .	. .	0.29 a	0.08 a
lsd	0.37	0.09			0.31	0.26
Pr>F	0.34	0.51			0.46	0.63
Fungicide applied to the lower leaf surface (translaminar) 24 hr after inoculation.						
625	0.33 a	0.00 a	. .	. .	0.38 a	0.17 a
1250	1.21 a	0.25 a	. .	. .	0.58 a	0.17 a
2500	0.58 a	0.17 a	. .	. .	0.67 a	0.17 a
lsd	1.09	0.31			0.71	0.43
Pr>F	0.22	0.23			0.64	1.00

[1]Disease scored on scale of 0 = no lesions to 3 = severe; phytotoxicity scored on scale of 0 – 5 where, 0 = no signs of damage 5 = severe tissue damage.

[2]Average values followed by the same letter were not significantly different at the 0.05 level using Least Significant Difference at P=0.05.

No systemic or translaminar activity was detected when azoxystrobin, CAY-1, or sampangine was applied 24 hrs after inoculation with *C. fragariae* (Pre, Table 1).

Field studies indicated that significant differences occurred in marketable yield of strawberries as a result of fungicide treatments. The highest marketable yield (17,071 kg/H) was recorded from plots receiving Pristine fungicide at the rate of 0.62 kg ai/H. The next highest yield (16,960 kg/H) was harvested from plants treated with Switch fungicide at the rate of 0.8 kg/H. Plants from the untreated control plots yielded more fruit than plants from four of the fungicide-treated plots. Plants treated with CAY-1 produced the lowest yield of cull fruit weighing 1,390 kg/H. compared with the highest yield of cull fruit weighing 5,816 kg/H produced on the plots treated with Pristine. CAY-1 produced high yields of diseased fruit weighing over 9,400 kg/H. Five fungicide treatments resulted in the production of marketable fruit at 60%: Switch® (69.8%), Pristine (68.51%), Cabrio® (65.5%), and the two CaptEvate® treatments (5.8 kg. and 3.9 kg, 65.8% and 62.7% marketable fruit, respectively). Most treatments that were combinations of two fungicides produced the highest marketable yields and lowest disease percentage.

The total number of berries with fruit rot symptoms and the number of berries with symptoms of anthracnose or stem end rot from the April 24 harvest time were significantly lower from plots treated with the fungicides Switch, Cabrio®, CaptEvate®, and Pristine® than from those receiving no fungicide treatment (Table 2). The most prevalent diseases in the Louisiana field study were anthracnose caused

Table 2. Fruit rot and field data from April 24, 2003 harvest of fungicide treated strawberry plots, Hammond, Louisiana.

Fungicide	Total Rots[1]	Anthracnose[2]	Stem end rot[2]	Plants Dead (%)[3]	Foliar disease[4]
Switch	1.5 f[5]	0.8 e	0.0 d	16.3 a	2.8 a
Cabrio	5.8 f	3.8 e	0.5 d	8.8 cde	2.6 ab
CaptEvate (High)	6.0 f	5.0 e	0.5 d	5.0 e	1.1 e
Pristine	7.0 ef	5.0 e	0.3 d	5.0 e	2.5 abc
CAY-1 (Half)	22.0 def	18.0 de	2.0 cd	15.6 ab	2.8 a
Captan	27.8 cdef	21.3 cde	3.0 cd	7.5 cde	2.0 bcd
Quadris	28.3 cdef	20.3 cde	2.8 cd	8.1 cde	1.9 cd
CaptEvate (Low)	29.0 cdef	23.3 bcde	2.5 cd	7.5 cde	1.8 de
Captan+Elevate	43.0 cdef	28.8 bcde	7.0 bcd	12.5 abc	2.0 bcd
CAY-1 (Full*	50.5 bcde	37.8 bcd	7.5 bcd	10.6 abcde	3.1 a
Scala	59.0 bcd	47.8 bcd	7.0 bcd	9.4 bcde	1.9 cd
Control	63.5 bcd	45.3 bcd	9.8 abc	5.6 de	2.8 a
Emerald	70.5 abc	50.3 Bc	7.8 abcd	5.6 de	2.9 a
Nova	87.8 ab	54.5 Ab	16.0 a	7.5 cde	2.0 bcd
Elevate	109.5a	84.5 A	14.3 ab	11.9 abcd	2.6 ab
LSD (0.05)	44.1	31.6	8.4	6.8	0.7

[1]Total number of fruit/20 plant plot with any disease symptom.
[2]Number of fruit with anthracnose fruit rot symptomn (*Colletotrichum acutatum*) or Stem-End Rot (*Gnomonia comari*).
[3]Percentage of plants/20 plant plot dead from anthracnose crown rot.
[4]Foliar disease scored on scale of 0 – 5, where 0 = no disease and 5 = plant defoliated due to foliar disease.
[5]Average values followed by the same letter were not significantly different at the 0.05 level using a Least Significant Difference.

by the fungus *C. acutatum* and stem end rot caused by the fungus *Gnomonia comari*. Stem end rot lesions were often invaded by secondary pathogens such as *Botrytis* and *Colletotrichum*. The combination fungicides such as Elevate® + Captan®, Pristine®, and Switch® were effective in controlling this disease complex. The untreated control plants and those treated with Emerald® had the most berries affected with stem end rot. The incidence of gray mold and anthracnose fruit rot was extremely low. Gray mold was controlled by the fungicides Scala®, Elevate® + Captan®, Pristine®, Switch®, Emerald®, and Elevate®.

4. CONCLUSIONS

Information gained from the examination of thousands of extracts and their associated pure compounds have culminated in the development of a variety of standardized operating protocols for natural product discovery that are currently being used in our laboratory. Successful discovery, evaluation, and development of natural product fungicides are totally dependent upon the availability of high quality miniaturized antifungal bioassays. Bioassay-directed screening of compounds and extracts is the initial step in the discovery process for new agrochemicals and pesticides.

Standardization of inoculum allows for meaningful comparison of growth inhibition between different fungal pathogens, test compounds, and experiments repeated in time. Bioautography provides a simple technique to visually follow antifungal components through the separation process. The 96-well microbioassay allows for the evaluation of microgram quantities, determination of dose-response relationships, and comparison of antifungal activity with fungicides with a known mode of action. Coupling bioautography techniques with the 96-well microbioassay provides us with a discovery protocol that combines the simple and visual nature of direct bioautography with the rapid, sensitive, and high throughput capabilities of a microtiter system.

The 96-well microbioassay is accurate and sensitive; as little as 0.1 μM amounts of test compound permit discrimination between germination and mycelial growth inhibitors and identification of fungicide resistant pathogens. The microbioassay utilizes a chemically defined liquid medium with a zwitterion buffer that limits chemical interaction with test compounds and controls for pH variations. This new standardized method provides high-throughput capability and the capacity to study chemical compounds in detail, to perform mode of action studies, and to determine fungicide resistance profiles for specific fungal pathogens.

Detached leaf assays are critical for establishing 'real world' activity prior to the field testing that agrochemical companies require before investing millions of dollars needed to develop a agrochemical. Subsequent efficacy testing in the greenhouse ultimately helps determine the potential usefulness of compounds as pest control agents. To maximize the detection of natural products, high-throughput bioassay techniques must target significant agricultural pests, include relevant commercial pesticide standards, and adhere to sound statistical principles.

While the use of SAR inducers is still in its infancy, these compounds will may be of use in specialized applications. However, our observations are such that the host plant must have a suitable 'metabolic engine' - found in many resistant cultivars - that is capable of being 'revved up.' Susceptible cultivars most often used in fungicide evaluations either produce a lower quantity of defense compound and or have a longer time lag between pathogen infection and plant symptom development than is the case in resistant cultivars. Systemic acquired resistance inducers have been marketed (Actigard, Messenger) with limited success and have never recouped their development costs. It thus appears unlikely that economically viable SAR products will be developed in the near future.

Even so, allelochemicals and other natural products represent potentially rich and new sources of agrochemical plant protectants. The identification of suitable natural products coupled with traditional synthesis chemistry should be an effective approach for optimizing pesiticide activity and associated chemical properties. New requirements for fungicides are stringent, and require that new products must be environmentally safe and efficacious at low rates. In addition they must posses low mammalian toxicity, operate using novel modes of action, and have a low to moderate risk of developing resistance in target pathogens.

5. REFERENCES

Agrios, G. N. How Pathogens Attack Plants. In: *Plant Pathology*. Academic Press, San Diego, 1997; 63-82.

Espinel-Ingrof, A., Kerkering, T. M. Spectrophotometric method of inoculum preparation for the *in vitro* susceptibility testing of filamentous fungi. Journal of Clinic Microbiology 1991; 29:393-394.

De Lucca, A. J., Bland, J. M.,Vigo, C. B., Cushion, M., Selitrennikoff, C. P., Peter, J., Walsh, T. J. CAY-1, a fungicidal saponin from *Capsicum* species. Med Mycol 2002; 40:131-137.

Farmer, E. E., Ryan, C. A. Octadecanoid precursors of jasmonic acid activate the synthesis of wound-inducible proteinase inhibitors. Plant Cell 1992; 4:129-134.

Gundlach, H., Muller, M. J., Kutchan, T. M., Zenk, M. H. Jasmonic acid is a signal transducer in elicitor-induced plant cell cultures. Proc Natl Acad Sci USA 1992; 89:2389-2393.

Homans, A. L., Fuchs, A. Direct bioautography on thin-layer chromatograms as a method for detecting fungitoxic substances. J Chromatogr 1970; 5(2):327-3299.

Kuhajek, J. M., Jeffers, S. N., Slattery, M., Wedge, D. E. A Rapid Microbioassay for Discovery of Novel Fungicides for *Phytophthora* spp. Phytopathology 2003; 93:46-53.

Nagle, D. G., Paul, V. J. Production of secondary metabolites by filamentous tropical marine cyanobacteria: ecological functions of the compounds. J Phycol 1999; 35 (Suppl. 607):1412-1421.

Meazza, G., Dayan, F. E., Wedge, D. E. Activity of activity quinones on *Colletotrichum* spp. J Agric Food Chem 2003; 51:3824-3828.

Moore, R. E. Cyclic peptides and depsipeptides from cyanobacteria: a review. J Ind Microbiol. 1996; 16:134-143.

Mueller, M. J., Brodschelm, W., Spannagl, E., Zenk, M. H. Signaling in the elicitation process is mediated through the octadecanoid pathway leading to jasmonic acid. Proc Natl Acad Sci USA 1993; 90:7490-7494.

Oku, H. In: Plant Pathogenesis and Disease Control. Lewis Publishers. Boca Raton, FL 1994; p.64.

Oliva, A., Meepagala, K. M. Wedge, D. E. Hale, A. L., Harries, D., Aliotta, G., Duke, S.O. Natural Fungicides from *Ruta graveolens* L. leaves, including a new quinolone alkaloid. J Agric Food Chem 2003; 51: 890-896

Pezzuto, J. M. Cancer chemopreventative agents: from plant materials to clinical intervention trials. In: A. D. Kinghorn, M. F. Balandrin (eds): Human Medicinal Agents from Plants. ACS Symposium Series No. 534, American Chemical Society, Washington DC, 1993; 205-215.

Roughi, A. M. Rediscovering natural products. Chemical and Engineering News. October 13, 2003; p. 77-91.

Salah, A. M., Bedir, E., Ngeh, T. J., Kahn, I. A., Wedge, D.E. Antifungal clerodane diterpenes from *Macaranga monandra* (L.) Muell. Et Arg. (Euphorbiaceae). J Agric Food Chem 2003; 51:7607-7610.

Tellez, M.R., Dayan, F.E., Schrader, K.K., Wedge, D.E., Duke, S.O. Composition and some biological activities of the essential oil of *Callicarpa americana* L. J Agric Food Chem 2000; 48:3008-3012.

Vick, B. A., Zimmerman, D. C. Biosynthesis of jasmonic acid by several plant species. Plant Physiol 1984; 75: 458-461.

Vincent, A., Dayan, F.E., Maas, J.L., Wedge, D.E. Detection and isolation of antifungal compounds in strawberry inhibitory to *Colletotrichum fragariae*. Adv Strawberry Res 1999; 18:28-36.

Wedge, D. E., Kuhajek, J. M. A microbioassay for fungicide discovery. SAAS Bull Biochem Biotech 1998; 11: 1-7.

Wedge, D. E., Camper, N. D. Connections between agrochemicals and pharmaceuticals. In: *Biologically Active Natural Products: Agrochemicals and Pharmaceuticals*. Cutler, H.G., Cutler, S. J. eds, ACS Symposium Series, American Chemical Society, Washington DC, 2000 .

Wedge. D. E., Nagle, D.G. A new 2D-TLC bioautography method for the discovery of novel antifungal agents to control plant pathogens. J Nat Prod 2000; 63:1050-1054.

Wedge, David E.; Nagle, Dale G. Preparation of sampangine and its analogs as fungicides. U.S. Pat. Appl. Publ. 12 pp. CODEN: USXXCO US 2004192721 A1 20040930 CAN 141:273006 AN 2004:803936, 2004.

Wedge, D.E., Curry, K.J., Boudreaux, J.E., Pace, P.F., Smith, B.J. *In vitro* sensitivity and resistance profiles of *Botrytis cinerea* isolates from Louisiana strawberry farms. Proceedings of the 5th North American Strawberry Conference, Strawberry Research to 2001; p78-81.

NANCY KOKALIS-BURELLE[1] and RODRIGO RODRÍGUEZ-KÁBANA[2]

ALLELOCHEMICALS AS BIOPESTICIDES FOR MANAGEMENT OF PLANT-PARASITIC NEMATODES

[1]Research Ecologist, USDA, Agricultural Research Service, U.S. Horticultural Research Lab, Fort Pierce, FL, 34945, USA, [2]Distinguished University Professor, Auburn University, Auburn, AL, 36849, USA.
Email:NBurelle@ushrl.ars.usda.gov

Abstract. Many allelopathic compounds in their native or processed forms have potential for development as viable components of plant-parasitic nematode management strategies. Allelochemicals have been identified that possess differing levels of activity against a wide range of plant-parasitic nematodes. In general, these compounds are less toxic to nontarget species, and less persistent in soil than chemical nematicides. Operative mechanisms for plant-parasitic nematode control with allelopathic compounds include nematicidal activity, nematostatic activity, and nematode behavior modification. Allelochemicals are sometimes produced in large quantities in plant material or as agricultural waste, making the use of rotation crops, cover crops, and organic amendments effective means for production and/or distribution of the active compounds. A greater understanding of the effects of soil microbes and environmental conditions on allelopathic compounds is necessary to improve their efficacy for control of parasitic nematodes. Use of allelochemicals for nematode control will require that growers know specifically what types and population levels of nematodes are present in their production fields. Development of improved production and incorporation methods for rotation and green manure crops, and appropriate application methods for processed allelochemical compounds, will also enhance the efficacy and consistency of these compounds for nematode control.

1. INTRODUCTION

For the purpose of this chapter, allelochemicals will be defined as plant metabolites or their products that are released into the environment through volatilization, exudation from roots, leaching from plants or plant residues, and decomposition of residues (Waller, 1985; Putnam and Tang, 1986; Einhellig,1995; Halbrendt, 1996). As such, allelochemicals are classified as biopesticides, which are defined as being derived from natural materials such as plants and microorganisms and include both substances that control pests (biochemical pesticides) and microorganisms that control pests (microbial pesticides). Biochemical and microbial pesticides are considered inherently less toxic than conventional pesticides and generally affect only the target pest and closely related organisms, in contrast to broad spectrum, conventional pesticides that may affect organisms as different as birds, insects, and mammals. Often biopesticides are effective in small quantities and decompose quickly, resulting in lower exposure and fewer pollution problems than conventional pesticides. Also, when used as a

Inderjit and K.G. Mukerji (eds.),
Allelochemicals: Biological Control of Plant Pathogens and Diseases, 15–29.

component of Integrated Pest Management (IPM) programs, biopesticides can decrease the use of conventional chemical pesticides (Halbrendt, 1996).

Increased restrictions on use and phase-out of chemical fumigants such as methyl bromide and other chemical nematicides for control of plant-parasitic nematodes make the discovery of target-specific, environmentally safe, naturally occurring biopesticide compounds that suppress nematode populations or modify nematode behavior increasingly important. For instance, allelochemicals that affect nematode chemotaxis could be invaluable in many different scenarios for nematode control and represents an area of research with both great needs for identification of active compounds, and great possibilities for their use. The production of chemical cues by plants and behavioral responses to these cues by nematodes are critical to successful host location and reproduction in nematodes, which are highly dependent on these chemotactic stimuli during many stages of their life cycles (Bridge, 1996; Huettel, 1986; Yasuhira et al., 1982).

In order to make effective use of nonchemical or biochemical pest control strategies such as allelochemicals for suppression of plant-parasitic nematodes, users need to be familiar with the types and quantity of nematodes present in their soil and determine the feasibility of growing a cover or rotation crop, using an organic amendment, or a formulated biopesticide. Identification of cash crops that produce compounds capable of reducing pathogenic nematode populations and that can be incorporated into existing production regimes is rare (Gardner et al., 1992; Mojtahedi et al., 1993a; Halbrendt, 1996). Some notable exceptions to this include commercial production of *Crotalaria*, mustard, African marigold, asparagus, and sesame in India (Bridge, 1996). The level to which growers will employ the use of cover crops or rotation crops is ultimately dependent on the economic feasibility of this method for nematode control. When determining the economic feasibility of this approach, the additional benefits that cover crops provide should be considered. These benefits include nitrogen fixation, soil stabilization, and weed management (Halbrendt, 1996).

Many plant constituents and metabolites have been investigated for activity against plant-parasitic nematodes. The conditions under which compounds are effective against nematodes vary with the compounds (Ferris and Zheng, 1999; Zasada and Ferris, 2004). These active compounds, or precursors of active compounds, can often be applied to soil as organic amendments, or refined and developed as biopesticide compounds.

Most allelochemicals are short-lived in soil as they are often easily metabolized or hydrolyzed, and may require that plants are actively growing and secreting them into the rhizosphere in order to be effective (Cheng, 1992). In some cases, breakdown products of allelochemicals are the active components against nematodes (Borek et al., 1995). Many such compounds are produced upon decomposition of plant material that can be useful when incorporated into soil as green manures (Brown and Morra, 1997; Mojtahedi et al., 1991; 1993a, b; Prot et al., 1992; Zasada and Ferris, 2004). In each of these instances, soil physical and chemical conditions, microbial populations, and environmental conditions influence the retention, transformation and transport of the allelochemicals (Cheng, 1992). Undoubtedly, these physical, microbiological, and environmental factors contribute to the inconsistency of nematode control observed in research trials on crop rotation and cover crop systems, biofumigation, and biochemical pesticides. It is likely that synthetic chemical nematicides will continue to fill short-term nematode control needs while research continues to improve nonchemical and biopestide-based approaches, which will eventually become the management strategies of choice (Thomas, 1996). In this chapter we will review examples of plants known to release allelochemicals into soil while actively growing, when incorporated into soil as green manures or organic amendments, and the direct application of purified allelochemicals or formulations of biopesticides for plant-parasitic nematode control.

2. ROTATION AND COVER CROPS

There are many examples of crop rotation sequences that passively suppress nematode populations which will not be reviewed here. Examples of active nematode suppression in crop rotation sequences are typically found with plant species that produce and excrete allelopathic compounds. These compounds then affect plant-parasitic nematodes in the rhizosphere either directly or indirectly by altering rhizosphere microbial populations (Halbrendt, 1996). For the purpose of this chapter, allelopathic

Table 1. Rotation/cover crops that actively suppress parasitic nematode populations in soil.

Common Name	*Scientific Name*	*References*
American joint vetch	*Aeschynomene* sp.	Rodriguez-Kabana et al., 1991a
Bahia grass	*Paspalum* spp.	Rodríguez-Kábana et al., 1994b
Castor bean	*Ricinus communis*	Rodriguez-Kabana et al., 1991b
Marigold	*Tagetes* spp.	Tyler, 1938
		Steiner, 1941
		Uhlenbroek and Bijloo, 1958, 1959
		Good et al., 1965
Hairy indigo	*Indigofera hirsuta*	Rodriguez-Kabana et al., 1988b
Horse bean	*Canavalia ensiformis*	Rodriguez-Kabana et al., 1992b
Partridge pea	*Cassia fasciculata*	Rodriguez-Kabana et al., 1991a
		Rodriguez-Kabana et al., 1995
Sesame	*Sesamum indicum*	Rodríguez-Kábana et al., 1994a
Showy clotalaria	*Crotalaria spectabilis*	Rodriguez-Kabana et al., 1992b
Sorghum-sudan grass	*S. bicolor* X *S. vulgare* var. *sudanense*	Kinloch and Dunavin, 1993
Sudan grass	*Sorghum vulgare* var. *sudanense*	Mojtahedi et al., 1993a
Sunn hemp	*Crotalaria juncea*	Sipes and Arakaki, 1997
		Robinson et al., 1998
		McSorley et al., 1999
		Wang et al., 2001
Velvet bean	*Mucuna deeringiana*	Rodriguez-Kabana et al., 1992a
		Weaver et al., 1993
		Taylor and Rodriguez-Kabana, 1999
		Vargas-Ayala and Rodriguez-Kabana, 2001
Vetch	*Vicia* spp.	Minton et al., 1966
		Minton and Donnelly, 1967

rotation and cover crops are considered to be those that are not typically incorporated into soil in order to realize their allelopathic potential. Green manure crops considered to be crops that are most effective for nematode control when incorporated into soil as organic amendments. However, there are many examples of effective rotation crops that are also incorporated into soil as green manure at the end of the growing-season.

Plants belonging to 57 families have been shown to possess nematicidal properties (Bridge, 1996). Crop rotation (crops planted in sequence) and intercropping (crops planted together) have both shown potential for reducing populations of parasitic nematodes. Cover crops are a type of rotation employed as an alternative to leaving land fallow, and are usually not cash crops. Many plants have been used as rotation or cover crops for plant-parasitic nematode control (Table 1).

The active allelopathic compounds differ with respect to each crop and, in many cases, the mode of action for nematode suppression and the effects on nematode chemotaxis have not been established. Several examples of rotation crops effective in actively suppressing populations or controlling parasitic nematodes are reviewed.

2.1. Marigold (Tagetes spp.*)*

Marigold was one of the first plants reported to be highly resistant to root-knot nematodes (*Meloidogyne* spp.) (Tyler, 1938). Soon after resistance was identified in marigold, it was reported that root-knot nematode larvae entered the roots but failed to develop to sexual maturity (Steiner, 1941). It was later discovered that *Tagetes* spp. produce nematotoxic compounds identified as alpha-terthienyls, which directly affect a wide range of nematodes (Uhlenbroek and Bijloo, 1958, 1959; Good et al., 1965). The evaluation of several marigold species as rotation or cover crops has shown *Tagetes minuta* to be highly resistant to both *Meloidogyne incognita* and *M. javanica*, and to perform well as a summer cover crop in southern Florida (McSorley, 1999). The more commonly used *T. erecta* and *T. patula* are often used as winter or spring bedding plants in Florida, and are not as tolerant of high temperatures as *T. minuta* (McSorley and Frederick, 1994; McSorley, 1999). Because *T. minuta* is more heat- tolerant, it has potential for use as a nematode suppressive summer cover crop in Florida during months when fields are often left fallow after fall and spring vegetable crop production cycles. Ploeg (1999) found that *T. erecta*, *T. patula*, *T. signata* and a hybrid *Tagetes* reduced galling by *M. incognita*, *M. javanica*, *M. arenaria*, and *M. hapla* in a tomato rotation study. However, all four *Meloidogyne* species reproduced on *T. signata* 'Tangerine Gem'. This indicates the potential for substantial variability in nematode suppression among similar cultivars of marigold.

There has been little research addressing the extent of allelopathic activity and the mechanism involved in nematode suppression with marigold. Sasanelli and DiVito (1991) found that aqueous leaf and root extracts and root leachates of two *Tagetes* sp. were nematostatic rather than nematicidal to eggs of the golden nematode *Globodera rostochiensis*. Emergence of juveniles that was suppressed or reduced in the presence of extracts and diffusates resumed when solutions were removed indicating that compounds found in solutions were nematostatic rather than nematicidal. More specific information is needed on the range of activity and mechanisms involved in nematode suppression with compounds found in marigold. Additional research in these areas would increase the potential for successful incorporation of marigold or its active allelopathic compounds into production systems for nematode control.

2.2. *Sorghum-sudangrass (Sorghum bicolor X S. sudanense)*

Sorghum has long been recognized for its allelopathic properties toward other plants (Guenzi and McCalla, 1966) and more recently to be suppressive to nematodes (Kinlock and Dunavin, 1993; Mojtahedi et al., 1993a). It was initially hypothesized that the nematicidal compound in sorghum-sudangrass green manures was hydrogen cyanide produced by hydrolysis of dhurrin in leaf tissue (Andewusi, 1990). However, Czarnota et al. (2003) examined root exudate production and composition of seven genetically diverse sorghum accessions, including two sorghum-sudangrass hybrids, and found that although variation occurred in exudate constituents among accessions, the predominant constituent in all exudates was the phenolic compound sorgoleone (Czarnota et al., 2003).

Suppression of parasitic nematodes in the field with sorghum-sudangrass has been inconsistent (MacGuidwin and Layne, 1995). It has been demonstrated that this crop is not effective for reducing populations of lesion nematodes (*Pratylenchus* sp.), an important plant-parasitic genus (MacGuidwin and Layne, 1995). Many studies have confirmed that sorghum-sudangrass is effective in reducing field populations of *Meloidogyne* spp., but that it cannot be recommended if stubby root nematode, *Paratrichodorus minor*, is present and of concern due to the high reproductive rates of *P. minor* on sorghum-sudangrass (McSorley and Gallaher, 1991; McSorley et al., 1994a; McSorley and Dickson, 1995). The selective nature of parasitic nematode control with this allelopathic crop serves as an example of why it is critical to know what nematode species are present in a location when employing cover or rotation crops for key nematode pests. Threshold levels for secondary parasitic nematodes should be established and susceptibility of potential cover crop known when employing these strategies.

2.3. *Sesame (Sesamum indicum)*

Sesame is an important seed and oil crop worldwide. In addition to being a poor host for root-knot nematodes (Rodriguez-Kabana et al., 1988a; 1989), sesame is known to produce several lignin compounds including sesamin and sesamolin which are antioxidants that function as insecticides and insecticidal synergists and are hypothesized to be the active allelopathic compounds in sesame (Bedigian and Harlan, 1986). Research has been conducted in the southern United States during the past 15 years to evaluate the potential for use of sesame as a profitable rotation crop for root-knot nematode control. *Meloidogyne arenaria* populations in peanut were reduced using a two-year sesame rotation followed by one year of peanut production (Rodriguez-Kabana et al., 1994a). Studies by Starr and Black (1995) confirm that sesame can be an effective rotation crop for control of *M. arenaria* and *M. incognita* but that it is not effective in controlling *M. javanica.* More work is needed to determine the identities of the active compounds found in sesame and the nature of their activity with regards to parasitic nematodes.

2.4. Velvetbean (Mucuna deeringiana)

Velvetbean (*Mucuna deeringiana*) is a legume commonly used as a cover crop, green manure, and forage in many subtropical and tropical regions (Allen and Allen, 1981). Velvetbean is known to produce L-DOPA (L-3,4-dihydroxyphenylalanine) which inhibits the growth of other plants and also has insecticidal properties (Fujii et al., 1991; Bell and Janzen, 1971). Vincente and Acosta (1987) demonstrated the antagonistic capabilities of velvetbean against plant-parasitic nematodes when used as a soil amendment. Vargas-Ayela and Rodriguez-Kabana (2001) found an increase in beneficial rhizosphere microorganisms including species of *Paecilomyces* and *Burkholderia* associated with reduction in disease caused by root-knot nematodes in velvetbean- soybean rotations compared to cowpea-soybean rotations. McSorley et al. (1994b) found velvetbean to be very successful in reducing populations of *M. arenaria*, *M. javanica*, and several races of *M. incognita*. Although the active compounds in velvetbean have been identified, little research has been performed to determine how to optimize the potential of this crop for nematode suppression.

2.5. Sunn hemp (Crotalaria juncea)

Sunn hemp is a legume that has many of the desirable qualities of a good rotation or cover crop. In addition to its ability to fix nitrogen, sunn hemp grows rapidly and produces a large amount of biomass, increases soil organic matter, sequesters carbon (Rotar and Joy, 1983), and suppresses many plant-parasitic nematodes (McSorley et al., 1999; Robinson et al., 1998; Wang et al., 2001). Sunn hemp also increases populations of important nematode-antagonistic fungi in soil (Quiroga-Madrigal et al., 1999; Rodríguez-Kábana and Kloepper, 1998; Wang et al., 2001). Populations of microbivorous nematodes, which are involved in soil nutrient cycling, have also been shown to increase in sunn hemp-planted soil (Venette et al., 1997; Wang et al., 2003). Sunn hemp is well suited for use as a cover crop in subtropical regions (McSorley, 1999) and did not increase population levels of *M. javanica* during a test in Hawaii (Sipes and Arakaki, 1997). Sunn hemp 'Tropic Sun' was highly resistant, but not immune, to the *Meloidogyne* spp. tested (McSorley, 1999). It has been suggested that elevated numbers of free-living nematodes, including bacteriovores and fungivores, may improve plant tolerance to parasitic nematodes by increasing plant vigor through more efficient nutrient cycling, or by increasing numbers of predatory and omnivorous nematodes that then feed on plant-parasitic nematodes (Wang et al., 2003).

3. GREEN MANURES AND ORGANIC AMENDMENTS

3.1. Brassicaceae

Glucosinolates are compounds produced by many members of the Brassicaceae plant family including mustard, rape, canola, and cabbage (Fahey et al., 2001). These compounds are β-D-thioglucosides with differing organic side chains, and can be aliphatic, aromatic, or indole forms. Enzymatic degradation of glucosinolates produces several types of compounds including isothyocyanate (ITC), a well known nematicide (Brown and Morra, 1997).

Glucosinolates are produced in various levels throughout the plant and to different degrees among plant species (Fahey et al., 2001). In studies by Potter et al. (1998) the nematicidal potential of *Brassica* spp. leaf and root tissue differed, with leaf tissue being far more toxic to *P. neglectus* than root tissue. However, leaf tissue toxicity was not correlated with either total glucosinolate content or with any individual glucosinolate, while the suppression achieved with the addition of root tissue was highly correlated with levels of 2-phenylethyl glucosinolate within roots (Potter et al., 1998). Zasada and Ferris (2004) performed studies where levels of glucosinolates in various *Brassica* spp. were determined and amendments containing equivalent amounts of active compounds were compared to determine their efficacy in controlling *M. javanica* and *Tylenchulus semipenetrans*. They determined that the degradation of brassicaceous amendments in soil resulted in a series of biological and chemical processes that differed among plant species. It was also determined that consistent nematode suppression could be achieved using brassicaceous amendments if the chemical composition of the organic material was considered and appropriate levels of biomass applied (Zasada and Ferris, 2004).

3.2. Neem (Azadirachta indica)

The neem tree is a tropical evergreen tree native to India whose biologically active properties have been recognized in Asia for centuries (Thakur et al., 1981). During the past 30 years, interest in the extensive variety of biologically active compounds produced by neem has increased. Various parts of the neem plant including leaves, seeds and bark, produce over 40 active diterpenoid, triterpenoid, limonoid, and flavonoid compounds, with a wide range of nematicidal activity (Bhatnagar and Goswami, 1987). The most well known and active compounds produced by neem are the azadirachtins (Thakur et al., 1981).

Nematode control has been achieved following the incorporation of neem products into soil, and their subsequent decomposition and release of nematicidal compounds (Stirling, 1991). Numerous studies have shown that leaves and oilcake of neem are effective in controlling *M. incognita* and increasing growth and yield of vegetable crops when used as a soil amendment (Akhtar, 1998; Bhatnagar and Goswami, 1987). This effect has also been reported to persist in successive crops (Akhtar and Alam, 1991; Akhtar and Mahmood, 1996). Neem based products have also shown nematicidal potential when applied as seed treatments (Akhtar and Mahmood, 1995; 1997) and bare-root treatments (Akhtar and Mahmood, 1993; 1994) leading some to conclude that compounds found in neem may act as inducers of resistance to some nematodes including *M. incognita* and *Rotylenchulus reniformis* (Siddiqui and Alam 1988). Other nematode-suppressive mechanisms of compounds derived from neem include antifeedent, repellent, deterrent, growth disruption, juvenile toxicant, and ovicidal properties (Akhtar, 1998). Testing of neem-based products and development of application techniques for plant-parasitic nematode control is increasing in western countries. There are currently several neem-based pesticides available in the United States for use on certain greenhouse and ornamental crops, with many more available for use in India as insecticides (Akhtar, 2000). A comprehensive review of the nematode suppressive potential of neem products is provided by Akhtar (2000).

4. ALLELOCHEMICALS AS BIOPESTICIDES

Extracts of many plants with anthelminthic or antimicrobial properties have been proven effective in reducing soil populations of plant-parasitic nematodes (Ferris and Zheng, 1999). Experiments which evaluated plant species documented in Chinese traditional medicine to be anthelminthic against plant-parasitic nematodes identified 153 aqueous plant extracts with activity against nematodes (Ferris and Zheng, 1999). Within a 24-hour exposure period, seventy-three of the extracts killed either juveniles of *M. javanica* or mixed developmental stages of *P. vulnus*, or both (Ferris and Zheng, 1999). Plants containing efficacious components included both annuals and perennials, which ranged in type from grasses and herbs to woody trees, representing 46 plant families (Ferris and Zheng, 1999). This research illustrates the tremendous potential for discovery of new active allelopathic compounds for plant-parasitic nematode control. In fact, many of the allelochemicals described below were isolated from crops observed to be nematode suppressive as rotation or green manure crops.

4.1. Glucosinolates

Glucosinolates are compounds primarily found in plants in the family Brassicaceae and are described previously in this chapter with respect to green manure crops. Enzymatic decomposition of glucosinolates in plant tissue occurs rapidly and is primarily attributed to microorganisms in soil (Fenwick et al., 1983). Products of glucosinolate degradation include organic cyanides and isothiocyanates which in addition to being evaluated as active compounds in green manures and organic amendments, have been studied for their their direct toxicity to nematodes as biochemical pesticides. Lazzeri et al. (1993) studied the direct effects of purified glucosinolates on second-stage juveniles of the sugar beet cyst nematode *Heterodera schachtii*. Compounds were isolated from seeds and plant tissue of brassicaceous hosts of the nematode. None of the glucosinolates tested in their native form were nematicidal. However, when exposed to the enzyme myrosinase, several compounds including sinigrin, gluconapin, glucotropeolin, glucode-hydroerucin, and the entire group of glucosinolate compounds extracted from rapeseed exhibited various levels of nematicidal activity depending on concentration and length of exposure (Lazzeri et al., 1993).

Later studies by Borek et al. (1995) investigated the persistence of glucosinolate-derived allyl isothiocyanate and allylnitrile in six soils. They found that the two compounds differed with respect to the temperature, moisture conditions, and soil physical conditions that effected their transformation in soil, and that both compounds dissipated from soil at relatively rapid rates. Donkin et al. (1995) studied the toxicity of glucosinolates and their enzymatic breakdown products to *Caenorhabditis elegans.* They found that allyl isothyocyanate, one of the decomposition products of the glucosinolate sinigrin, was three times more toxic to the nematode *C. elegans* than corresponding glucosinolate itself.

4.2. Benzaldehyde (benzoic aldehyde)

Benzaldehyde occurs in seeds of bitter almond (*Prunus dulcis*). It is found naturally in several cyanogenic glucosides and is used in food and fragrances for its almond-like aroma and flavor (Harborne and Baxter, 1993). The value of benzaldehyde as a fungicide is well established (Flor, 1926), with the nematicidal activity more recently demonstrated. Benzaldehyde reduced populations of *M. incognita* in field microplots with no phytotoxicity to cotton at 0.18 -2.14 ml/kg soil (Bauske et al., 1994). The combination of chitin and benzaldehyde added to peat-based potting mix improved tomato transplant growth and reduced galling by *M. incognita* in greenhouse trials (Kokalis-Burelle et al., 2002). When tested *in vitro* against *M. incognita* eggs, benzaldehyde was 100% effective as an ovicide for this species of root-knot nematode (Kokalis-Burelle et al., 2002).

The direct effect of benzaldehyde on *C. elegans* chemotaxis kinetics was analyzed by Nuttley et al. (2001). An initial attractive response to 100% benzaldehyde was reported, followed by a strong aversion to the chemical. They determined this behavior to be mediated by two genetically separable response pathways. Initially, upon exposure, the attraction response dominates but eventually gives way to a repulsive response. Oka (2001) found that with juveniles of *M. javanica*, immobilization and hatching inhibition *in vitro* were greater with benzaldehyde and furfural than with several other essential oils. Benzaldehyde and furfural also reduced galling on tomato in pot experiments where other aldehydes were not effective (Oka, 2001).

The effects of benzaldehyde combined with several organic amendments on soil microbial populations and plant-parasitic and nonparasitic nematodes were investigated by Chavarría-Carvajal et al. (2001). They found that benzaldehyde combined with most organic amendments reduced damage from parasitic nematodes and selected for predominantly gram-positive rhizosphere bacteria. When benzaldehyde was combined with root-knot nematode egg parasitizing isolates of the fungus *Fusarium solani,* increasing rates of benzaldehyde in soil reduced nematode penetration and infection of the host plant, and resulted in increased parasitism of *M. javanica* females by the fungus (Siddiqui and Shaukat, 2003). However, the increasing rates of benzaldehyde resulted in lower egg parasitism by the fungus (Siddiqui and Shaukat, 2003).

4.3. Furfural

The biological activity of furfural (also known as 2-furancarboxaldehyde, furaldehyde; 2-furanaldehyde, 2-furfuraldehyde, fural, furfurol) has also been recognized for decades (Flor, 1926). Furfural is the aldehyde of pyromucic acid and has properties similar to those of benzaldehyde. Furfural is a derivative of furan and is prepared commercially by dehydration of pentose sugars obtained from sugarcane, cornstalks and corncobs, husks of oat and peanut, and other agricultural waste products (Harborne and Baxter, 1993). Commercially available products for disease and nematode control are available in several countries including the United States. These products include Multiguard FFA (furfural (75%) + allyl isothiocynate (25%), Harborchem, Cranford, New Jersey), and Multiguard Protect (furfural (50%) + metham sodium (50%), Harborchem, Cranford, New Jersey).

Rajendran et al. (2003) reported improved plant growth and reductions in soil populations of *M. arenaria* and *R. reniformis* in groundnut with formulations of furfural compared to an untreated control, with no effect on free-living nematodes in soil. Spaull (1997) also found that free-living nematodes were relatively unaffected by furfural application while parasitic nematodes differed in their susceptibility with species of *Paratichodorus* and *Xiphinema* being more susceptible than *Helicotylenchus* and *Tylenchorhynchus*. Sipes (1997) found furfural to be as effective as 1,3-D for reduction of preplant soil nematode populations in pineapple production. Rodríguez-Kábana et al. (1993) found that furfural was an effective nematicide against *M. arenaria*, *M. incognita*, *Heterodera glycines*, and *Pratylenchus* spp. on squash, okra, and soybean. Furfural reduced populations of *M. incognita* in field microplots with no phytotoxicity to cotton at 0.18 -2.14 ml/kg soil (Bauske et al., 1994).

4.4. Thymol

Thymol (isopropyl-*m*-cresol) is a volatile, phenolic monoterpene produced by several plants including thyme (*Thymus vulgaris* L.) (Baerheim Svendsen and Scheffer, 1985). Thymol has well-known antiseptic, antifungal, and anthelminthic properties (Wilson et al., 1977) and is also used for food and fragrance applications (Bauer et al., 1990).

Research by Soler-Serratosa et al. (1996) using combinations of thymol and benzaldehyde for root-knot and cyst nematode control on soybeans showed that both compounds exhibited wide spectrum nematicidal activity with *Meloidogyne* spp. and Dorylaimid nematodes being more sensitive than cyst nematode and nonparasitic nematodes (Soler-Serratosa et al., 1996). In addition to the direct toxicity of these compounds to nematodes, it was hyopothesized that stimulation of beneficial microflora by the compounds or their products, altered host response, and a deleterious physico-chemical environment may all contribute to reduced gall formation (Soler-Seratosa et al., 1996).

4.5. Citral

Citral, is the aldehyde of geraniol, and occurs in the volatile oils of lemon grass, lemon, orange, limetta, and pimento (Harborne and Baxter, 1993). The flavor of lemon oil is largely due to its citral content, and the pure aldehyde may be used to increase the flavoring power of commercial samples of that oil (Harborne and Baxter, 1993). In research trials evaluating the nematicidal potential of citral compared to other allelopathic chemicals, citral was less nematicidal against *M. incognita* juveniles, and more phytotoxic to tomato than benzaldehyde *in vitro,* and when added to a peat-based potting mix (Kokalis-Burelle et al., 2002). When tested *in vitro* against root-knot nematode eggs, citral reduced egg viability by 80%, but also decreased tomato seed germination and growth in greenhouse trials (Kokalis-Burelle et al., 2002). However, when evaluated in soil, citral reduced populations of root-knot nematode juveniles, galling on roots, and increased cotton growth when applied at 0.1 -0.5 ml/kg soil in the greenhouse, and at 0.18 -2.14 ml/kg soil in field microplots with no phytotoxicity to cotton (Bauske et al., 1994). This difference in phytotoxicity may be due to the difference in host plants tested or to the presence of microorganisms in soil compared to the relatively uncontaminated conditions that occur *in vitro* and in potting media.

5. CONCLUSIONS

The use of allelopathic compounds in either their native, degraded, or processed forms for plant-parasitic nematode management is receiving increased attention as agricultural producers face increasing restrictions on chemical biocides and nematicides. Allelochemicals are classified as biochemical pesticides and have been shown to possess various levels of activity against a wide range of plant-parasitic nematodes, while exhibiting reduced toxicity to nontarget species and reduced persistence in soil. The fact that some allelochemicals can be produced in large quantities in plant material and incorporated into soil as green manures or organic amendments increases their potential for use as components of nematode management strategies. The wide array of plant families that produce nematicidal or nematistatic compounds provides almost unlimited research opportunities for discovery of novel compounds. Determination of the mechanisms responsible for nematode suppression with allelopathic compounds also represents a research area with exciting possibilities. In order to improve the nematicidal and nematistatic efficacy achieved with allelopathic compounds, a greater understanding of the effects of soil microbiology, soil properties and environmental conditions on the active compounds is necessary. Currently available application methods need to be refined and new and improved methods developed to enhance the performance of allelochemicals for nematode control. Further investigation is also needed to develop compatible companion applications of biochemical pesticides and biological control agents, to determine the effects of allelochemicals on nematode nervous systems, and how they act as nematode attractants and/or repellents. This research area has enormous potential for discovery of new compounds, for elucidating the direct effects of known compounds on nematode behavior, and for the development of new products to serve as components in multifaceted approaches to nematode management in the post methyl bromide era.

6. REFERENCES

Adewusi, S.R.A. Turnover of dhurrin in green sorghum seedlings. Plant Physiol 1990; 94:1219-1224.

Akhtar, M. Biological control of plant-parasitic nematodes by neem products in agricultural soils. Appl Soil Ecol 1998; 7:219-223.

Akhtar, M. Nematicidal potential of the neem tree *Azadirachta indica* (A. Juss). Integ Pest Manag Rev 2000; 5:57-66.

Akhtar, M., Alam, M.M. Integrated control of plant-parasitic nematodes on potato with organic amendments, nematicide, and mixed cropping with mustard. Nematol Medit 1991; 19:169-171.

Akhtar, M., Mahmood, I. Control of plant-parasitic nematodes with 'Nimin' and some plant oils by bare root-dip treatment. Nematol Medit 1993; 21:89-92.

Akhtar, M., Mahmood, I. Control of root-knot nematode by bare-root dip in undecomposed and decomposed extracts of neem cake and leaf. Nematol Medit 1994; 22:55-57.

Akhtar, M., Mahmood, I. Control of root-knot nematode *Meloidogyne incognita* in tomato plant by seed-coating with Achook and neem oil. Integ Pest Control 1995; 37: 86-87

Akhtar, M., Mahmood, I. Organic soil amendments in relation to nematode management with particular reference to India. Integ Pest Manag Reviews 1996; 1:201-215.

Akhtar, M., Mahmood, I. Control of root-knot nematode *Meloidogyne incognita* in tomato plants by seed coating with Suneem and neem oil. J Pesticide Sci 1997; 22:37-38.

Allen, O.N., Allen, E.K. *The Leguminosae, a source book of characteristics, uses and nodulation.* The University of Wisconsin Press, Madison, WI, 1981; p. 446.

Baerheim Svendsen, A., Scheffer, J.J.C. *Essential oils and aromatic plants.* Martinus Nijhoff/Dr. W. Junk Publishers, Dordrecht, The Netherlands, 1985.

Bauer, K., Garbe, D., Surburg, H. *Common fragrance and flavor materials. Preparation, properties, and Uses.* VCH Verlagsgesellschraft, Weinheim, Germany, 1990.

Bauske, E.M., Rodríguez-Kábana, R., Estaun, V., Kloepper, J.W., Robertson, D.G., Weaver, C.F., King, P.S. Management of *Meloidogyne incognita* on cotton by use of botanical aromatic compounds. Nematropica 1994; 24:143-150.

Bedigian, D., Harlan, J.R. Evidence for cultivation of sesame in the ancient world. Econ Bot 1986; 40:137-154.

Bell, E.A., Janzen, D.H. Medical and ecological considerations of L-Dopa and 5 HTP in seeds. Nature 1971; 229:136-137.

Bhatnagar, A., Goswami, B.K. Comparative efficacy of neem and groundnut oilcakes with aldicarb against *Meloidogyne incognita* in tomato. Rev Nematol 1987; 10:461-471.

Borek, V., Morra, M.J., Brown, P.D., McCaffrey, J.P. Transformation of the glucosinolate-derived allelochemicals allyl isothiocyanate and allylnitrile in soil. J of Agricult and Food Chem 1995; 43:1935-1940.

Bridge, J. Nematode management in sustainable and subsistence agriculture. Ann Rev Phytopathol 1996; 34:201-225.

Brown, P.D., Morra, M.J. Control of soilborne plant pests using glucosinolate-containing plants. Adv in Agron 1997; 61:167-231.

Chavarría-Carvajal, J.A., Rodríguez-Kábana, R., Kloepper, J.W., Morgan-Jones, G. Changes in populations of microorganisms associated with organic amendments and benzaldehyde to control plant parasitic nematodes. Nematropica 2001; 31:165-180.

Cheng, H.H. A conceptual framework for assessing allelochemicals in the soil environment. In: *Allelopathy, basic and applied aspects.* S. J. H. Rizvi and V. Rizvi, eds. Chapman and Hall, NY, 1992; pp. 21-29.

Czarnota, M.A., Rimando, A.M., Weston, L.A. Evaluation of root exudates of seven sorghum accessions. J Chem Ecol 2003; 29: 2073-2083.

Donkin, S.G., Eiteman, M.A., Williams, P.L. Toxicity of glucosinolates and their enzymatic decomposition products to *Caenorhabditis elegans.* J Nematol 1995; 27:258-262.

Einhellig, F.A. *Allelopathy:* Current status and future goals. In: *Allelopathy, organisms, processes, and applications*. Inderjit, K.M.M. Dakshini, F.A. Einhellig. eds. Washington, DC: American Chemical Society. 1995; pp. 1-24.

Fahey, J.W., Zalemann, A.T., Talalay, P. The chemical diversity and distribution of glucosinolates and isothiocyanates among plants. Phytochemistry 2001; 56: 5-51.

Fenwick, G.R., Heaney, R.K., Mullin, W.J. Glucosinolates and their breakdown products in food and plants. Crit Rev Food Sci Nutrit 1983; 18:123-201.

Ferris H., Zheng, L. Plant sources of Chinese herbal remedies: Effects on *Pratylenchus vulnus* and *Meloidogyne javanica*. J Nematol 1999; 31:241-263.

Flor, H.H. Fungicidal activity of furfural. Iowa State College J Science 1926; 1:199-227.

Fujii, Y., Shibuya, T., Yasuda, T. L-3,4-Dihydroxyphenylalanine as an allelochemical candidate from *Mucuna puriens* (L.) DC. var. *utilis*. Agricult Biolog Chem 1991; 55:617-618.

Gardner, J., Casswell-Chen, E.P., Westerdahl, B., Anderson, C., Lanini, T. The influence of *Raphanus sativus*, *Sinapis alba*, and *Phacelia tanacetifolia* on California populations of *Heterodera schactii*. J Nematol 1992; 24:591.

Good, J.M., Minton, N.A., Jaworski, C.A. Relative susceptibility of selected cover crops and coastal bermudagrass to plant nematodes. Phytopathol 1965; 55:1026-1030.

http://www.epa.gov/pesticides/biopesticides/

Guenzi, W.D., McCalla, I.N. Phenolic acids in oats, wheat, sorghum, and corn residues and their phytotoxicity. Agronomy J 1966; 58: 303-304.

Halbrendt, J.M. Allelopathy in the management of plant parasitic nematodes. J Nematol 1996; 28:8-14.

Harborne, J.B., Baxter, H. *Phytochemical Dictionary. A handbook of bioactive compounds from plants.* Taylor and Francis, London, 1993.

Huettel, R.N. Chemical Communicators in Nematodes. J Nematol 1986; 18:3-8.

Kinloch, R.A., Dunivan, L.S. Summer cropping effects on the abundance of *Meloidogyne arenaria* race 2 and subsequent soybean yield. Suppl J Nematol 1993; 25:806-808.

Kokalis-Burelle, N., Martinez-Ochoa, N., Rodríguez-Kabana, R., Kloepper, J.W. Development of multi-component transplant mixes for suppression of *Meloidogyne incognita* on tomato (*Lycopersicon esculentum*). J Nematol 2002; 34:362-369.

Lazzeri, L., Tacconi, R., Palmieri, S. *In vitro* activity of some glucosinolates and their reaction products toward a population of the nematode *Heterodera schachtii*. J Agricult Food Chem 1993; 41:825-829.

MacGuidwin, A.E., Layne, T.L. Response of nematode communities to sudangrass and sorghum-sudangrass hybrids grown as green manure crops. Suppl J Nematol 1995; 27:609-616.

McSorley, R. Host suitability of potential cover crops for root-knot nematodes. Suppl J Nematol 1999; 31:619-623.

McSorley, R., Dickson, D.W. Effect of tropical rotation crops on *Meloidogyne incognita* and other plant-parasitic nematodes. Suppl J Nematol 1995; 27:535-544.

McSorley, R., Frederick, J.J. Response of some common annual bedding plants to three species of *Meloidogyne*. Suppl J Nematol 1994; 26:773–777.

McSorley, R., Gallaher, R. Nematode changes and forage yields of six corn and sorghum cultivars. J Nematol 1991; 23:673-677.

McSorley, R., Dickson, D.W., de Brito, J.A. Host status of selected tropical rotation crops to four populations of root-knot nematodes. Nematropica 1994a; 24:45-53.

McSorley, R., Dickson, D.W., de Brito, J.A., Hochmuth, R.C. Tropical Rotation Crops Influence Nematode Densities and Vegetable Yields. J Nematol 1994b; 26:308-314.

McSorley, R., Ozores-Hampton, M., Stansly, P.A., Conner, J.M. Nematode management, soil fertility, and yield in organic vegetable production. Nematropica 1999; 29:205–213.

Minton, N.A., Donnelly, E.D. Additional *Vicia* species resistant to root-knot nematodes. Plant Dis Report 1967; 51:614-616.

Minton, N.A., Donnelly, E.D., Shepherd, R.L. Reaction of *Vicia* species and F hybrids from *V. sativa* x *V. angustifolia* to five root knot nematode species. Phytopathol 1966; 56:102^{5}-107.

Mojtahedi, H., Santo, G.S., Hang, A.N., Wilson, J.W. Suppression of root-knot nematodes with selected rapeseed cultivars as green manures. J Nematol 1991; 23:170-174.

Mojtahedi, H., Santo, G.S., Ingham, R.E. Suppression of Meloidogyne chitwoodi with sudangrass cultivars as green manure. J Nematol 1993a; 25: 303-311.

Mojtahedi, H., Santo, G.S., Wilson, J.H., Hang, A.N. Managing Meloidogyne chitwoodi on potato with rapeseed as green manure. Plant Dis 1993b; 77:42-46.

Nuttley, W.M., Harbinder, S., van der Kooy, D. Regulation of distinct attractive and aversive mechanisms mediating benzaldehyde chemotaxis in *Caenorhabditis elegans*. Learning Memory 2001; 8:170-181.

Oka, Y. Nematicidal activity of essential oil components against the root-knot nematode *Meloidogyne javanica*. Nematology 2001; 3: 159-164.

Ploeg, A.T. Greenhouse studies on the effect of marigolds (*Tagetes* spp.) on four *Meloidogyne* species. J Nematol 1999; 31:62-69.

Potter, M.J., Davies, K., Rathjen, A.J. Suppressive impact of glucosinolates in Brassica vegetative tissues on root lesion nematode *Pratylenchus neglectus*. J Chem Ecol 1998; 24:67-80.

Prot, J.C., Soriano, I.R.S., Mantias, D.M., Savary, S. Use of green manure crops in control of *Hirschmanniella mucronata* and *H. oryzae* in irrigated rice. J Nematol 1992; 24:127-132.

Putnam, A.R., Tang, C. Allelopathy: State of the science. *In The science of allelopathy*. A.R. Putnam & C. Tang, eds. New York: John Wiley, 1986, pp. 1-19.

Quiroga-Madrigal, R., Rodríquez-Kábana, R., Robertson, D.G., Weaver, C.F., King, P.S. Nematode populations and enzymatic activity in rhizospheres of tropical legumes in Auburn, Alabama. Nematropica 1999; 29:129 (Abstr.).

Rajendran, G., Ramakrishnan, S., Subramanian, S. Cropguard – a botanical nematicide for the management of *Meloidogyne arenaria* and *Rotylenchulus reniformis* in groundnut. In: *Proceedings of National Symposium on Biodiversity and Management of Nematodes in Cropping Systems for Sustainable Agriculture*, Jaipur, India, 2003, pp. 122-125.

Reeves, D.W., Mansoer, Z., Wood, C.W. Suitability of sunn hemp as an alternative legume cover crop. In: Proceedings of the new technology and conservation tillage, vol. 96-07. Jackson, TN: University of Tennessee Agricultural Experiment Station, 1996, pp. 125-130.

Robinson, A.F., Cook, C.G., Bridges, A.C. Comparative reproduction of *Rotylenchulus reniformis* and *Meloidogyne incognita* race 3 on kenaf and sunn hemp grown in rotation with cotton. Nematropica 1998; 28:143 (Abstr.).

Rodríguez-Kábana, R., King, P.S., Robertson, D.G, Weaver, C.F. Potential of crops uncommon to Alabama for management of root-knot and soybean cyst nematodes. Suppl J Nematol 1988a; 20:116-120.

Rodríguez-Kábana, R., Robertson, D.G., Wells, L., Young, R.W. Hairy indigo for the management of *Meloidogyne arenaria* in peanut. Nematropica 1988b; 18:137-142.

Rodríguez-Kábana, R., Robertson, D.G., Wells, L., King, P.S., Weaver, C.F. Crops uncommon to Alabama for the management of *Meloidogyne arenaria* in peanut. Suppl J Nematol 1989; 21:712-716.

Rodríguez-Kábana, R., Robertson, D.G., King, P.S., Wells, L. American jointvetch and partridge pea for the management of *Meloidogyne arenaria* in peanut. Nematropica 1991a; 21:97-103.

Rodríguez-Kábana, R., Robertson, D.G., Weaver, C.F., Wells, L. Rotations of bahiagrass and castorbean with peanut for the management of *Meloidogyne arenaria*. Suppl J Nematol 1991b; 23:658-661.

Rodríguez-Kábana, R., Kloepper, J.W., Robertson, D.G., Wells, L.W. Velvetbean for the management of root-knot and southern blight in peanut. Nematropica 1992a; 22:75-80.

Rodríguez-Kábana, R., Pinochet, J., Robertson, D.G., Weaver, C.F., King, P.S. Horsebean (*Canavalia ensiformis*) and crotalaria (*Crotalaria spectabilis*) for the management of *Meloidogyne* spp. Nematropica 1992b; 22:29-35.

Rodríguez-Kábana, R., Kloepper, J.W., Weaver, C.F., Robertson, D.G. Control of plant parasitic nematodes with furfural – a naturally occurring fumigant. Nematropica 1993; 26:63-73.

Rodríguez-Kábana, R., Kokalis-Burelle, N., Robertson, D.G., Wells, L.W. Evaluation of sesame for control of *Meloidogyne arenaria* and *Sclerotium rolfsii* in peanut. Nematropica, 1994a; 24:55-61.

Rodríguez-Kábana, R., Kokalis-Burelle, N., Robertson, D.G., Wells, L., King, P.S. Rotations with Coastal Bermudagrass, Cotton, and Bahiagrass for the Management of Root-Knot Nematode and Southern Blight in Peanut. Suppl J Nematol 1994b; 26:665-668.

Rodríguez-Kábana, R., Kokalis-Burelle, N., Robertson, D.G., Weaver, C.F., Wells, L. Effects of partridge pea-peanut rotations on populations of *Meloidogyne arenaria*, incidence of *Sclerotium rolfsii*, and yield of peanut. Nematropica 1995; 25:27-34.

Rodríguez-Kábana, R., Kloepper, J.W. Cropping systems and the enhancement of microbial activities antagonistic to nematodes. Nematropica 1998; 28:144 (Abstr.).

Rotar, P.P., Joy, R.J. 'Tropic Sun' sunn hemp. *Crotalaria juncea* L. Research Extension Series 036, University of Hawaii, Honolulu, 1983.

Sasanelli, N., M. DiVito. The effect of *Tagetes* spp. extracts on the hatching of an Italian population of *Globodera rostochiensis*. Nematol Medit 1991; 19:135-137.

Siddiqui, M.A., Alam, M.M. Control of root-knot and reniform nematodes by bare-root dip in leaf extract or margosa and persian lilac. *Zeitsch.* Pflazenkh Pflazen Sch 1988; 95:138-142.

Siddiqui, I.A., Shaukat, S.S. Factors influencing the effectiveness of non-pathogenic *Fusarium solani* strain Fs5 in the suppression of root-knot nematode in tomato. Phytopathol. Mediterran, 2003; 42:17-26.

Sipes, B.S. Pre-plant and post-plant pesticides for nematode control in pineapple. Acta Horticult 1997; 425:457-464.

Sipes, B.S., Arakaki, A.S. Root-knot nematode management in dryland taro with tropical cover crops. Suppl J Nematol 1997; 29:721–724.

Soler-Serratosa, A., Kokalis-Burelle, N., Rodriguez-Kabana, R., Weaver, C.F., King, P.S. Allelochemicals for control of plant-parasitic nematodes. 1. *In vivo* nematicidal efficacy of thymol and thymol/benzaldehyde combinations. Nematropica 1996; 26: 57-71.

Spaull, V.W. On the use of furfural to control nematodes in sugarcane. *Proceedings of the Annual Congress – South African Sugar Technologists Association* 1997; 71:96.

Starr, J.L., Black, M.C. Reproduction of *Meloidogyne arenaria, M. incognita*, and *M. javanica* on sesame. Suppl J Nematol 1995; 27: 624-627.

Steiner, G. Proc. Proc Biol Soc Washington 1941; 54:31-34.

Stirling, G.R. *Biological Control of Plant-Parasitic Nematodes. Progress, Problems, and Prospects.* CAB International, Wallingford, UK, 1991; pp. 212.

Taylor, C.R., Rodriguez-Kabana, R. Population dynamics and crop yield effects of nematodes and white mold in peanut, cotton and velvetbean. Agricult *Syst* 1999; 59:177-191.

Thakur, R.S., Singh, S.B., Goswami, A. *Azadirachta indica* A. Juss. – A Review. CROMAP 1981; 3:135-140.

Thomas, W.B. Methyl bromide: Effective pest management tool and environmental threat. J Nematology 1996; 28:286-289.

Tyler, J. Proceeding of the root-knot nematode conference, Atlanta, GA, Feb. 4, 1938. Plant Dis Report Suppl 1938; 109:133-151.

Uhlenbroek, J.H., Bijloo, J.D. Investigations on nematicides. I. Isolation and structure of a nematicidal principle occurring in Tagetes roots. Recuceil des Travaux Chimiques des Pays-Bas 1958; 77:1004-1008.

Uhlenbroek, J.H., Bijloo, J.D. Investigations on nematicides. II. Structure of a second nematicidal principle isolated from Tagetes roots. Recuceil des Travaux Chimiques des Pays-Bas 1959; 78:382-390.

Vargas-Ayala, R., Rodriguez-Kabana, R. Bioremediative management of soybean nematode population densities in crop rotations with velvetbean, cowpea, and winter crops. Nematropica 2001; 31:37-46.

Venette, R.C., Monstafa, F.A.M., Ferris, H. Trophic interactions between bacterial-feeding nematodes in plant rhizospheres and the nematophagus fungus *Hirsutella rhossiliensis* to suppress *Heterodera schachtii.* Plant Soil 1997; 191:213–223.

Vincente, N.E., Acosta, N. Effects of *Mucuna deeringiana* on *Meloidogyne incognita.* Nematropica 1987; 17:99-102.

Waller, G.R. *Allelochemicals: Role in agriculture and forestry*. Washington, DC: American Chemical Society, 1985.

Wang, K.H., Sipes, B.S., Schmitt, D.P. Suppression of *Rotylenchulus reniformis* by *Crotalaria juncea, Brassica napus,* and *Target erecta.* Nematropica 2001; 31:237–251.

Wang, K.H., McSorley, R., Gallaher, R.N. Effect of *Crotalaria juncea* amendment on nematode communities in soil with different agricultural histories J Nematol 2003; 35:294–301.

Weaver, D.B., Rodríguez-Kábana, R., Carden, E.L. Velvetbean in rotation with soybean for management of *Heterodera glycines* and *Meloidogyne arenaria.* Suppl J Nematol 1993; 25:809-813.

Wilson, C.O., Grisvold, O., Doerge, R.F. *Textbook of organic medicinal and pharmaceutical chemistry*. J.B. Lippincott Co., Philadelphia, PA, USA, 1977.

Yasuhira, T., Nagase, A., Kuwahara, Y., Sugawara, R. Aggregation of *Bursaphelenchus lignicolus* (Nematoda: Aphelenchoididae) to several compounds containing oleyl group. Appl Entomol & Zool 1982; 17:46-51.

Zasada, I.A., Ferris, H. Nematode suppression with brassicaceous amendments: application based upon glucosinolate profiles. Soil Biol Biochem 2004; 36:1017-1024.

ANA LUISA ANAYA

ALLELOPATHIC ORGANISMS AND MOLECULES: PROMISING BIOREGULATORS FOR THE CONTROL OF PLANT DISEASES, WEEDS, AND OTHER PESTS

Laboratorio de Alelopatía, Instituto de Ecología, Universidad Nacional Autónoma de México. Circuito Exterior, Ciudad Universitaria. 04510 México, D.F.
Email: alanaya@miranda.ecologia.unam.mx

Abstract. Increasing attention has been given to the role and potential of allelopathy as a management strategy for crop protection against weeds and other pests. Incorporating allelopathy into natural and agricultural management systems may reduce the use of herbicides, insecticides, and other pesticides, reducing environment/soil pollution and diminish autotoxicity hazards. There is a great demand for compounds with selective toxicity that can be readily degraded by either the plant or by the soil microorganisms. In addition, plant, microorganisms, other soil organisms and insects can produce allelochemicals which provide new strategies for maintaining and increasing agricultural production in the future.

1. INTRODUCTION
AGRICULTURE AND PEST MANAGEMENT SYSTEMS

Agriculture is one of the world's largest industries. On a worldwide basis, more people are in some way involved in agriculture than in all other occupations combined. Agriculture is also United States' largest industry. This country produces more food than any other nation in the world and is the world's largest exporter of agricultural products. According to the 2002 survey from the United States Department of Agriculture (USDA) National Agricultural Statistics Service, there are more than 941 million acres used for farming in the U.S. with the average farm size being 436 acres.

Agriculture in the United States is becoming more productive. In 1935, there were 6.8 million farms in the United States, and the average farmer produced enough food to feed 20 people. In 2002, the number of farms was estimated to be 2.16 million, and the average U.S. farmer produced enough food each year to feed more than 100 people. In addition to providing an abundant food supply for domestic markets, crops from nearly one-third of U.S. farm acreage are exported to overseas customers (IFIC 2004).

Pest problems and their management vary widely throughout the world, based on economical resources, cultural techniques, climate, soil types, and many other conditions. As a result, chemical pest control has won a central place in modern

Inderjit and K.G. Mukerji (eds.),
Allelochemicals: Biological Control of Plant Pathogens and Diseases, 31–78.

agriculture, contributing to the dramatic increase in crop yields achieved in recent decades for most major field, fruit, and vegetable crops.

Farmers must contend with approximately 80,000 plant diseases, 30,000 species of weeds, 1,000 species of nematodes, and more than 10,000 species of insects. Today, national and international agricultural organizations estimate that as much as 45 percent of the world's crops continue to be lost to these types of hazards. In the United States alone, about $20 billion worth of crops (one-tenth of production) are lost each year (IFIC 2004).

The word "pesticides" refers to a broad class of crop protection chemicals, including four major groups: insecticides, rodenticides, herbicides, and fungicides. All pesticides must be toxic, or poisonous, to kill the pests they are intended to control. Because pesticides are toxic, they are potentially hazardous to humans and animals. Therefore, people who use pesticides or regularly come in contact with them must understand the relative toxicity and potential health effects of the products they use (Pesticide Education Program. Penn State's College of Agricultural Sciences, 2004). Some pesticides (administered at extremely high dosages) have been found to cause cancer in laboratory animals (Agri 21FAO).

Weeds represent the most serious threat to the sustainability and profitability of agricultural production around the world considering these facts:

- Weeds represent the most economically serious pest complex reducing world food and fiber production.
- Controlling weeds costs the United States economy more than $15 billion annually, surpassing the combined cost of controlling disease and insect pests. U.S. herbicide expenditures account for 85% of all pesticide purchases and 62% of the total amount of active ingredient applied.
- Nearly all (greater than 95%) of the corn and soybean acreage in the U.S. receives herbicide applications. In developing countries, the cost of weed control in terms of labor and loss of yields is even greater, proportionally, than in the U.S. Worldwide herbicide purchases ($16.9 billion in 1997) constitute 58% of all pesticides bought and 53% of the amount of active ingredient applied (1 billion kg a.i.) worldwide.

Crop yield losses from weed competition can be substantial. The degree of loss depends on, crop and weed species present; timing and duration of competitive interactions; and resource availability (Agri 21 FAO*, IFIC 2004**).

Worldwide, the competitive effect of weeds causes a 10% loss in agricultural production. Yield losses in rice and other grass crops in West Africa have been reported to range from 28-100% if weeds such as witchweed (*Striga hermonthica*) –a parasitic weed– are not controlled; the greatest reductions occur on nutrient-poor soils. Left unchecked, weeds cause dramatic reductions in food production that eventually can

---

** Agri 21 FAO: Agriculture 21. IPM and Weed Management. Food and Agriculture Organization of the United Nations (FAO). Agriculture Department. 2004. http://www.fao.org/WAICENT/FAOINFO/AGRICULT/Default.htm.*

*** IFIC 2004 : International Food Information Council. 2004. Agriculture & Food Production. Background on Agriculture & Food Production. http://www.ific.org/food/agriculture/index.cfm.*

destabilize economic and social systems. Hence, there is an urgent need to develop and refine weed management strategies in crop production that are effective, safe, and economically viable.

Regardless of cropping system or agricultural region of the world, effective weed management strategies are needed continually to maintain crop yields and crop quality as well as to reduce the negative impact of weeds in future years. This ongoing quest to optimize weed management strategies is largely because of the ability of agricultural weeds to adapt to many of our most important crop production systems (Agri 21 FAO, IFIC 2004).

Integrated Pest Management (IPM) is a system that works in partnership with nature to produce foods efficiently (Upadhyay et al., 1996). The concept began in U.S.A. in the 1950s, and recently resurfaced in popularity. Although many definitions of IPM have been advanced, two elements are critical:

- using multiple control tactics;
- integrating a knowledge of pest biology into the management system.

IPM involves the carefully managed use of an array of pest control techniques including biological, cultural, and appropriate chemical methods to achieve the best results with the least disruption of the environment. With IPM, growers are adopting less chemically intensive methods of farming, which may include pest-resistant plant varieties, adjustments in planting times, low tillage, and other non-chemical techniques. The objectives of IPM are:

- Appreciate the importance of controlling weeds within an integrated pest management program.
- Understand the key biological differences that make IPM for weeds more difficult than IPM for insects.
- Become familiar with integrated weed management (IWM) strategies.

The United States Department of Agriculture (USDA) proposed national standards for organic farming and handling in 1997.The federal regulations for organic standards were finalized in 2001 and began full implementation in 2002. Generally, organic food is produced by farmers who emphasize the use of renewable resources and the conservation of soil and water to enhance environmental quality for future generations (IFIC 2004, Agri 21 FAO).

Barberi (2002), assessed that despite the serious threat which weeds offer to organic crop production, relatively little attention has so far been paid to research on weed management in organic agriculture, an issue that is often approached from a reductionist perspective. Compared with conventional agriculture, in organic agriculture the effects of cultural practices (e.g. fertilization and direct weed control) on crop-weed interactions usually manifest themselves more slowly. Weed management should be tackled in an extended time domain and needs deep integration with the other cultural practices, aiming to optimize the whole cropping system rather than weed control *per se*. Many organic farmers are aware that successful weed management implies putting into practice the concept of maximum diversification of their cropping system. However, this task is often difficult to achieve, because practical solutions have to pass through local filters (soil and climate conditions, availability of and accessibility to external inputs -seeds, crop cultivars, machinery, etc.) and socio-economic constraints (market, tenure status, attitude towards entrepreneurial risk, etc.). The use of cover crops and organic amendments, via the promotion of diversity in insect, fungal, bacterial or mychorrhyzal communities, may alter antagonist or competitive effects to the benefit of crops and to the detriment of weeds. Once factors driving these effects are better understood, it might be possible to use this knowledge to improve organic weed management systems locally. It would also be helpful to find indicators of "functional biodiversity", where weed species abundance is assessed on the role that they have in the agroecosystem (e.g. strong /weak competitors, promoters of the presence of beneficial arthropods, etc.). Management of allelopathy is another potential tool in the arsenal of the organic farmer (Barberi, 2002). In the United States, the rate of increase of organic growers was estimated at 12% in 2000. However, many producers are reluctant to undertake the organic transition because of uncertainty of how organic production will affect weed population dynamics and management. The organic transition has a profound impact on the agroecosystem. Changes in soil physical and chemical properties during the transition often impact indirectly insect, disease, and weed dynamics. Greater weed species richness is usually found in organic farms but total weed density and biomass are often smaller under the organic system compared with the conventional system. The improved weed suppression of organic agriculture is probably the result of combined effects of several factors including weed seed predation by soil microorganisms, seedling predation by phytophagus insects, and the physical and allelopathic effects of cover crops (Ngouajio and McGiffen, 2002).

2. ALLELOPATHY

Increasing attention has been given to the role and potential of allelopathy as a management strategy for crop protection against weeds and other pests. Incorporating allelopathy into natural and agricultural management systems may reduce the use of herbicides, fungicides, nematicides, and insecticides, cause less pollution and diminish autotoxicity hazards. There is a great demand for compounds with selective toxicity that can be readily degraded by either the plant or by the soil microorganisms. Plant, microorganisms, other soil organisms and insects can produce allelochemicals which provide new strategies for maintaining and increasing agricultural production in the future. Compounds with allelopathic activity may provide novel chemistry for the synthesis of herbicides, insecticides, nematicides, and fungicides that are not based on the persistent petroleum derived compounds which are such a public health concern (Waller and Chou, 1989; Waller, 1999).

Several crops (some of which can be used as cover crops) have been proved to release allelopathic compounds in the soil (Jimenez-Osornio and Gliessman, 1987; Blum et al., 1997; Inderjit and Keating, 1999; Anaya, 1999), many of which have been chemically characterized (Pereda-Miranda et al., 1996; Inderjit, 1996; Seigler, 1996; Waller et al., 1999). The idea of exploiting these compounds as natural herbicides is therefore very attractive (Putnam, 1988; Weston, 1996; Duke et al., 2000).

However, the large majority of the studies carried out on this topic have referred to reductionist trials carried out in controlled environments, often with the only aim to extract and characterize allelochemicals or, at the most, to test the effect of these compounds on the germination of selected sensitive species in bioassays. In the case of crop-weed interactions, absolute evidence of the occurrence of allelopathy in the field is difficult to obtain, mainly because allelopathic effects are difficult to disentangle from resource competition and other biotic effects (Weidenhamer, 1996; Inderjit and del Moral, 1997). Additionally, the production and release of allelochemicals depend largely upon environmental conditions, usually being higher when plants are under stress, e.g. extreme temperatures, drought, soil nutrient deficiency, high pest incidence (Einhellig, 1987); also, the range and concentration of chemicals that a given species can produce can vary with environment conditions (Anaya, 1999). Other effects that need to be examined are allelopathy-mediated weed-weed, weed-crop and crop-following (or companion) crop interactions. It is therefore questionable whether allelopathy management *per se* would ever represent a consistently effective weed management tool; however, a better understanding of allelopathic occurrence in field situations, and of how it is influenced by cultural practices, would make it possible to include allelopathic crops in organic cropping systems and use them as a complementary tactic in a weed management strategy (Barberi, 2002).

The extracts of many dominant plants in Taiwan, such as *Delonix regia, Digitaria decumbens, Leucaena leucocephala*, and *Vitex negundo*, contain allelopathic compounds, including phenolic acids, alkaloids, and flavonoids that can be used as natural herbicides, fungicides, etc. which are less disruptive of the global ecosystem than are synthetic agrochemicals (Chou, 1995). Many important crops, such as rice, sugarcane, and mungbean, are affected by their own toxic exudates or by phytotoxins produced when their residues decompose in the soil. For example, in Taiwan the yield of the second annual rice crop is typically 25% lower than that of the first, due to phytotoxins produced during the fallowing period between crops. Autointoxication can be minimized by eliminating, or preventing the formation of the phytotoxins through field treatments such as crop rotation, water draining, water flooding, and the polymerization of phytotoxic phenolics into a humic complex (Chou, 1995).

Wetland soils provide anoxia-tolerant plants with access to ample light, water, and nutrients. Intense competition, involving chemical strategies, ensues among the plants. The roots of wetland plants are prime targets for root-eating pests, and the wetland rhizosphere is an ideal environment for many other organisms and communities because it provides water, oxygen, organic food, and physical protection. Consequently, the rhizosphere of wetland plants is densely populated by many specialized organisms, which considerably influence its biogeochemical functioning. The roots protect themselves against pests and control their rhizosphere organisms by bioactive chemicals, which often also have medicinal properties. Anaerobic metabolites, alkaloids, phenolics, terpenoids, and steroids are bioactive chemicals abundant in roots and rhizospheres in wetlands. Bioactivities include allelopathy, growth regulation, extraorganismal enzymatic activities, metal manipulation by phytosiderophores and phytochelatines, various pest-control effects, and poisoning. Complex biological-biochemical interactions among roots, rhizosphere organisms, and the rhizosphere solution determine the overall biogeochemical processes in the wetland rhizosphere and in the vegetated wetlands. In order to comprehend how wetlands really function and to understand these interactions it is necessary to implement long-term collaborative research (Neori et al., 2000). We can find promising allelochemicals and useful interactions in the rich biodiversity of these particular ecoystems, but without doubt, in all type of ecosystems.

3. CHEMISTRY OF ALLELOPATHY

As we know, plant and microbial compounds are continuously analyzed as potential sources of herbicides, pesticides, and pharmaceuticals because they provide a diversity of carbon skeletons and there has been success in that a number of compounds have shown biological activity. The same bioassays and techniques to reveal mechanisms of action apply to the search for herbicides as in the study of allelopathy. Certainly there is overlap in goals and compounds studied, but there also is a difference in that the starting point in the 'herbicide search' might be any natural product, as opposed to one identified with allelopathy. Most inhibitors of plants are secondary compounds that have their origin in either the shikimate or acetate pathways, or they are compounds having skeletal components from both of these origins (Einhellig, 2002). Waller et al. (1999) listed over twenty classes of secondary metabolites that are produced, stored, and released into the rhizosphere where they have biological activity as well as undergo microbial transformation and degradation. Einhellig (2002) concluded that the 14 categories suggested by Rice (1984) are sufficiently broad to still retain validity: water-soluble organic acids, straight-chain alcohols, aliphatic aldehydes and ketones, unsatured lactones, long-chain fatty acids and polyacetylenes, naphthoquinones, anthraquinones and complex quinones, gallic acids and polyphloroglucinols, cinnamic acids, coumarins, flavonoids, tannins, terpenoids, and steroids, amino acids, and purines and nucleosides.

In this chapter some of the main compounds and studies associated with allelopathy will be mentioned.

Terpenoids and phenolics are the most common compounds involved in allelopathic interactions. Terpenoids are the largest group of plant chemicals (15,000-20,000) with a common biosynthethic origin. The terpenoid pathway generates great structural diversity and complexity of compounds, thus generating enormous potential for mediating ecological interactions (Duke, 1991; Langenheim, 1994). Terpenoids may produce effects on seeds and soil microbiota through volatilization, leaching from plants, or decomposition of plant debris. These interactions can significantly affect community and ecosystem properties, although studies of plant-plant chemical interactions have often been controversial because of difficulty in unambiguously demonstrating interference by chemical inhibition rather than through resource competition or other mechanisms (Harper, 1977).

3.1. Terpenoids and sesquiterpene lactones

Vokou et al. (2003) compared the potential allelopathic activity of 47 monoterpenoids of different chemical groups, by estimating their effect on seed germination and subsequent growth of *Lactuca sativa* seedlings. Apart from individual compounds, eleven pairs at different proportions were also tested. As a group, the hydrocarbons, except for (+)-3-carene, were the least inhibitory. Of the oxygenated compounds, the least inhibitory were the acetates; whenever the free hydroxyl group of an alcohol turned into a carboxyl group, the activity of the resulting ester was markedly lower (against both germination and seedling growth). Twenty-four compounds were extremely active against seedling growth (inhibiting it by more than 85%), but only five against seed germination. The compounds that were most active against both processes belonged to the groups of ketones and alcohols; they were terpinen-4-ol, dihydrocarvone, and two carvone stereoisomers. These authors used a model to investigate whether compounds acted independently when applied in pairs. The combined effect varied. In half of the cases, it followed the pattern expected under the assumption of independence; in the rest, either synergistic or antagonistic interactions were found in both germination and elongation. However, even in cases of synergistic interactions, the level of inhibition was not comparable to that of a single extremely active compound, unless such a compound already participated in the combination.

The effect of the sesquiterpene cacalol and extracts (water and petroleum ether) derived from the roots of *Psacalium decompositum* (Asteraceae) on the germination and radicle growth of two plants, *Amaranthus hypochondriacus* (Amaranthaceae) and *Echinochloa crus-galli* (Poaceae), and the radial growth of four phytopathogenic fungi was described (Anaya et al., 1996). The activity of two cacalol derivatives (methyl cacalol and cacalol acetate) was also investigated. Germination of *A. hypochondriacus* was inhibited by almost all the treatments. The extracts and cacalol produced a significant inhibition of radicle growth of *A. hypochondriacus* and *E. crus-galli.* Cacalol acetate showed a specific inhibition on *E. crus-galli,* and methyl cacalol inhibited significantly the growth of *A. hypochondriacus.* In general, antifungal activity depended upon the target fungi and the concentration of each treatment. Cacalol had also effects on the morphology and coloration of the fungal mycelium. The bioactivity shown by the extracts of *Psacalium decompositum* on the tested seeds and fungi is mainly due to their content in cacalol.

The allelochemical potential of *Callicarpa acuminata* (Verbenaceae) was investigated using a biodirected fractionation study as part of a long-term project to search for bioactive compounds among the rich biodiversity of plant communities in the Ecological Reserve El Eden, Quintana Roo, Mexico. Aqueous leachate, chloroform-methanol extract, and chromatographic fractions of the leaves of the plant species inhibited the root growth of *Amaranthus hypochondriacus*, *Echinochloa crus-galli*, and tomato (23% , 59%, and 70% respectively). Some of these treatments caused a moderate inhibition of the radial growth of two phytopathogenic fungi, *Helminthosporium longirostratum* and *Alternaria solani* (18% to 31%). The chloroform-methanol (1:1) extract prepared from the leaves rendered five compounds:

isopimaric acid, a mixture of two diterpenols: sandaracopimaradien-19-ol and akhdarenol, α-amyrin, and the flavone salvigenin. The phytotoxicity exhibited by several fractions and the full extract almost disappeared when pure compounds were evaluated on the test plants, suggesting a synergistic or additive effect. Akhdarenol, α-amyrin and isopimaric acid methyl ether had antifeedant effects on *Leptinotarsa decemlineata.* Alpha-amyrin was most toxic to this insect. No correlation was found between antifeedant and toxic effects on this insect, suggesting that different modes of action were involved. All the test compounds were cytotoxic to insect Sf9 cells while salvigenin, akhdarenol, and isopimaric acid also affected mammalian Chinese Hamster Ovary (CHO) cells. Alpha-amyrin showed the strongest selectivity against insect cells (Anaya et al., 2003). In this study the authors emphasized that allelochemicals involved in allelopathic interactions often have multiple functionality.

Sesquiterpene lactones (SL) occur in over 15 plant families, predominantly in the Asteraceae, and represent with about 3,500 naturally occurring compounds, one of the largest groups of natural products. It has been demonstrated that some sesquiterpene lactones exhibit a broad spectrum of biological activities including phytotoxic and plant growth regulatory properties, cytotoxicity and antitumor properties, antimicrobial, insecticidal, molluscicidal and antimalarial activity (Fischer, 1986). Phytotoxic terpenoids and their possible involvement in allelopathy were covered in reviews on mono- and sesquiterpenes (Evenari, 1949; Fischer, 1986, 1991, 1994) and biological activities of SL were reviewed by Stevens and Merrill (1985), Picman (1986) and Elakovich (1988). Seedlings of *Ambrosia cumanensis* are inhibited by leachates of the adult plants and residues in soils. Some SL of this species have been implicated in this autototoxic mechanism (Anaya and del Amo, 1978). In the same way, parthenin and coronopilin of *Parthenium hysterophorus* also exhibited autotoxicity toward seedlings and older plants, this fact possibly reveal a mechanism of intraspecific population regulation (Picman and Picman, 1984). Axivalin and tomentosin from the seeds of *Iva axillaris* were inhibitory toward the germination and growth of *Abutilon theophrasti* (velvetleaf) (Spencer et al., 1984). The germacranolide-type SL represented by dihydrotartridin B significantly inhibited the root growth of *Brassica rapa* var. *pervidis* (Sashida et al., 1983). The α-methylene-γ-lactone group is present in many of the isolated natural sesquiterpene lactones, and has been proposed as one of the factors which can determine their allelopathic activity, in particular, as well as their biological activity in general. The different spatial arrangements that a molecule of SL can adopt is the other factor that has been related with the potential allelopathic activity of this type of secondary compounds (Macias et al., 1992). Data of several studies on the allelopathic potential of SL clearly demonstrated that they can selectively promote or inhibit germination or growth at concentrations as low as 1 μM. It is reasonable to assume that rain washes transport SL from the source plant or decomposing litter into the soil where they can reach significant concentration levels. In the case of isoalantolactone, it has been demonstrated that it can persist in mineral and organic soil for 3 months, supporting the assumption that SL play a significant role in allelopathic interactions in the environment (del Amo and Anaya, 1976; Stevens and Merril, 1985; Picman, 1986; Fischer, 1991).

Dehydrozaluzanin C, a natural SL, is a weak plant growth inhibitor with an I_{50} (or IC_{50}, the concentration required to inhibit plant growth 50 %) of about 0.5 mM for lettuce root growth. It also causes rapid plasma membrane leakage in cucumber cotyledon discs. Dehydrozaluzanin C is more active at 50 μM than the same concentration of the herbicide acifluorfen. Symptoms include plasmolysis and the disruption of membrane integrity is not light dependent. Reversal of its effects on root growth was obtained with treatment by various amino acids, with histidine and glycine providing ca. 40% reversion. The strong reversal effect obtained with reduced glutathione is due to cross-reactivity with DHZ and the formation of mono- and di-adducts. Photosynthetic, respiratory and mitotic processes, as well as NADH oxidase activity appear to be unaffected by this compound. Dehydrozaluzanin C exerts its effects on plants through two different mechanisms, only one of which is related to the disruption of plasma membrane function (Galindo et al., 1999).

A structure-activity study to evaluate the effect of the trans,trans-germacranolide SL lactones costunolide, parthenolide, and their 1,10-epoxy and 11,13-dihydro derivatives (in a range of 100-0.001 μM) on the growth and germination of several mono and dicotyledon target species was carried out by Macias et al. (1999). These compounds appear to have more selective effects on the radicle growth of monocotyledons. Certain factors such as the presence of nucleophile-acceptor groups and their accessibility enhance the inhibitory activity. The levels of radicle inhibition obtained with some compounds on wheat are totally comparable to those of commercial herbicide Logran and allow proposing them as lead compounds. In addition, a structure-activity study to evaluate the effect of 17 guaianolide SL (in a range of 100-0.001 μM) on the growth and germination of several mono and dicotyledon target species was also performed by Macias et al. (2002a). These compounds appear to have deeper effects on the growth of either monocots or dicots than the previously tested germacranolides. Otherwise, the lactone group seems to be necessary for the activity, though it does not necessarily need to be unsaturated. However, the presence of a second and easily accessible unsaturated carbonyl system greatly enhances the inhibitory activity. Lipophilicity and the stereochemistry of the possible anchoring sites are also crucial factors for the activity.

The dichloromethane extract of dried leaves of *Helianthus annuus* has yielded, in addition to the known SL annuolide E and leptocarpin, and the sesquiterpenes heliannuols A,C,D,F,G,H,1, the new bisnorsesquiterpene, annuionone E, and the new sesquiterpenes heliannuol L, helibisabonol A and helibisabonol B. Structural elucidation was based on extensive spectral (one and two-dimensional NMR experiments) and theoretical studies. The sesquiterpenes heliannuol A and helibisabonol A and the SL leptocarpin inhibited the growth of etiolated wheat coleoptiles (Macias et al., 2002b). In addition to (+)-, (-)- and (±)-heliannuol E, growth-inhibitory activities of five synthetic chromanes and four tetrahydrobenzo[b]oxepins were examined against oat and cress. All heliannuol E isomers exhibited similar biological activities against cress, whereas when tested against oat roots, the unnatural optical isomer (+) showed no inhibitory activity. Four brominated chromans and two tetrahydrobenzo[b]oxepins derivatives also showed apparent inhibition against both cress and oat (Doi et al., 2004).

The tremendous impact of parasitic plants on world agriculture has prompted much research aimed at preventing infestation. *Orobanche* and *Striga* spp. are two examples of parasitic weeds that represent a serious threat to agriculture in large parts of the world. The life cycle of these parasitic weeds is closely regulated by the presence of their hosts, and secondary metabolites that are produced by host plants play an important role in this interaction. A special interest has been arising on those host-produced stimulants that induce the germination of parasite seeds. Three classes of compounds have been described that have germination-stimulating activity: dihydrosorgoleone, the strigolactones and SL. Keyes et al. (2001) suggest that dihydrosorgoleone is the active stimulant in the root exudates of sorghum and other monocotyledonous hosts. However, Butler et al. (1995) and Wigchert et al. (1999) suggest that dihydrosorgoleone is less likely to be the germination stimulant *in vivo* because of its low water solubility, and because no correlation between its production and the germination of *Striga* has been found. To date, there is no definite proof that the germination of parasitic weed seeds in the field is induced by one single signal compound or class of compounds (and indeed such proof will be hard to obtain) (Bouwmeester et al., 2003). The capacity of SL, which share some structural features with the strigolactones, to induce the germination of *S. asiatica* has been reported (Fischer et al., 1989, 1990). In addition, a decade after the results of Fischer studies, Francisco Macías and his group (Pérez de Luque et al., 2000; Galindo et al., 2002) performed some studies of the structure-activity relationship (SAR) directed to evaluate the effect of several SL as germination stimulants of three *Orobanche* spp. (*O. cumana, O. crenata,* and *O. ramosa*). Results are compared with those obtained in the same bioassay with an internal standard, the synthetic analogue of strigol GR-24. A high specificity in the germination activity of SL on the sunflower parasite *O. cumana* has been observed, and a relationship between such activity and the high sunflower SL content is postulated. Molecular properties of the natural and synthetic germination stimulants (GR-24, GR-7, and Nijmegen-1) and SL have been studied using MMX and PM3 calculations. Consequently, comparative studies among all of them and their activities have been made. SL tested present similarities in molecular properties such as the volume of the molecule and the spatial disposition of the carbon backbone to the natural germination stimulant orobanchol. These properties could be related to their biological activity. Considering that the sun-flower–*O. cumana* interaction is highly specific and that sunflower contains many SL, it is tempting to speculate that *O. cumana* has evolved to respond to sesquiterpene lactones (and not or less to strigolactones) (Bouwmeester et al., 2003).

3.2. Phenolics

In relation with phenolics, Inderjit et al. (1997) conducted a study to understand the effects of certain phenolics, terpenoides, and their equimolar mixture through agar gel and soil growth bioassays and their recovery from soils. The eight compounds selected for this study were p-hydroxybenzoic acid, ferulic acid, umbelliferone, catechin, emodin, 1,8-cineole, carvone, and betulin. Lettuce (*Lactuca sativa* L.) was used as test species for agar gel and soil growth bioassays. Root and shoot growth of lettuce was inhibited for all the above except emodin and catechin. However, in soils treated with different phenolics and terpenoids, only root growth of lettuce was inhibited, whereas shoot growth was promoted. Recovery of p-hydroxybenzoic acid and umbelliferone was higher in unautoclaved soils, while that of catechin was lower.

Nava-Rodriguez et al. (in press) observed the *in vitro* effects of aqueous leachates from fresh and dry, flowering and vegetative stage of *Phaseolus* species, faba bean, alfalfa, vetch, maize, and squash, and weed species on the root growth of selected crop and weeds, as well as on two strains of *Rhizobium leguminosarum* biovar *phaseoli* (CPMex1 and Tlaxcala). Most of the specimens were collected in a traditional agricultural drained field ("Camellon") in Tlaxcala, Mexico where maize, beans, squash, alfalfa, faba-beans, and vetch are cultivated in mixed or rotation crops. Significant effects of leachates from fresh vegetative and flowering cultivated plants and weeds were predominantly stimulatory on the growth of tested crops, being the leachates from fresh aerial parts of alfalfa and pinto bean the most stimulatory. Nevertheless, aqueous leachates from fresh and dry cultivated legumes (vegetative and flowering) inhibited the growth of weeds. In contrast, the aqueous leachates from the dry aerial part of almost all plants resulted inhibitory on the root growth of the test crops, except maize. Aqueous leachates were also evaluated on the growth of two strains of *Rhizobium leguminosarum* biovar *phaseoli.* Leachates from some of the tested crops significantly stimulated the growth of both *Rhizobium* strains. The aqueous leachates from fresh aerial parts of the weeds *Simsia amplexicaulis* and *Tradescantia crassifolia* significantly inhibited the growth of CPMex1 *Rhizobium* strain. On the other hand, the aqueous leachates from fresh roots of these same weeds inhibited the growth of the Tlaxcala strain. In preliminary chemical tests using thin layer chromatography (TLC), phenolics were detected in dry aerial parts of vegetative alfalfa, pinto bean, and vetch, and dry aerial part of flowering faba bean suggesting the role of these compounds in the allelopathic effects of these legumes.

Nilsson et al. (1998) reported on the temporal variation of phenolics and a dihydrostilbene, batatasin III, in *Empetrum hermaphroditum* leaves. These authors reported that first year shoots produced higher levels of phenolics than older tissues. High phenolic concentration was maintained through the second year, but it declined afterwards. However, the phytotoxicity of *E. hermaphroditum* extracts was related more to batatasin III than phenolics.

Hyder et al. (2002) performed a study focused on the presence and distribution of secondary phenolic compounds found within creosotebush (*Larrea tridentata*). Total phenolics, condensed tannins and nordihydroguaiaretic acid (NDGA) were measured in nine categories of tissue within creosotebush. Total phenolic and condensed tannin concentrations were determined using colorimetric methods while NDGA content was determined with high performance liquid chromatography (HPLC). Phenolics were present throughout the plant with the highest concentrations in green stems (40.8 mg/g), leaves (36.2 mg/g), and roots (mean for all root categories=28.6 mg/g).

Condensed tannins were found in all tissues with highest concentrations in flowers (1.7 mg/g), seeds (1.1 mg/g), and roots less than 5 mm in diameter (1.1 mg/g). Flowers, leaves, green stems and small woody stems (<5 mm in diameter) all contained NDGA with highest concentrations in leaves (38.3 mg/g) and green stems (32.5 mg/g).

Another study conducted by Singh et al. (2003a) assessed the phytotoxicity of *Ageratum conyzoides*, a weed of cultivated areas, to the growth and establishment of wheat (*Triticum aestivum*). The lengths of the radicle and coleoptile and the seedling dry weight of wheat were significantly reduced when wheat was grown in field soil previously infested with *A. conyzoides*, compared to control soil collected from an area devoid of this weed. Even extracts prepared from *A. conyzoides* soil were inhibitory, indicating the presence of some water-soluble phytotoxins in the soil. To determine the possible contribution of the weed in releasing these phytotoxins, growth studies involving leaf residues and their extracts and amended soils (prepared by incorporating leaf residues and residue extracts) were also performed on wheat. With all treatments, an inhibitory effect of *A. conyzoides* was found, compared to respective controls. A significant amount of water-soluble toxic phenolics was found to be present in the soil infested with *A. conyzoides*, leaf residues and the amended soils. The amount of phenolics correlated well with growth performance in the respective treatments (see chapter 11).

Aqueous leachates of roots of the perennial weed *Pluchea lanceolata*, its root-incorporated soil and rhizosphere soil, interfered with the seedling growth of certain plant species. The soils from the rhizosphere zone of this plant had significantly higher total phenolics and HPLC analysis revealed that phenolic fractions represented by retention times of 1.6, 1.9, 2.5 (simple phenol, chlorogenic acid and phloroglucinol respectively), 3.7 and 4.3 min were contributed by roots of the weed to the soil. The phenolic fraction represented by the retention time 3.3 (formononetin 7-O-glucoside) was detected in the weed's rhizosphere soils and not in the root-incorporated soils. UV spectral studies established the presence of phloroglucinol, simple phenol, chlorogenic acid, formononetin 7-O-glucoside, and methylated coumarins in the root leachate, which affect the seedling growth of mustard (*Brassica juncea*) (Inderjit and Dakshini, 1994).

The effects of five phenolic compounds, catechol, protocatechuic, p-coumaric, p-hydroxybenzoic, ferulic acids and their mixture were studied on pH, organic matter, organic-nitrogen, total phenolic content and certain inorganic ions of forest mineral soils (Ae and B horizons). The A- and B-horizon soils, were amended with 10^{-4} M concentration of each phenolic compound and their mixture. In general, soil properties were affected by phenolics amendement. However, soils amended with catechol did not influence any of the soil characteristics. Contents of organic matter, nitrogen and phosphate were lower in soils amended with different phenolic compounds compared to the unamended control soil (Inderjit and Mallik, 1997).

Low molecular weight phenolic compounds have been identified in fresh leaves and in soils in which leaves of five varieties of *Capsicum annuum* were decomposing. Six phenolic compounds were tested in laboratory bioassays for their allelopathic effects on germination and seedling growth of six weeds. Ferulic acid, gallic acid, p-coumaric acid, p-hydroxybenzoic acid, vanillic acid, and p-vanillin were bioassayed in concentrations of 10, 1, 0.1, and 0.01 mM. Equimolar mixtures containing all these phenolics were prepared at the final total concentration of 10, 1, 0.1, and 0.01 mM to test for possible interactive effects. *Chenopodium album, Plantago lanceolata, Amaranthus retroflexus, Solanum nigrum, Cirsium* sp. and *Rumex crispus* were the selected target weeds. The highest concentration of the compounds inhibited the germination of all these weeds, but lower concentrations had no effect or were stimulatory. However, effects varied with the weed species, the concentration of the compound tested and the compound itself. In assays with the mixture of phenolics some additive effects were found (Reigosa et al., 1999).

Reversible sorption of phenolic acids by soils may provide some protection to phenolic acids from microbial degradation. In the absence of microbes, reversible sorption 35 days after addition of 0.5-3 mu mol/g of ferulic acid or p-coumaric acid was 8-14% in Cecil A(p) horizon and 31-38% in Cecil B-t horizon soil materials. The reversibly sorbed/solution ratios (r/s) for ferulic acid or p-coumaric acid ranged from 0.12 to 0.25 in A(p) and 0.65 to 0.85 in B-t horizon soil materials. When microbes were introduced, the r/s ratio for both the A(p) and B-t horizon soil materials increased over time up to 5 and 2, respectively, thereby indicating a more rapid utilization of solution phenolic acids over reversibly sorbed phenolic acids. The increase in r/s ratio and the overall microbial utilization of ferulic acid and/or p-coumaric acid were much more rapid in A(p) than in B-t horizon soil materials. Reversible sorption, however, provided protection of phenolic acids from microbial utilization for only very short periods of time. Differential soil fixation, microbial production of benzoic acids (e.g., vanillic acid and p-hydroxybenzoic acid) from cinnamic acids (e.g., ferulic acid and p-coumaric acid, respectively), and the subsequent differential utilization of cinnamic and benzoic acids by soil microbes indicated that these processes can substantially influence the magnitude and duration of the phytoxicity of individual phenolic acids (Blum, 1998).

Soil solution concentrations of allelopathic agents (e.g., phenolic acids) estimated by soil extractions differ with extraction procedure and the activities of the various soil sinks (e.g., microbes, clays, organic matter). This led to the hypothesis that root uptake of phenolic acids is a better estimator of dose than soil solution concentrations based on soil extracts. This hypothesis was tested by determining the inhibition of net phosphorus uptake of cucumber seedlings treated for 5 hr with ferulic acid in whole-root and split-root nutrient culture systems. Experiments were conducted with II ferulic acid concentrations ranging from 0 to 1 mM, phosphorus concentrations of 0.25, 0.5, or 1 mM, and solution pH values of 4.5, 5.5, or 6.5 applied when cucumber seedlings were 9, 12, or 15 days old. The uptake or initial solution concentration of ferulic acid was regressed on ferulic acid inhibition of net phosphorus uptake. Attempts were made to design experiments that would break the collinearity between ferulic acid uptake and phosphorus uptake. The original hypothesis was rejected because the initial ferulic acid solution concentrations surrounding seedling roots were more frequently and consistently related to the inhibition of net phosphorus uptake than to ferulic acid uptake by these roots. The data suggest that root contact, not uptake, is responsible for the inhibitory activity of phenolic acids (Lehman and Blum, 1999).

Bulk-soil and rhizosphere bacteria are thought to exert considerable influence over the types and concentrations of phytotoxins, including phenolic acids that reach a root surface. Induction and/or selection of phenolic acid-utilizing (PAU) bacteria within the bulk-soil and rhizosphere have been observed when soils are enriched with individual phenolic acids at concentrations greater than or equal to 0.25 μmol/g soil. However, since field soils frequently contain individual phenolic acids at concentrations well below 0.1 μmol/g soil, the actual importance of such induction and/or selection remains uncertain. Common bacteriological techniques (e.g., isolation on selective media, and plate dilution frequency technique) were used to demonstrate in Cecil Ap soil systems: (i) that PAU bacterial communities in the bulk soil and the rhizosphere of cucumber seedlings were induced and/or selected by mixtures composed of individual phenolic acids at concentrations well below 0.25 μmol/g soil; (ii) that readily available carbon sources other than phenolic acids, such as glucose, did not modify induction and/or selection of PAU bacteria; (iii) that the resulting bacterial communities readily utilize mixtures of phenolic acids as a carbon source; and (iv) that depending on conditions (e.g., initial PAU bacterial populations, and phenolic acid concentration) there were significant inverse relationships between PAU bacteria in the rhizosphere of cucumber seedlings and absolute rates of leaf expansion and/or shoot biomass. The decline in seedling growth could not be attributed to resource competition (e.g., nitrogen) between the seedlings and the PAU bacteria in these studies. The induced and/or selected rhizosphere PAU bacteria, however, reduced the magnitude of growth inhibition by phenolic acid mixtures. For a 0.6 μmol/g soil equimolar phenolic acid mixture composed of ρ-coumaric acid, ferulic acid, ρ-hydroxybenzoic acid, and vanillic acid, modeling indicated that an increase of 500% in rhizosphere PAU bacteria would lead to an approximate 5% decrease (e.g., 20-25%) in inhibition of absolute rates of leaf expansion (Blum et al., 2000).

Allelopathy due to humus phenolics is a cause of natural regeneration deficiency in subalpine Norway spruce (*Picea abies*) forests. If inhibition of spruce germination and seedling growth due to allelochemicals is generally accepted, in contrast there is a lack of knowledge about phenolic effects on mycorrhizal fungi. Thus, Souto et al. (2000) tested the effects of a humic solution and its naturally occurring phenolics on the growth and respiration of two mycorrhizal fungi: *Hymenoscyphus ericae* (symbiont of *Vaccinium myrtillus*, the main allelochemical-producing plant) and *Hebeloma crustuliniforme* (symbiont of *P. abies*, the target plant). Growth and respiration of *H. crustuliniforme* were inhibited by growth medium with the original humic solution (-6% and -30%), respectively, whereas the same humic solution did not affect growth but decreased respiration of *H. ericae* (-55%). When naturally occurring phenolics (same chemicals and concentrations in the original humic solution) were added to the growth medium, growth of *H. crustuliniforme* was not affected, whereas that of *H. ericae* significantly increased (+10%). These authors concluded that *H. ericae* is better adapted to the allelopathic constraints of this forest soil than *H. crustuliniforme* and that the dominance of *V. myrtillus* among understory species could be explained in this way.

Inderjit and Duke (2003) mentioned that the best evidence for allelopathy should include some understanding of natural concentrations and rates of allelochemicals. For example, (±)-catechin has been isolated from *Centaurea maculosa* (Bais et al., 2002), an invasive species in North America for which other lines of evidence suggest root allelopathy (Ridenour and Callaway, 2001). The more common enantiomer, (+)-catechin, has anti-bacterial functions, whereas (–)-catechin has strong allelopathic effects on other plants. (±)-catechin is harmless to *C. maculosa*, but has negative effects on other species at concentrations of ˜100 mg L^{-1}. (±)-catechin is exuded from *C. maculosa* roots creating concentrations from 83.2 to 185 mg L^{-1} in aqueous solutions. Importantly, Bais et al. (2002) found (±)-catechin in extracts from natural soils in fields containing *C. maculosa* in concentrations as far higher than the minimum required dose, ranging from 291.6 to 389.8 μg cm^{-3}.

In allelopathy studies a central goal is to isolate, identify, and characterize allelochemicals from the soil. However, since it is essentially impossible to simulate exact field conditions, experiments must be designed with conditions resembling those found in natural systems. Inderjit (1996) argued that allelopathic potential of phenolics can be appreciated only when we have a good understanding of i) species responses to phenolic allelochemicals, ii) methods for extraction and isolation of active phenolic allelochemicals, and iii) how abiotic and biotic factors affect phenolic toxicity.

Duke et al. (2003) summarized the recent research of the Agricultural Research Service of United States Department of Agriculture on the use of natural products to manage pests. They discussed some studies on the use of both phytochemicals and diatomaceous earth to manage insect pests. Chemically characterized compounds, such as a saponin from pepper (*Capsicum frutescens* L), benzaldehyde, chitosan and 2-deixy-D-glucose are being studied as natural fungicides. Resin glycosides for pathogen resistance in sweet potato and residues of semitropical leguminous plants for nematode control are also under investigation. Bioassay-guided isolation of compounds with potential use as herbicides or herbicide leads is underway at several locations. New natural phytotoxin molecular target sites (asparagine synthetase and fructose-1,6-bisphosphate aldolase) have been discovered. Weed control in sweet potato and rice by allelopathy is under investigation. Molecular approaches to enhance allelopathy in sorghum are also being undertaken. The genes for polyketide synthases involved in production of pesticidal polyketide compounds in fungi are found to provide clues for pesticide discovery. Gene expression profiles in response to fungicides and herbicides are being generated as tools to understand more fully the mode of action and to rapidly determine the molecular target site of new, natural fungicides and herbicides.

Research on the chemical basis for allelopathy has often been hindered by the complexity of plant and soil matrices, making it difficult to track active compounds. Recent improvements in the cost and capabilities of bench-top chromatography-mass spectrometry instruments make these tools more powerful and more widely available to assist with molecular studies conducted in today's expanding field. Such instrumental techniques are herein recommended as economically efficient means of advancing the rigor of allelopathy research and assisting the development of a better understanding of the chemical basis for the allelopathy phenomenon (Haig, 2001).

4. THE MODE OF ACTION OF ALLELOCHEMICALS

Allelopathic chemicals alter plant growth and development by a multiplicity of actions on physiological processes because there are hundreds of different structures and many of the compounds have several phytotoxic effects. Whole plants bioassays and physiological tests designed to use a small quantity of compound are keys to strategies for elucidating mechanisms of action. Insight into the action of the responsible chemicals is critical to a more complete explanation of allelopathy and to its application for improving crop production. Unfortunately, we still find that linkages between inhibition of growth and the corresponding physiological mechanism are elusive. A major part of the problem is the array and diversity of allelochemicals. Several hundred different compounds have been identified, and we speculate that many others will be eventually being discovered. Most instances of allelopathic inhibition are the result of the simultaneous action of several compounds, and often these include compounds whose chemistry is divergent (Einhellig, 2002). The visible evidence in bioassays is that many phenolics, quinones, sesquiterpene lactones, alkaloids, and others alter root morphology. Although the cell membrane is an early interface with allelochemicals, relatively little attention has been given to membrane-related effects and their molecular targets.

Effects of a single allelochemical or their mixtures on physiological processes include disruption of membrane permeability (Galindo, et al., 1999), ion uptake (Yu and Matsui, 1997), inhibition of electron transport in both the photosynthesis and respiratory chains (Calera et al., 1995a; Abrahim, et al., 2000), alteration of enzymatic activity (Politycka, 1999, Romagni, et al., 2000), and inhibition of cell division (Cruz-Ortega et al., 1988; Anaya and Pelayo-Benavides, 1997). In other examples, bioactivity-directed fractionation of the methanol extract of the roots of *Ratibida mexicana* resulted in the isolation of two bioactive sesquiterpene lactones, iso*allo*alantolactone and elema-1,3,11-trien-8,12-olide. Both compounds caused a significant inhibition of the radicle growth of *Amaranthus hypochondriacus* and *Echinochloa crus-galli*, exerted moderate cytotoxicity activity against three different solid tumour cell lines and inhibited the radial growth of three phytopathogenic fungi. Iso*allo*alantolactone also caused the inhibition of ATP synthesis, proton uptake, and electron transport (basal, phosphorylating and uncoupled) from water to methylviologen therefore acting as a Hill's reaction inhibitor.The lactone inhibited only photosystem II (Calera et al., 1995b). Other compounds studied by Calera et al., (1995c) were the resin glycoside mixture from *Ipomoea tricolor*. They tested its effect on seedling growth and plasma membrane H^+-ATPase activity in *Echinochloa crus-galli*. The resin glycoside mixture as well as Tricolorin A, the main compound in the mixture, inhibited the activity of the plasma membrane ATPase. In other studies, Cruz-Ortega et al. (1998) observed plasma membrane disruption in root tips and plasmolized cells in the peripheral zone of beans and bottle gourd roots treated with the aqueous leachate of *S. deppei* suggesting that the allelopathics of this plant alter some membrane processes.

4-phenyl coumarins isolated from *Exostema caribaeum* and *Hintonia latiflora* (Rubiaceae) and some semisynthetic derivatives acted as uncouplers in spinach chloroplasts. The glycoside 5-O-β-D-glucopyranosyl-7-methoxy-3',4'-dihidroxy-4-phenylcoumarin, 5,7,3',4'-tetrahydroxy-4-phenyl-coumarin, and 7-methoxy-5,3',4'-trihydroxy-4-phenylcoumarin inhibited ATP synthesis and proton uptake. On the other hand, basal and phosphorylating electron transport were enhanced by these compounds. The light-activated Mg^{2+}-ATPase was slighted stimulated by the last two coumarins. In addition, at alkaline pH compound 5,7,3',4'-tetrahydroxy-4-phenyl-coumarin stimulated the basal electron flow from water to methylviologen, but at the pH range from 6 to 7.5 the coumarin did not have any enhancing effect. This last compound, which possesses four free phenolic hydroxyl groups, was the most active uncoupler agent. Probably, the phenolate anions may be the active form responsible for the uncoupling behavior of 4-phenylcoumarins (Calera et al., 1996).

Low molecular weight phenolic compounds were identified in two soils with different vegetative cover, *Fagus sylvatica* and *Pinus laricio*, and were tested at different concentrations on seed germination of *Pinus laricio*, and on respiratory and oxidative pentose phosphate pathway enzymes involved in the first steps of seed germination. There are marked differences in the phenolic acid composition of the two investigated soils. All the phenolic compounds bioassayed inhibited seed germination and those extracted from *Pinus laricio* soil were particularly inhibitory. Inhibition of germination of seeds is strongly correlated to the inhibition of the activities of enzymes of glycolysis and the oxidative pentose phosphate pathway (Muscolo et al., 2001).

Seven-day-old seedlings of cucumber (*Cucumis sativus* cv. *Wisconsin)* were treated with 0.1 mM solutions of cinnamic acid (ferulic and p-coumaric acids) and benzoic acid (hydroxybenzoic and vanillic acids) derivatives as stressors. The content of free and glucosylated soluble phenols and the activity of phenylalanine ammonia-lyase (E.C.4.3.1.5), phenol-beta-glucosyltransferase (E.C.2.4.1.35.), and beta-glucosidase (E.C.3.2.1.21.) in seedling roots as well as their length and fresh weight were examined. Changes in glucosylated phenolic content and phenol-beta-glucosyltranspherase activity were observed under the influence of all phenolics applied. Treatment with ferulic and p-coumaric acids stimulated the increase of phenylalanine ammonia-lyase and beta-glucosidase activity and slightly inhibited cucumber root growth (Politycka, 1998).

Environmental stresses (biotic and abiotic), including allelochemicals have been shown to induce the synthesis of new proteins in plants. These proteins might have evolutionary value for survival under adverse environmental situations. Romero-Romero et al. (2002) tested the effect of the mixture of toxic allelochemicals from the aqueous leachates from *Sicyos deppei, Acacia sedillense, Sebastiania adenophora*, and *Lantana camara* on the radicle growth and cytoplasmic protein synthesis patterns of *Zea mays* (maize), *Phaseolus vulgaris* (bean), *Cucurbita pepo* (squash), and *Lycopersicon esculentum* (tomato). In general, high, medium and low molecular weight cytoplasmic proteins were affected by the different aqueous leachates. Crop plant responses were diverse, but in general, an increase in protein synthesis was observed in the treated-roots. Maize was the least affected, but both the radicle growth and also the protein pattern of tomato were severely inhibited by all allelopathic plants. The changes observed on protein expression may indicate a biochemical alteration at the cellular level of the tested crop plants.

Roshchina (2001) discussed the molecular-cellular basis of pollen allelopathy, related to possible chemosensory mechanisms. The phenomenon consists of a series of events, viz., a) excretion of signalling and regulatory substances from donor cell (pollens, pistil stigma); b) recognition of specific signal-stimulus from plant excretions by acceptor cell (pollen or pistil stigma); c) transmission of chemical information within the acceptor cell (pollen); and d) development of characteristic response in acceptor cell. The processes occur in growth, development and normal fertilization. In the first stage of interactions, allelochemicals are excreted, which act as chemical signals, growth regulators and modulators of cellular metabolism, etc. The allelochemicals, acting on fertilization may be, nitrogen-containing substances (acetylcholine, histamine, serotonin, dopamine, noradrenaline), phenols [(flavonoids: quercetin, kaempferol, rutin), aromatic acids (benzoic, gallic, vanillic)], terpenoids (monoterpenes: citral, linalool, cymol), sesquiterpene lactones: azulene and proazulenes (desacetylinulicine, inulicine, ledol, artemisinine, grosshemine, gaillardine and austricine), and polyacetylenes (capilline) found in flower excretions. These compounds were tested *in vitro* and *in vivo* on pollen germination of *Hippeastrum hybridum*. Nitrogenous compounds stimulate the growth of pollen tube, whereas, their antagonists blocked normal fertilization and thus fruits or seeds did not form. Terpenoids act on pollen germination and their stimulatory and inhibitory effects (block fruit formation) depend on their concentration. These effects of terpenoids on pollen germination are through chemosignalling and possible steps are: a) spreading of information in pollen secretions e.g. in olfactory slime; b) binding with special sensors or receptors in plasmalemma; and c) transfer of stimulus within the pollen cell to nucleus, where spermia appear and a pollen tube starts to grow. Moving from donor cell, allelochemicals penetrate the wall of acceptor cell either a) directly (without any changes in protoplasmic membrane); or b) after conversions [interaction with foreign substance of low or high-molecular weight (enzymes and protectory proteins) secreted from donor cells. or compounds of acceptor cell]. Often the second case includes free radical processes. The transmission of information within cell is third stage which includes participation of secondary messengers (cyclic AMP and GMP, inositol triphosphate, Ca ions) and some related enzymatic systems. The final transmission occurs in membranes of cellular organelles, which respond to information received through changes in enzymatic activity and metabolism. At cellular level, in pollen and pistil it may be active excretion, changes in the autofluorescence and membrane permeability, regulation of alternative pathways in respiration and photosynthesis and switching on free radical processes.

The phenyl propanoid pathway (PPP) can be stimulated as demonstrated by Randhir et al. (2004) in mung bean sprouts through the pentose phosphate and shikimate pathways, by natural elicitors such as fish protein hydrolysates (FPH), lactoferrin (LF) and oregano extract (OE). Elicitation significantly improved the phenolic, antioxidant and antimicrobial properties of mung bean sprouts. The optimal elicitor concentrations were 1 ml/l FPH, 250 ppm LF and 1 ml/l OE for the highest phenolic content that was approximately 20, 35 and 18% higher than control, respectively, on day 1 of dark germination. The antioxidant activity estimated by P-carotene assay in mung bean sprouts was highest on day 1 of germination for all treatments and control. In general, higher antioxidant activity was observed in the elicited sprouts compared with control. In the case of 1,1-diphenyl-2-picrythydrazyl (DPPH) assay the antioxidant activity for all treatments and control was highest on day 2. Among the different elicitor treatments, OE elicited mung bean sprouts showed the highest antioxidant activity of 49% DPPH inhibition on day 2. This increased activity correlates with high guaiacol peroxidase (GPX) activity indicating that the polymerizing phenolics required during lignification with growth have antioxidant function. For all elicitor treatments a higher glucose-6-phosphate dehydrogenase (G6PDH) activity was observed during early germination following the high phenolic content. This is due to the general mobilization of carbohydrates to the growing sprouts in response to elicitation. In general the GPX activity steadily increased with germination for treatments and control. The higher phenolics produced on day 1 was utilized for GPX-mediated polymerization to form polymeric phenolics and lignin required during germination. The late stage polymerization linked to GPX activity preceded stimulation of G6PDH. This indicated that as phenolics were polymerized by GPX in late stages, G6PDH linked precursors such as NADPH (2) and sugar phosphates were being made available. Antimicrobial activity against *Helicobacter pylori* was observed in the mung bean sprout extract from control, LF and OE treatments from the day 1 stage. Both the LF and OE elicited extracts showed high antimicrobial activity, which correlated to high antioxidant activity on day 1. The higher antimicrobial activity was also observed with the higher stimulation of G6PDH and GPX activity during early stages of germination. This leads to the hypothesis that enhanced mobilization of carbohydrates (as indicated by G6PDH activity on days 2) and 4), enhanced polymerization of simple phenols (as indicated by GPX activity on day 3) contributed to high antioxidant activity producing intermediary metabolites (day 2).

5. WEED MANAGEMENT

Bhowmik and Inderjit (2003) examined some considerable efforts in designing alternative weed management strategies due to increase in the number of herbicide-resistant weeds and environmental concerns in the use of synthetic herbicides. The conventional synthetic herbicides are becoming less and less effective against the resistant weed biotypes. These authors discussed the role of allelopathic cover crops/ crop residues, natural compounds, and allelopathic crop cultivars in natural weed management, and gave numerous examples of employing crop residues, cover crops, and allelopathic crop cultivars in weed management. They concluded that although we cannot eliminate the use of herbicides, their use can be reduced by exploiting allelopathy as an alternate weed management tool for crop production against weeds and other pests.

The use of allelopathy for controlling weeds could be either through directly utilizing natural allelopathic interactions, particularly of crop plants, or by using allelochemicals as natural herbicides. In the former case, a number of crop plants with allelopathic potential can be used as cover, smother, and green manure crops for managing weeds by making desired manipulations in the cultural practices and cropping patterns. These can be suitably rotated or intercropped with main crops to manage the target weeds (including parasitic ones) selectively. Even the crop mulch/ residues can also give desirable benefits. The allelochemicals present in the higher plants as well as in the microbes can be directly used for weed management along with the management of some herbicides (Singh et al., 2003b).

Singh et al. (2003b) also mentioned that the bioefficacy of allelochemicals can be enhanced by structural changes or the synthesis of chemical analogues based on them. Further, in order to enhance the potential of allelopathic crops, several improvements can be made with the use of biotechnology or genomics and proteomics. In this context either the production of allelochemicals can be enhanced or the transgenics with foreign genes encoding for a particular weed-suppressing allelochemical could be produced. These authors comment that in the former, both conventional breeding and molecular genetical techniques are useful. However, with conventional breeding being slow and difficult, more emphasis is laid on the use of modern techniques such as molecular markers and the selection aided by them. Although the progress in this regard is slow, nevertheless some promising results are coming and more are expected in future. In this sense, is important to point out that the potential use of transgenic plants and other genetically modified organisms (GMO's) with such or other proposal, cause a strong controversial with the principles of organic agriculture defined and established by the International Federation of Organic Agriculture Movements (IFOAM) founded with the aim to promote an agriculture that is ecologically, economically, and socially sustainable. IFOAM is opposed to genetic engineering in agriculture, in view of the unprecedented danger it represents for the entire biosphere and the particular economic and environmental risks it poses for organic producers (IFOAM, 2002)[*].

Using a soil bioassay technique, Conkling et al. (2002) assessed seedling growth and incidence of disease of wild mustard (*Brassica kaber*) and sweet corn (*Zea mays*) in soil from field plots that received either of two treatments: incorporated red clover (*Trifolium pratense*) residue plus application of compost ('amended soil'), or application of ammonium nitrate fertilizer ('unamended soil'). Soils were analyzed for percent moisture, dissolved organic carbon, conductivity, phenolics, and nutrient content. A trend toward greater incidence of *Pythium* spp. infection of wild mustard seedlings grown in amended soil was observed during the first 40 days after

* *IFOAM, 2002: International Federation of Organic Agriculture Movements (IFOAM). Position on Genetic Engineering and Genetically Modified Organisms. http://www.ifoam.org/pospap/ge_position_0205.html 2002.*

incorporation (DAI) of red clover and compost, with significant differences (α= 0.05) at two out of four sampling dates in 1997, and four out of four sampling dates in 1998. Incidence of *Pythium* infection was 10-70% greater in the amended soil treatment during that period. Asymptomatic wild mustard seedlings grown in amended soil were also on average 2.5 cm shorter (α= 0.05) at 5 DAI than those grown in unamended soil in one year out of two. Concentration of phenolic compounds in soil solution was correlated with decreased shoot and root growth (r = 0.50, 0.28, respectively) and increased incidence of disease (r = 0.48) in wild mustard seedlings in one year out of two. Dissolved organic carbon concentration was correlated with increased disease in wild mustard seedlings in both years (r = 0.51, 0.33, respectively). Growth of corn seedlings did not differ between the two soil treatments, suggesting that red clover green manure and compost may selectively reduce density and competitive ability of wild mustard in the field. Bioassay results corresponded well with emergence and shoot weight results from a related field study, indicating that this technique may be useful for screening potential soil treatments prior to field studies.

Anaya et al. (1987) applied leaves of *Alnus firmifolia, Berula erecta* and *Juncus* sp., as green manures in corn fields with corn, bean, and squash grown using traditional techniques. The growth of weeds during the crop period was decreased by the presence of the green manures. At the same time, stimulation of bean root nodulation by *Rhizobium* was obtained with these particular green manure species, nodulation was also increased in plots with abundant weed growth. These results suggest that the presence of different secondary metabolites liberated by these green manures and by some living weeds in the field plots increase the ability of *Rhizobium* sp. to infect bean roots.

Weed control by rye, crimson clover, subterranean clover, and hairy vetch cover crops was evaluated in no-tillage corn during 1992 and 1993 at two North Carolina locations (Yenish, et al., 1996). Weed biomass reduction was similar with rye, crimson clover, and subterranean clover treatments, ranging between 19 and 95% less biomass than a conventional tillage treatment without cover. Weed biomass reduction using hairy vetch or no cover in a notillage system was similar averaging between 0 and 49%, but less than other covers approximately 45 and 90 d after planting. Weed biomass was eliminated or nearly eliminated in all cover systems with pre- plus post-herbicide treatments. Weed species present varied greatly between years and locations, but were predominantly common lambsquarters, smooth pigweed, redroot pigweed, and broadleaf signalgrass. Corn grain yield was greatest using pre-herbicides or pre-plus post-herbicides, averaging between 16 to 100% greater than the nontreated control across all cover treatments depending on the year and location.

Studies were conducted by Burgos and Talbert (1996) at the Main Agricultural Experiment Station in Fayetteville and the Vegetable Substation in Kibler, Arkansas, in 1992 and 1993 on the same plots to evaluate weed suppression by winter cover crops alone or in combination with reduced herbicide rates in no-till sweet corn and to evaluate cover crop effects on growth and yield of sweet corn. Plots seeded to rye plus hairy vetch, rye, or wheat had at least 50% fewer early season weeds than hairy vetch alone or no cover crop. None of the cover crops reduced population of yellow nutsedge. Without herbicides, hairy vetch did not suppress weeds 8 wk after cover crop desiccation. Half rates of atrazine and metolachlor (1.1 + 1.1 kg al ha (-1)) reduced total weed density more effectively in no cover crop than in hairy vetch. Half rates of atrazine and metolachlor controlled redroot pigweed, Palmer amaranth, and goosegrass regardless of cover crop. Full rates of atrazine and metolachlor [2.2 + 2.2 kg al ha (-1)] were needed to control large crabgrass in hairy vetch. Control of yellow nutsedge in hairy vetch was marginal even with full herbicide rates- Yellow nutsedge population increased and control with herbicides declined the second year, particularly with half rates of atrazine and metolachlor. All cover crops except hairy vetch alone reduced emergence, height, and yield of sweet corn. Sweet corn yields from half rates of atrazine and metolachlor equalled the full rates regardless of cover crops.

It is presently not known what effect wheat root residues have in regulating dicotyledonous (dicot) weed emergence in no-till management systems. Past research has focused almost entirely on the role of shoot residues, while the role of root residues in weed control has been essentially ignored. A field study was designed by Blum et al. (2002) to determine the respective effects of wheat shoot and root residues in regulating the emergence of three dicotyledonous weed species (morning-glory, pigweed and prickly sida). Glyphosate-desiccated wheat plots and fallow plots were surface seeded with morning-glory, pigweed and prickly sida during the spring of 1996 and 1997. Weed seedling emergence was determined for two months during each experimental period in plots with or without wheat shoot and/or root residues. The resulting data suggested that: a) the closer desiccation of the wheat cover crop occurred to the initial emergence of pigweed seedlings, the lower the emergence of that weed, b) the effects of wheat shoot and/or root residues on dicot weed seedling emergence vary considerably for the different weed species ranging from stimulation to inhibition and c) the role of root residues appear to be much more important to regulating weed emergence than that of surface shoot residues, Differences in soil moisture and temperature associated with the presence or absence of wheat residues could not be used to explain the observed treatment effects.

The growth of four summer season crops, namely *Cyamopsis tetragonoloba, Sorghum vulgare, Pennisetum americanum* and *Zea mays*, in fields with or without residues of the preceding sunflower crop was poor. Crop density, weight of seed or grain and total yield were significantly lower in sunflower fields than in the control fields (i.e. those without previous sunflower crops). Growth in terms of plant height and biomass was drastically reduced after 60 days. The effect was more pronounced in the fields where sunflower residues were allowed to decompose than in those where residues were completely removed. The soil collected from sunflower fields (both with and without residues) was found to be rich in phenolics, which in a laboratory bioassay were found to be phytotoxic. The reduced growth and yield of crops can be attributed to the release of phytotoxic phenolics from decomposing sunflower residues (Batish, et al., 2002).

John and Narwal (2003) assessed that *Leucaena leucocephala* is the most productive and versatile multipurpose legume tree in tropical agriculture and has several uses, thus called ‘miracle tree’. It is a popular choice for intercropping with annuals in hedgerow or alley cropping systems. Its allelopathic effects on oil cereals, pulses (peas and beans), oilseeds, vegetables, fodder crops, weeds, trees etc. are reviewed in this paper. The foliage and pods of *Leucaena* contain the toxic amino acid mimosine [beta-N-(3-hydroxy-4-pyridone)-alpha-aminopropionic acid] and many other phytotoxic compounds. The toxic effects of mimosine oil plants and physiology of its action also are discussed. The future areas identified for research in *Leucaena* are: (a) studies on its allelopathic compatibility with different crops to identify sustainable agroforestry systems (b) investigations to overcome its adverse allelopathic effects and mimosine toxicity and (c) possibility of using the allelopathic compounds in *Leucaena* as natural herbicides.

Marigold (*Tagetes erecta*) is another multipurpose crop with ceremonial, ornamental, medical and pharmaceutical uses, and reported antimicrobial properties. Gómez-Rodríguez et al. (2003) evaluated the effect of marigold intercropped with tomato (*Lycopersicon esculentum*) on *Alternaria solani* conidia germination *in vitro*, on conidial density and tomato leaf damage *in vivo*, as well as microclimatic changes, compared to tomato intercropped with pigweed (*Amaranthus hypochondriacus*) and monocropped tomato. They found that intercropping with marigold induced a significant ($P < 0{:}05$) reduction in tomato early blight caused by *A. solani*, by means of three different mechanisms. One was the allelopathic effect of marigold on *A. solani* conidia germination, as it was shown *in vitro* conditions; while pigweed did not have any of this inhibitory effect in conidia germination. The second way was by altering the microclimatic conditions around the canopy, particularly by reducing the number of hours per day with relative humidity 92%, thus diminishing conidial development. The third mechanism was to provide a physical barrier against conidia spreading. When intercroppped with tomato, pigweed plants worked also as a physical barrier and promoted reductions in the maximum relative humidity surrounding the canopy, but to a lesser extent than marigold.

The allelopathic properties of unburnt (UR) and burnt (BR) residues of *Parthenium hysterophorus* towards the growth of two winter crops-radish and chickpea were investigated (Singh et al., 2003c). The extracts prepared from both UR and BR were toxic to the seedling length and dry weight of the test crops, those from BR in particular. The difference was attributed to the highly alkaline nature of the extracts prepared from BR. Growth studies conducted in soil amended with UR and BR extracts and residues also revealed phytotoxic effects towards test crops, UR being more active than BR unlike crude extracts. These effects were attributed to the presence of phenolics rather than to any significant change in pH or conductivity.

Weston and Duke (2003) focused a review on a variety of weed and crop species that establish some form of potent allelopathic interference, either with other crops or weeds, in agricultural settings, in the managed landscape, or in naturalized settings. They remarked that recent research suggests that allelopathic properties can render one species more invasive to native species and thus potentially detrimental to both agricultural and naturalized settings. In contrast, allelopathic crops offer strong potential for the development of cultivars that are more highly weed suppressive in managed settings. Both environmental and genotypic effects impact allelochemical production and release over time. A new challenge that exists for future plant scientists is to generate additional information on allelochemical mechanisms of release, selectivity and persistence, mode of action, and genetic regulation. In this manner, it is possible to further protect plant biodiversity and enhance weed management strategies in a variety of ecosystems.

Ohno et al. (2000) based on previous studies that suggested phenolics from legume green manures may contribute to weed control through allelopathy, investigated if red clover (*Trifolium pratense*) residue amended field soils expressed phytotoxicity to a weed species, wild mustard (*Sinapis arvensis*). Field plots involving incorporation treatments of wheat (*Triticum aestivum*) stubble or wheat stubble plus 2530 kg ha(-1) red clover residue, were sampled at -12, 8, 21, 30, 41, 63, and 100 days after residue incorporation (DAI). Soil-water extracts (1:1, m:v) were analyzed for plant nutrients and phenolic content. Phytotoxicity of the extracts was measured using a laboratory wild mustard bioassay. There was a 20% reduction of radicle growth in the green manure treatment in comparison with the wheat stubble treatment, but only at the first sample date after residue incorporation (8 DAI). The radicle growth reduction had the highest correlation with the concentration of soluble phenolics in the soil, water extracts. Bioassays using aqueous extracts of the clover shoots and roots alone predicted a radicle growth reduction of 18% for the quantity of clover amendment rate used in the field plots. The close agreement of the predicted and observed root growth reduction at 8 DAI further supports clover residue as the source of the phytotoxicity.

The allelopathic influence of sweet potato cultivar 'Regal' on purple nutsedge was compared to the influence on yellow nutsedge under controlled conditions. Purple nutsedge shoot dry weight, total shoot length and tuber numbers were significantly lower than the controls (47, 36, and 19% inhibition, respectively). The influence on the same parameters for yellow nutsedge (35, 21, and 43% inhibition, respectively) was not significantly different from purple nutsedge. Sweet potato shoot dry weight was inhibited by purple and yellow nutsedge by 42% and 45%, respectively. The major allelopathic substance from 'Regal' root periderm tissue was isolated and tested *in vitro* on the two sedges. The I_{50}'s for shoot growth, root number, and root length were 118, 62, and 44 μg/ml, respectively, for yellow nutsedge. The I_{50}'s for root number and root length were 91 and 85 μg/ml, respectively, for purple nutsedge and the I_{50} for shoot growth could not be calculated (Peterson and Harrison, 1995). These allelopathic substances, the resin glycosides mixture extracted from the periderm tissue of storage roots from sweet potato, *Ipomoea batatas*, was bioassayed for effects on survival, development, and fecundity of the diamondback moth, *Plutella xylostella*. The resin glycoside was incorporated into an artificial diet and fed to *P. xylostella* larvae. First instars were placed individually into snap-top centrifuge vials containing artificial diet with one of six concentrations of resin glycoside material (0.00. 0.25, 0.50, 1.00, 1.50, and 2.00 μg/ml). Each replication consisted of 10 individuals per concentration, and the experiment was repeated 13 times. Vials were incubated at 25°C and a photoperiod of 14:10 (L:D) h in a growth chamber. After 6 d, surviving larvae were weighed and their sex determined, then returned to their vials. Later, surviving pupae were weighed and incubated at 25°C until moths emerged. Females were fed, mated with males from the laboratory colony, and allowed to lay eggs on aluminum foil strips. Lifetime fecundity (eggs/female) was measured. There were highly significant negative correlations between resin glycoside levels and survival and between glycoside levels and larval weight after 0 d of feeding. For larvae that lived at least 6 d, there was no additional mortality that could be attributed to the resin glycoside material. However, there was a significant positive correlation between glycoside dosages and developmental time of larvae (measured as days until pupation). Lifetime fecundity also was negatively affected at sublethal doses. Resin glycosides may contribute to the resistance in sweet potato breeding lines to soil insect pests (Jackson and Peterson, 2000). It is important to consider that the use of allelopathic crops or plant residues in agricultural management will inevitably affect other crop pests, for example insect populations.

The total resin glycoside content in the periderm of 37 sweetpotato cultivars and breeding clones was measured by HPLC and varied greatly among the clones, the highest content was 10.02 % of the periderm dry weight and the lowest was 0.05 %. Insect damage ratings of the clones and their periderm resin glycoside content were negatively correlated and all clones with high resin glycoside content exhibited moderate or low injury from insects. Resin glycosides extracted from 'Regal' periderm and incorporated into potato dextrose agar medium were inhibitory to the growth of four fungal species of sweetpotato roots; however, these fungi exhibited variable response. These observations provide evidence that sweetpotato resin glycosides contribute to the insect and disease resistance in the roots of some sweetpotato-clones (Harrison et al., 2003).

Barazani and Friedman (2001) discussed the impact of allelopathic, nonpathogenic bacteria on plant growth in natural and agricultural ecosystems. In some natural ecosystems, evidence supports the view that in the vicinity of some allelopathically active perennials (e.g., *Adenostoma fasciculatum*, California), in addition to allelochemicals leached from the shrub's canopy, accumulation of phytotoxic bacteria or other allelopathic microorganisms amplify retardation of annuals. In agricultural ecosystems allelopathic bacteria may evolve in areas where a single crop is grown successively, and the resulting yield decline cannot be restored by application of minerals. Transfer of soils from areas where crop suppression had been recorded into an unaffected area induced crop retardation without readily apparent symptoms of plant disease. Susceptibility of higher plants: to deleterious rhizobacteria is often manifested in sandy or so-called skeletal soils. The allelopathic effect may occur directly through the release of allelochemicals by a bacterium that affects susceptible plant(s) or indirectly through the suppression of an essential symbiont. The process is affected by nutritional and other environmental conditions; some may control bacterial density and the rate of production of allelochemicals. Allelopathic nonpathogenic bacteria include a wide range of genera and secrete a diverse group of plant growth-mediating allelochemicals. Although a limited number of plant growth-promoting bacterial allelochemicals have been identified, a considerable number of highly diversified growth-inhibiting allelochemicals have been isolated and characterized. Some species may produce more than one allelochemical; for example, three different phytotoxins, geldanamycin, nigericin, and hydanthocidin, were isolated from *Streptomyces hygroscopicus*. Efforts to introduce naturally produced allelochemicals as plant growth-regulating agents in agriculture have yielded two commercial herbicides, phosphinothricin, a product of *Streptomyces viridochromogenes*, and bialaphos from *S. hygroscopicus*. Both herbicides have the same mechanism of action. Many species of allelopathic bacteria that affect growth of higher plants are not plant specific, but some do exhibit specificity; for example, dicotyledonous plants were more susceptible to *Pseudomonas putida* than were monocotyledons. Differential susceptibility of higher plants to allelopathic bacteria was noted also in much lower taxonomical categories, at the subspecies level, in different cultivars of wheat, or of lettuce. Therefore, when test plants are employed to evaluate bacterial allelopathy, final evaluation must include those species that are assumed to be suppressed in nature. The release of allelochemicals from plant residues in plots of 'continuous crop cultivation' or from allelopathic living plants may induce the development of specific allelopathic bacteria.

Striga hermonthica is an obligate root-parasitic flowering plant that severely threatens cereal production in sub-Saharan Africa. A potential biological control option for reduction of crop yield-loss within the season of application is the use of soil-borne antagonists of *S. hermonthica* seed. A study was made (Ahonsi et al., 2002) with the aim to select soil-borne fluorescent pseudomonad strains capable of suppressing germination of *S. hermonthica* seeds and consequently reducing parasitism and damage to maize. An *in vitro* screening procedure was developed and was used to evaluate 460 fluorescent pseudomonad isolates from naturally suppressive soils. This resulted in the identification of 15 *Pseudomonas fluorescens/P. putida* isolates that significantly inhibited germination of *S. hermonthica* seeds. In a pot experiment using steam-sterilized soil, there was a significant reduction in the number of *S. hermonthica* plants on maize grown from seeds that were inoculated with any of the 15 bacterial isolates. Inoculation of maize seed with six of these isolates resulted not only in a reduced number of S. *hermonthica* plants, but also in an increased maize shoot biomass compared with the check. When soils inoculated with these bacterial isolates were left dried for 5 weeks after maize harvest and then planted with a second maize crop, no reduction in *S. hermonthica* parasitism was observed. This suggested that the bacteria did not persist in the soil after the first crop of maize. These results suggest that saprophytic fluorescent pseudomonads have potential for biological control of *S. hermonthica* in maize and that periodic application of bacteria, perhaps through seed treatment, may be necessary for sustained control.

Chittapur *et al.* (2001) asserted that integrated weed management systems involving catch and trap crops are needed to reduce herbicide use in agriculture and to help to control parasitic weed growth. The effective catch crops viz., fodder millet (*Panicum miliaceum*), sorghum (*Sorghum bicolor*), corn (*Zea mays*), sudangrass (*Sorghum sudanense*) have been identified for the management of *Striga asiatica*, and the cowpea (*Vigna catjang*) for *S. gesnerioides*. Cotton (*Gossypium* spp.), soybean (*Glycine max*) and peanut (*Arachis hypogaea*) are important trap crops. Intercropping of soybean or peanut with sorghum effectively controls *S. hermonthica*. Flax

(*Linum usitatissimum*) is a useful trap crop for *Orobanche ramosa, O. cernua, O. crenata*, and *O. aegyptica*. In India, sunnhemp (*Crotolaria juncea*), blackgram (*Phaseolus mungo*), greengram (*Phaseolus aureus*) and sesame (*Sesamum indicum*) have shown good potential for *Orobanche* control. Rotation of trap crop reduces the population of *Orobanche* and 3 to 4 years long rotation of catch/trap crops provides its effective control. *Sorghum*/maize/paddy (*Oryza sativa*)-tobacco (*Nicotiana tabacum*) rotation reduces the infestation and weed biomass of *Orobanche*. Relay cropping of tobacco in capsicum (*Capsicum annuum*), onion (*Allium cepa*) and peanut also reduces the incidence of *Orobanche*.

Nagabhushana *et al.* (2001) remarked that no matter how one may define sustainable agriculture, use of soil-conserving cropping practices, less synthetic herbicide inputs and better weed control would be compatible components. Previously, these components were considered incompatible, since it was widely believed that soil-conserving practices required increased pesticide use, including herbicides. However, it has been shown that environmental and ecological differences between the no-till and conventional tillage can enhance the control of certain weed species in no-till cropping systems. With proper choice and manipulation of cover crops and residues, it is often possible to reduce the herbicides use. Thus, in eliminating tillage, by utilizing the surface mulch and allelochemicals leached from a killed cover crop and using most effective herbicides when needed, weed management has become much more effective in no-till. In North Carolina, these authors have grown soybean (*Glycine max*), tobacco (*Nicotiana tabaccum*), corn (*Zea mays*), sorghum (*Sorghum bicolor*) and sunflower (*Helianthus annuus*) in killed heavy mulches of rye (*Secale cereale*) without herbicides, other than a non-selective one to kill the rye. Early-season control of broadleaf weeds such as sicklepod (*Cassia obtusifolia*), morningglory spp. (*Ipomoea* spp.), cocklebur (*Xanthium strumarium*), prickly sida (*Sida spinosa*), common purslane (*Portulaca oleracea*) and pigweed spp. (*Amaranthus* spp.) has been 80 to 95%. Rye is the most weed suppressing cover crop among several small grains and subterranean clover (*Trifolium subterraneum*) and crimson clover (*Trifolium incarnatum*) the most suppressive legumes. This approach will still enhance agricultural sustainability because: (a) productive top-soil will be conserved, (b) herbicide use (especially preemergence herbicides) can be reduced and (c) herbicides for cover crop kill and postemergence selective herbicides, even if used, have little potential for environmental contamination.

Staman *et al.* (2001) stated that in order to demonstrate that allelopathic interactions are occurring, one must, among other things, demonstrate that putative phytotoxins move from plant residues on or in the soil, the source, through the bulk soil to the root surface, a sink, by way of the rhizosphere. These authors hypothesized that the incorporation of phytotoxic plant residues into the soil would result in a simultaneous inhibition of seedling growth and a stimulation of the rhizosphere bacterial community that could utilize the putative phytotoxins as a carbon source. If true and consistently expressed, such a relationship would provide a means of establishing the transfer of phytotoxins from residue in the soil to the rhizosphere of a sensitive species under field conditions, presently, direct evidence for such transfer is lacking. To test this hypothesis, cucumber seedlings were grown in soil containing various concentrations of wheat or sunflower tissue. Both tissue types contain phenolic acids, which have been implicated as allelopathic phytotoxins. The level of phytotoxicity of the plant tissues was determined by the inhibition of pigweed seedling emergence and cucumber seedling leaf area expansion. The stimulation of cucumber seedling rhizosphere bacterial communities was determined by the plate dilution frequency technique using a medium containing phenolic acids as the sole carbon source. When sunflower tissue was incorporated into autoclaved soil (to reduce the initial microbial populations), a simultaneous inhibition of cucumber seedling growth and stimulation of the community of phenolic acid utilizing rhizosphere bacteria occurred. Thus, it was possible to observe simultaneous inhibition of cucumber seedlings and stimulation of phenolic acid utilizing rhizosphere bacteria, and therefore provide indirect evidence of phenolic acid transfer from plant residues in the soil to the root surface. However, the simultaneous responses were not sufficiently consistent to be used as a field screening tool but were dependent upon the levels of phenolic acids and the bulk soil and rhizosphere microbial populations present in the soil. It is possible that this screening procedure may be useful for phytotoxins that are more unique than phenolic acids. Such an inverse relationship between phytotoxicity and the response of rhizosphere bacterial populations was also observed by Blum et al. (2000), and such interactions provide indirect evidence for the transfer of allelochemicals from the plant root to the rhizosphere.

In relation with resistance of weeds to herbicides, Duke *et al.* (2000) mentioned that new mechanisms of action for herbicides are highly desirable to fight evolution of resistance in weeds, to create or exploit unique market niches, and to cope with new regulatory legislation. Comparison of the known molecular target sites of synthetic herbicides and natural phytotoxins reveals that there is little redundancy. Comparatively little effort has been expended on determination of the sites of action of phytotoxins from natural sources, suggesting that intensive study of these molecules will reveal many more novel mechanisms of action. These authors gave some examples of natural products that inhibit unexploited steps in the amino acid, nucleic acid, and other biosynthetic pathways: AAL-toxin, hydantocidin, and various plant-derived terpenoids.

Natural products have not been utilized as extensively for weed management as they have been for insect and plant pathogen management, but there are several notable successes such as glufosinate and the natural product-derived triketone herbicides. The molecular target sites of these compounds are often unique. Strategies for the discovery of these materials and compounds are outlined by Duke *et al.* (2002a). Numerous examples of individual phytotoxins and crude preparations with weed management potential are provided by these authors. They described an example of research to find a natural product solution of a unique pest management problem (blue-green algae in aquaculture), and mentioned the two fundamental approaches to the use of natural products for weed management: i) as a herbicide or a lead for a synthetic herbicide and ii) use in allelopathic crops or cover crops (Duke et al., 2002b).

As it was mentioned, crops may be genetically engineered for weed management purposes by making them more resistant to herbicides or by improving their ability to interfere with competing weeds. Transgenes for bromoxynil, glyphosate, and glufosinate resistance are found in commercially available crops. Other herbicide resistance genes are in development. Glyphosate-resistant crops have had a profound effect on weed management practices in North America, reducing the cost of weed management, while improving flexibility and efficacy. In general, transgenic, herbicide-resistant crops have reduced the environmental impact of weed management because the herbicides with which they are used are generally more environmentally benign and have increased the adoption of reduced-tillage agriculture. Crops could be given an advantage over weeds by making them more competitive or altering their capacity to produce phytotoxins (allelopathy). Strategies for producing allelopathic crops by biotechnology are relatively complex and usually involve multiple genes. One can choose to enhance production of allelochemicals already present in a crop or to impart the production of new compounds. The first strategy involves identification of the allelochemical(s), determination of their respective enzymes and the genes that encode them, and, the use of genetic engineering to enhance production of the compound(s). The latter strategy would alter existing biochemical pathways by inserting transgenes to produce new allelochemicals (Duke et al., 2002c).(See controversy between organic agriculture and biotechnology - *Control of Weeds and Management of Agroecosystems*, Pag. 18, first paragraph)

More sophisticated techniques will be used to search for alternative to herbicides in agroecosystems. The use of winter cover crops is beneficial to agriculture. Stanislaus and Cheng (2002) tried to design a cover crop that self-destructs in response to an environmental cue, thereby eliminating the use of herbicides and tillage to remove the cover crop in late spring. Here, this novel concept is tested in a model system. The onset of summer brings with it elevated temperatures. Using this as the environmental cue, a self-destruction cassette was designed and tested in tobacco. A heat-shock-responsive promoter was used to direct expression of the ribonuclease Barnase. Because Barnase is extremely toxic to cells, it was necessary to coexpress its inhibitor, Barstar, whose expression was under the control of the CaMV 35S promoter. The wild-type and two mutated Barnase genes, one missense and one translation attenuated, were tested. The results indicated that the translation-attenuated version of the Barnase gene was most effective in causing heat shock regulated plant death. Analysis of the T-2 progeny of a transgenic plant carrying this Barnase mutant showed that the Barnase gene expression was sixfold higher in heat-shock-treated plants compared with untreated plants. This level of Barnase gene expression was sufficient to kill transgenic plants.

Many advances in disciplines such as chemistry, biochemistry, plant breeding, genetics, engineering, and others have been applied in a positive manner to improve knowledge in weed science. The emerging field of genomics is likely to have a similar positive effect on our understanding of weeds and their management in various plant agriculture systems. Genomics involves the large-scale use of molecular techniques for identification and functional analysis of complete or nearly complete genomic complements of genes. Commercial application of genomics has already occurred for improvement in certain crop input and output traits, including improved quality characteristics and herbicide and insect resistance. Additional commercial applications of genomics in weed science will be identification of genes involved in crop ability. Genes controlling early crop root emergence, rapid early-season leaf and root development for fast canopy closure, production of allelochemicals for natural weed control, identification of novel herbicide target sites, resistance mechanisms, and genes for protecting crops against specific herbicides can and will be identified. Successful crop improvement in these areas using the tools of genomics will dramatically affect weed-crop interactions and improve crop yields while reducing weed problems. In relation to improved basic knowledge of weeds and the resulting ability to improve our weed management techniques, genomics will offer the weed science community many new and exciting research opportunities. Scientists will be able to determine the genetic composition of weed populations and how it changes over time in relation to agricultural practices, Identification of genes contributing to weediness, perennial growth habit, herbicide resistance, seed and vegetative structure dormancy, plant architecture and morphology, plant reproductive characters (outcrossing and hybridization, introgression), and allelopathy will be identified and utilized with high-throughput DNA sequencing and other genomics-based technologies. Using genomics to improve our understanding of weed biology by determining which genes function to affect the fitness, competitiveness, and adaptation of weeds in agricultural environments will allow the development of improved management strategies. Information is provided concerning the current state of molecular research in various areas of weed science and specific genomic research currently being conducted at Purdue University using transfer DNA (TDNA) activation tagging to generate large populations of mutated plants that can be screened for genes of importance to weed science (Weller et al., 2001).

6. SUPPRESSIVE SOILS

When soils are characterized by a very low level of disease development even though a virulent pathogen and susceptible host are present, they are known as suppressive soils. Biotic and abiotic elements of the soil environment contribute to suppressiveness, however most defined systems have identified biological elements as primary factors in disease suppression. Many soils possess similarities with regard to microorganisms involved in disease suppression, while other attributes are unique to specific pathogen-suppressive soil systems. The organisms' operative in pathogen suppression does so via diverse mechanisms including competition for nutrients, antibiosis and induction of host resistance (Mazzola, 2002). Non-pathogenic *Fusarium* spp. and fluorescent *Pseudomonas* spp. play a critical role in naturally occurring soils that are suppressive to *Fusarium* wilt. Suppression of take-all of wheat, caused by *Gaeumannomyces graminis* var. *tritici*, is induced in soil after continuous wheat monoculture and is attributed, in part, to selection of fluorescent pseudomonads with capacity to produce the antibiotic 2,4-diacetylphloroglucinol. Cultivation of orchard soils with specific wheat varieties induces suppressiveness to *Rhizoctonia* root rot of apple caused by *Rhizoctonia solani* AG 5. Wheat cultivars that stimulate disease suppression enhance populations of specific fluorescent pseudomonad genotypes with antagonistic activity toward this pathogen. Methods that transform resident microbial communities in a manner which induces natural soil suppressiveness have potential as components of environmentally sustainable systems for management of soilborne plant pathogens (Mazzola, 2002).

Actually, agricultural soils suppressive to soilborne plant pathogens occur worldwide, and for several of these soils the biological basis of suppressiveness has been described. Two classical types of suppressiveness are known. General suppression owes its activity to the total microbial biomass in soil and is not transferable between soils. Specific suppression owes its activity to the effects of individual or select groups of microorganisms and is transferable. The microbial basis of specific suppression to four diseases, *Fusarium* wilts, potato scab, apple replant disease, and take-all, is discussed by Weller et al. (2002). One of the best-described examples occurs in take-all decline soils. In Washington State, take-all decline results from the buildup of fluorescent *Pseudomonas* spp. that produce the antifungal metabolite 2,4-diacetyl-phloroglucinol. Producers of this metabolite may have a broader role in disease-suppressive soils worldwide. By coupling molecular technologies with traditional approaches used in plant pathology and microbiology, it is possible to dissect the microbial composition and complex interactions in suppressive soils.

In three of 12 soils obtained from agricultural fields in California, population density development of *Meloidogyne incognita* under susceptible tomato was significantly suppressed when compared to identical but methyl iodide (MI)-fumigated, *M. incognita* re-infested soils. When the 12 soils were infested with second-stage juveniles (J2) of *M. incognita* and the juveniles were extracted after 3 days, significantly fewer J2 were recovered from 9 of the 12 non-treated soils than from the MI-fumigated equivalents. In one of the 12 soils, infestation 3 weeks before planting resulted in lower nematode population densities than infestation at planting in both MI-fumigated and non-treated soil. The combination of infestation 3 weeks before planting with infestation at planting did not alter the occurrence or degree of root-knot nematode suppressiveness (Pyrowolakis et al., 2002).

Yin et al. (2004) established that for suppressive soils that have a biological nature, one of the first steps in understanding them is to identify the organisms contributing to this phenomenon. They presented a new approach for identifying microorganisms involved in soil suppressiveness. This strategy identifies microorganisms that fill a niche similar to that of the pathogen by utilizing substrate use assays in soil. To demonstrate this approach, they examined an avocado grove where a *Phytophthora cinnamomi* epidemic created soils in which the pathogen could not be detected with baiting techniques, a characteristic common to many soils with suppressiveness against *P. cinnamomi*. Substrate utilization assays were used to identify rRNA genes (rDNA) from bacteria that rapidly grew in response to amino acids known to attract *P. cinnamomi* zoospores. Six bacterial rDNA intergenic sequences were prevalent in the epidemic soils but uncommon in the non-epidemic soils. These sequences belonged to bacteria related to *Bacillus mycoides, Renibacterium salmoninarum*, and *Streptococcus pneumoniae*. We hypothesize that bacteria such as these, which respond to the same environmental cues that trigger root infection by the pathogen, will occupy a niche similar to that of the pathogen and contribute to suppressiveness through mechanisms such as nutrient competition and antibiosis.

A similar experimental approach was developed by Borneman et al. (2004) for identifying microorganisms involved in specified functions such as pathogen suppressiveness. In this approach, it was postulated that the microorganisms involved in pathogen suppressiveness could be discovered by identifying those organisms whose populations positively correlate with high levels of suppressiveness. The approach has three phases. The first phase is to identify bacterial and fungal rRNA genes (rDNA) from soils possessing various levels of suppressiveness. Ribosomal DNA sequences that are more abundant in the highly suppressive soils than in the less suppressive soils are considered candidate sequences. A method termed oligonucleotide fingerprinting of rRNA genes (OFRG) is used to obtain extensive analysis of microbial community composition. The second phase of this experimental approach is to verify the results obtained from phase one using quantitative PCR. Here, selective PCR primers for each of the candidate rDNA sequences are designed. These primers are then used to determine the relative amounts of the candidate sequences in soils possessing various levels of suppressiveness produced by several different methods such as mixing various quantities of suppressive and fumigation-induced non-suppressive soil, biocidal treatments and temperature treatments. In phase three, the organisms that consistently correlate with suppressiveness are isolated and amended to non-suppressive soils to assess their abilities to produce suppressiveness. The utility of this experimental approach was demonstrated by using it to identify microorganisms involved in suppressiveness against the plant-parasitic nematode, *Heterodera schachtii*. This general experimental approach should also be useful for the identifying microorganisms involved in functions other then pathogen suppressiveness.

Kloepper et al. (1999) discussed concepts and examples of how naturally occurring bacteria (plant-associated bacteria residing in the rhizosphere, phyllosphere, and inside tissues of healthy plants –endophytic), and introduced bacteria may contribute to management of soilborne and foliar diseases. Some introduced rhizobacteria have been found to enhance plant defences, leading to systemic protection against foliar pathogens upon seed or root-treatments with the rhizobacteria. In these cases, introduction of the rhizobacteria results in reduced damage to multiple pathogens, including viruses, fungi and bacteria. An alternative strategy to the introduction of specific antagonists is the augmentation of existing antagonists in the root environment. This augmentation may result from the use of specific organic, amendments, such as chitin, which stimulate populations of antagonists, thereby inducing suppressiveness. Intercropping or crop rotation with some tropical legumes, including velvetbean (*Mucuna deeringiana*), lead to management of phytoparasitic nematodes, partly through stimulation of antagonistic microorganisms. Some biorational nematicides, such as specific botanical aromatic compounds, also appear to induce suppressiveness through alterations in the soil microbial community.

Single isolates of bacterial endophytes, obtained from the nematode antagonistic plant species African (*Tagetes erecta*) and French (*T. patula*) marigold, were introduced into potatoes (*Solanum tuberosum*). Several bacterial species possessed activity against root-lesion nematodes (*Pratylenchus penetrans*) in soils around the root zone of potatoes, namely: *Microbacterium esteraromaticum, Tsukamurella paurometabolum*, isolate TP6, *Pseudomonas chlororaphis, Kocuria varians* and *K. kristinae*. Of these, *M. esteraromaticum* and *K. varians* depressed the population densities of root-lesion nematodes without incurring any yield penalty (tuber wet weight). No significant differences were found in the total numbers of *P. penetrans* nematodes, rhabditid nematodes or 'other' parasitic nematode species within the root tissues of bacterized potato plants compared to the unbacterized check. Overall, tuber fresh weights and tuber number were equal to or significantly lower ($P < 0.05$) in bacterized plants than their unbacterized counterpart (Sturz and Kimpinski, 2004). The authors of this study conclude that endoroot bacteria from *Tagetes* spp. can play a role in nematode suppression through the attenuation of nematode proliferation, and proposed that these nematode control properties are capable of transfer to other crops in a rotation as a beneficial 'residual' microflora – a form of beneficial microbial allelopathy.

In relation with this same type of study, Hallman et al. (1998) performed a greenhouse experiments with cotton and cucumber to determine the effects of inoculation of the parasitic nematode *Meloidogyne incognita* on population dynamics of indigenous bacterial endophytes and introduced endophytic bacterial strains JM22 (*Enterobacter asburiae*) and 89B-61 (*Pseuedomonas fluorescens*) applied as seed treatments. Internal communities of endophytic bacteria in roots were generally largest in the presence of *M. incognita*. Recovery of JM22 from cucumber roots was positively, but not significantly, associated with soilborne nematode inoculum size, except at 2 weeks after inoculation. The internal populations of 89B-61 applied to seed also increased with nematode applications. The diversity of indigenous bacterial endophytes changed within 7 d after *M. incognita* inoculation. Species richness and diversity of endophytic bacteria were slightly, but not significantly, greater for nematode-infested plants than for non-infested plants. *Alcaligenes piechaudii* and *Burkholderia pickettii* occurred only in nematode-infested plants, whereas *Bievundimonas vesicularis* was mainly isolated from nematode-free plants. *Agrobacterium radiobacter* and *Pseudomonas* spp. were the most common taxa found in both treatments, accounting for a total of 41% and 37% of the community for non-inoculated and inoculated plants, respectively. JM22 colonized cotton roots internally and was also found in high numbers on the root surface around nematode penetration sites and on root galls where the root tissue had been disruptured due to gall enlargement. Single cells of JM22 were attached to the cuticle of *M. incognita* juveniles. Sturz et al. (2000) assesses that endophytic bacteria and *M. incognita* form complex associations and an understanding of these associations will aid efforts to develop and manage microbial communities of endophytic bacteria for practical use as biocontrol agents against plant-parasitic nematodes and soil-borne pests and pathogens.

In addition, Postma et al. (2003) found that compost amended soil has also been found to be suppressive against plant diseases in various cropping systems. The level and reproducibility of disease suppressive properties of compost might be increased by the addition of antagonists. In this study, the establishment and suppressive activity of two fungal antagonists of soil-borne diseases was evaluated after their inoculation in potting soil and in compost produced from different types of organic waste and at different maturation stages. The fungal antagonists *Verticillium biguttatum*, a mycoparasite of *Rhizoctonia solani*, and a non-pathogenic isolate of *Fusarium oxysporum* antagonistic to *Fusarium* wilt, survived at high levels (10^3-10^5 CFU g^{-1}) after 3 months incubation at room temperature in green waste compost and in potting soil. Their populations faded-out in the organic household waste compost, especially in the matured product. In bioassays with *R. solani* on sugar beet and potato, the disease suppressiveness of compost increased or was similar after enrichment with *V. biguttatum*. The largest effects, however, were present in potting soil, which was very conducive for the disease as well as the antagonist. Similar results were found in the bioassay with *F. oxysporum* in carnation where enrichment with the antagonistic *F. oxysporum* had a positive or neutral effect. Postma et al. (2003) foresee great potential for the application of antagonists in agriculture and horticulture through enrichment of compost or potting soil with antagonists or other beneficial micro-organisms.

All soils are suppressive to phytonematodes to some degree. The degree of suppressiveness to them or other soilborne pathogens in a soil can be enhanced not only by infesting soil with selected microorganisms, but by the use of appropriate cropping systems and the application to soil of specific organic amendments or chemical compounds. Conducive cropping systems such as monoculture can reduce soil suppressiveness to the point where the soil is not resistant to plant parasitic nematodes (Wang et al., 2002).

Inorganic fertilizers containing ammoniacal nitrogen or formulations releasing this form of N in the soil are most effective for suppressing nematode populations. Anhydrous ammonia has been shown to reduce soil populations of *Tylenchorhynchus claytoni, Helicotylenchus dihystera*, and *Heterodera glycines*. The rates required to obtain significant suppression of nematode populations are generally in excess of 150 kg N/ha. Urea also suppresses several nematode species, including *Meloidogyne* spp., when applied at rates above 300 kg N/ha. Additional available carbon must be provided with urea to permit soil microorganisms to metabolize excess N and avoid phytotoxic effects. There is a direct relation between the amount of "protein" N in organic amendments and their effectiveness as nematode population suppressants. Most nematicidal amendments are oil cakes, or animal excrements containing 2-7% (w/w) N; these materials are effective at rates of 4-10 t/ha. Organic soil amendments containing mucopolysaccharides (e.g., mycelial wastes, chitinous matter) are also effective nematode suppressants (Rodriguez-Kábana, 1986).

Vargas-Ayala and Rodriguez-Kábana (2001) established a field microplot trial to evaluate nematode population dynamics in a rotation program utilizing nematode-suppressive and non-suppressive legumes, and nematode-host and nonhost grass species. The rotation treatments consisted of velvetbean (*Mucuna deeringiana*) or cowpea (*Vigna unguiculata*) during the first year, followed in winter by oat (*Avena sativa*), wheat (*Triticum aestivum*), rye (*Secale cereale*), rye grass (*Lolium* sp.), clover (*Trifolium* sp.), hairy vetch (*Vicia villosa*), lupine (*Lupinus* sp.) or fallow. Rotation in the second and third year consisted of soybean (*Glycine max*). Results showed that velvetbean had a generally suppressive effect on populations of root-knot (*Meloidogyne incognita*), cyst (*Heterodera glycines*), and stunt (*Tylenchorhynchus claytoni*) nematodes in soil and roots. It had little effect on populations of *Helicotylenchus dihystera*. Velvetbean rotations with winter grass species were also effective in reducing nematode population densities in soil. Soybean yields were positively correlated with velvetbean in rotations with winter grass species. High populations of *M. incognita* were negatively correlated with soybean yields. The use of velvetbean as a rotation crop assures reduction of important plant-parasitic nematodes in soil and an improvement in soybean yield.

Wang et al. (2002) made an extensive review on the use of *Crotalaria* spp. (Fabaceae) as a suppressor of agricultural pests, particularly nematodes. These authors summarized the knowledge of the efficacy of *Crotalaria* spp. for plant-parasitic nematode management, described the mechanisms of nematode suppression, and outline prospects for using this crop effectively. They mentioned that *Crotalaria* is a poor host to many plant-parasitic nematodes including *Meloidogyne* spp., *Rotylenchulus reniformis*, *Radopholus similis*, *Belonolaimus longicaudatus*, and *Heterodera glycines*. It is also a poor or non-host to a large group of other pests and pathogens. Besides, *Crotalaria* is competitive with weeds without becoming a weed, grows vigorously to provide good ground coverage for soil erosion control, fixes nitrogen, and is a green manure. However, most *Crotalaria* species are susceptible to *Pratylenchus* spp., *Helicotylenchus* sp., *Scutellonema* sp. and *Criconemella* spp.

Crotalaria species are used as preplant cover crops, intercrops, or soil amendments. When used as cover crops, *Crotalaria* spp. reduces plant-parasitic nematode populations by: i) acting as a nonhost or a poor host, ii) producing allelochemicals that are toxic or inhibitory, iii) providing a niche for antagonistic flora and fauna, and iv) trapping the nematode.

A non-host to a nematode species is a plant in which the nematode fails to reproduce. A plant is considered as resistant to nematodes when these fail to live inside the host or early dead in the host; they decreased the production of eggs; or their growth or development are inhibited by the plant (Wang et al., 2002).

Allelopathic effects of the plant against nematodes were described as a mechanism of suppression of nematodes. Soler-Serratosa et al. (1996) evaluated the nematicidal activity of thymol, a phenolic monoterpene present in the essential oils of several plant families. Thymol was added to soil at rates 25-250 ppm. Initial and final population densities of *Meloidogyne arenaria, Heterodera glycines, Paratrichodorus minor*, and Dorylaimoid nematodes, as well as disease incidence, declined sharply with increased dosages of thymol. Thymol was also applied at 0, 50, 100, and 150 ppm to soil in combination with 0, 50, and 100 ppm benzaldehyde, an aromatic aldehyde present in nature as a moiety of plant cyanogenic glycosides. Combinations in which benzaldehyde was applied at 100 ppm showed synergistic effects in suppressing initial and final soil populations of *M. arenaria* and *H. glycines*. Significant reductions in root galling and cyst formation in soybean were attributable to thymol at $\geq$ 50 ppm.

Hallmann and Sikora (1996) confirmed that endophytic fungi isolated from the cortical tissue of surface sterilized tomato roots collected from field plots produced secondary metabolites in nutrition broth that were highly toxic to *Meloidogyne incognita*. Especially strains of *Fusarium oxysporum* were highly active with 13 of 15 strains producing culture filtrates toxic to nematodes. They investigated also the mechanism of action of the toxic metabolites produced by the non-pathogenic *F. oxysporum* strain 162 with proven biological control of *M. incognita* in pot experiments. These metabolites reduced *M. incognita* mobility within 10 min of exposure. After 60 min, 98% of juveniles were inactivated. Fifty percent of juveniles with exposure of 5 h were dead, and 24 h exposure resulted in 100% mortality. In a bioassay with lettuce seedlings metabolite concentrations >100 mg/l reduced the number of *M. incognita* juveniles on the roots comparing to the water control. The *F. oxysporum* toxins were highly effective towards sedentary parasites and less effective towards migratory endoparasites. Non-parasitic nematodes were not influenced at all. Metabolites of strain 162 also reduced significantly the growth of *Phytophthora cactorum, Pythium ultimum* and *Rhizoctonia solani in vitro.*

Three out of 15 bacterial strains preselected for antagonistic activity in different pathosystems showed biocontrol activity towards *Meloidogyne incognita* on lettuce and tomato as described by Hoffmann-Hergarten et al. (1998). They found that seed treatment with the rhizobacteria *Pseudomonas* sp. W34 or *Bacillus cereus* S18 resulted in significant reductions in root galling and enhanced seedling biomass. The yield response of *M. incognita*-infested tomato was tested in a long-term pot experiment using three antagonistic bacteria, i.e., *Pseudomonas* sp. W34, *Bacillus cereus* S18 and *Bacillus subtilis* VM1-32. Significant reduction in *M. incognita* gall index was observed within 18 weeks after inoculation with all three bacterial strains. *B. cereus* S18 caused a 9 % yield increase when compared with the nematode control and thereby compensated for the yield loss due to nematode infection. Early maturity of fruits on *M. incognita*-infested tomato plants after inoculation of *B. cereus* S18 was observed when compared with both the nematode and the untreated control.

Vargas-Ayala et al. (2000) hypothesized that the induction of soil suppressiveness to plant parasitic nematodes that occurs following planting of velvetbean (*Mucuna deeringiana*) is associated with the development of an antagonistic microflora in soils and rhizospheres. They performed a crop rotation study in microplots, consisting of three crop cycles. Cycle 1 involved planting of either velvetbean or cowpea (*Vigna unguiculata*) in the first spring. Cycle 2 during the next fall and winter was fallow or cover-cropped with wheat *(Triticum aestivum)* or crimson clover (*Trifolium incarnatum*). Cycle 3 the next spring was soybean (*Glycine max*). Rhizosphere fungal populations were significantly smaller on velvetbean than on cowpea at the end of cycle 1. The use of velvetbean in cycle 1 significantly decreased rhizosphere bacterial populations on crops in cycle 2, compared to treatments which had cowpea in cycle 1. Velvetbean also influenced bacterial diversity, generally increasing frequency of bacilli, *Arthrobacter* spp. and *Burkholderia cepacia*, while reducing fluorescent pseudomonads. Some of these effects persisted through cycle 3. Fungal diversity was influenced in cycle 1 by velvetbean; however, effects generally did not persist through cycles 2 and 3. The results indicate that the use of velvetbean in a cropping system alters the microbial communities of the rhizosphere and soil, and they are consistent with the hypothesis that the resulting control of nematodes results from induction of soil suppressiveness.

6.1. Resistence through natural compounds

Sometimes, natural compounds that confer resistance to a plant against nematode infestation could be of foreign origin (Reitz et al., 2000). Recent studies have shown that living and heat-killed cells of the rhizobacterium *Rhizobium etli* strain G12 induce in potato roots systemic resistance to infection by the potato cyst nematode *Globodera pallida*. To better understand the mechanisms of induced resistance, Reitz et al. (2000) focused on identifying the inducing agent. Since heat-stable bacterial surface carbohydrates such as exopolysaccharides (EPS) and lipopolysaccharides (LPS) are essential for recognition in the symbiotic interaction between *Rhizobium* and legumes, their role in the *R. etli*-potato interaction was studied. EPS and LPS were extracted from bacterial cultures, applied to potato roots, and tested for activity as an inducer of plant resistance to the plant-parasitic nematode. Whereas EPS did not affect *G. pallida* infection, LPS reduced nematode infection significantly in concentrations as low as 1 and 0.1 mg ml(-1). Split-root experiments, guaranteeing a spatial separation of inducing agent and challenging pathogen, showed that soil treatments of one half of the root system with LPS resulted in a highly significant (up to 37%) systemic induced reduction of *G. pallida* infection of potato roots in the other half. The results clearly showed that LPS of *R. etli* G12 act as the inducing agent of systemic resistance in potato roots.

In relation with *Crotalaria* spp. secondary metabolites, it is well known that these plants produce pyrrolizidine alkaloids and monocrotaline which have high vertebrate toxicity and could potentially be toxic to nematodes; but it is possible also that the low C/N ratio of *Crotalaria* may also contribute to its allelopathic effect against nematodes. Materials with very low C/N or high content of ammonia will either result in plasmolysis of nematodes, or proliferation of nematophagous fungi due to the release of NH4+-N (Rich and Rahi, 1995).

6.2. Soil Amendments

A study was conducted to determine the effects of combinations of organic amendments and benzaldehyde on plant-parasitic and non-parasitic nematode populations, soil microbial activity, and plant growth (Chavarria-Carvajal et al., 2001). Pine bark, velvetbean and kudzu were applied to soil at rates of 30 g/kg and paper waste at 40 g/kg alone and in combination with benzaldehyde (300 mul/kg), for control of plant-parasitic nematodes. Pre-plant and post-harvest soil and soybean root samples were analyzed, and the number of parasitic and non-parasitic nematodes associated with soil and roots were determined. Soil samples were taken at 0, 2, and 10 weeks after treatment to determine population densities of bacteria and fungi. Treatment effects on microbial composition of the soybean rhizosphere were also determined by identifying microorganisms. Bacteria strains were identified rising fatty acid analysis, and fungus identification was done rising standard morphological measurements and appropriate taxonomic keys. Results showed that most amendments alone or in combination with benzaldehyde reduced damage from plant parasitic nematodes. Benzaldehyde applied alone or in combination with the amendments exerted a selective action on the activity and composition of microbial populations in the soybean rhizosphere. In control soils the bacterial flora was predominantly Gram-negative, while in soils amended with velvetbean or kudzu alone or with benzaldehyde. Gram-positive bacteria were dominant. Mycoflora promoted by the different amendments or combinations with benzaldehyde included species of *Aspergillus, Myrothecium, Penicillium,* and *Trichoderma.*

Calvet et al. (2001) evaluated the survival of two species of plant parasitic nematodes: the root-lesion nematode *Pratylenchus brachyurus*, and the root-knot nematode *Meloidogyne javanica*, in saturated atmospheres of 12 natural chemical compounds. The infectivity of two isolates of arbuscular mycorrhizal fungi: *Glomus mosseae* and *Glomus intraradices*, under identical experimental conditions, was also determined. All the compounds tested exerted a highly significant control against *M. javanica* and among them, benzaldehyde, salicilaldehyde, borneol, p-anisaldehyde and cinnamaldehyde caused a mortality rate above 50% over *P. brachyurus*. The infectivity of *G. intraradices* was inhibited by cinnamaldehyde, salicilaldehyde, thymol, carvacrol, p-anisaldehyde, and benzaldehyde, while only cinnamaldehyde and thymol significantly inhibited mycorrhizal colonization by *G. mosseae*.

When soybean plant responses to *Meloidogyne incognita* infestation were compared to resistant (Bryan) and susceptible (Brim) cultivars at 0, 1, 3, 10, 20, and 34 days after infestation, Qiu and collaborators (1997) observed that the resistant cultivar had higher chitinase activity than the susceptible cultivar at every sample time beginning at the third day. Results from isoelectric focusing gel electrophoresis analyses indicated that three acidic chitinase isozymes with isoelectric points (pIs) of 4.8, 4.4, and 4.2 accumulated to a greater extent in the resistant compared to the susceptible cultivar following challenge. SDS-PAGE analysis of root proteins revealed that two proteins with molecular weights of approximately 31 and 46 kD accumulated more rapidly and to a higher level in the resistant than in the susceptible cultivar. Additionally, three major protein bands (33, 22, and 20 kD) with chitinase activity were detected with a modified SDS-PAGE analysis in which glycolchitin was added into the gel matrix. These results indicate that higher chitinase activity and early induction of specific chitinase isozymes may be associated with resistance to root-knot nematode in soybean.

Antagonists, most likely favored by selected cover crops, include mainly fungal egg parasites, trapping fungi, endoparasitic fungi, fungal parasites of females, endomycorrhizal fungi, planthealth promoting rhizobacteria, and obligate bacterial parasites. There are several hypotheses on how cover crops can enhance nematode-antagonistic activities. A series of ecological events may be involved. The decomposing organic material is a significant event because the bacteria which proliferate after organic matter incorporation become a food base for microbiovorous nematodes. In turn, these nematodes serve as a food source for nematophagous fungi (Wang et al., 2002). Leguminous crops enhance nematophagous fungi better than other crops. Rootknot symptoms were reduced more by alfalfa amendments in a 4-year microplot test than by chemical fertilization of plots (Mankau, 1968). Microplots amended with alfalfa meal increased nematode-trapping fungal activity of *Drechmeria coniospora* (Van Den Boogert *et al.*, 1994). Pea enhanced the densities and species diversity of nematode-trapping fungi more than white mustard or barley. In addition, formation of conidial traps of nematode-trapping fungi was more prevalent in the pea rhizosphere than in root-free soil (Persmark and Nordbring-Hertz, 1997; Persmark and Jansson, 1997).

Being a legume, *Crotalaria juncea* has characteristics that may make the crop useful for nematode antagonism. Plant exudates from *Crotalaria* spp. were selective for microbial species antagonistic to phytopathogenic fungi and nematodes (Rodriguez-Kábana and Kloepper, 1998). The changes in soil enzymatic activity was investigated by Chavarría and Rodriguez-Kábana (1998) when they incorporated four organic amendments (velvetbean, kudzu, pine bark, and urea-N) to the soil to evaluate their effects on the root-knot nematode (*Meloidogyne incognita*). The amendments were applied to nematode-infested soil at rates of 0 to 5% and placed in pots planted with 'Davis' soybean (*Glycine max*). The number of *M. incognita* juveniles and nonparasitic nematodes associated with the soil and root tissues were determined after 8 weeks. Soil samples were taken at 0, 2, and 10 weeks after amendment application for determination of soil enzyme activities. Most organic amendments were effective in reducing root galling and root-knot nematodes and increasing populations of non-parasitic nematodes. Catalase and esterase were sharply increased by most rates of velvetbean, kudzu, and pine bark. Application of velvetbean, kudzu, and urea to soil stimulated urease activity in proportion to the amendment rates. Results suggest that complex modes of action operating in amended soils are responsible for suppression of *M. incognita*.

In relation with nematode-trapping fungi, a major group of nematode antagonists, they can be enhanced by incorporation of residues of *C. juncea*. These fungi have been categorized into two groups: parasitic and saprophytic. The saprophytic group consists of predators characterized by sticky three-dimensional networks and non-spontaneous trap formation. These fungi have a saprophytic and a predatory (trap formation) phase. In the presence of nematodes, or even exudates and homogenates of nematodes, trap formation is induced. The parasitic group consists of nematode-trapping fungi that form constricting rings, adhesive knobs, or adhesive branches. These fungi form traps spontaneously, and thus are more effective trappers (Wang, 2000).

Among these two groups of nematode trapping fungi, the population densities of parasitic fungi are more likely to be enhanced by organic matter due to the rich microbial flora and fauna. The nematode trapping by these fungi are not nematode species- or trophic group specific, therefore the enhancement of nematode-trapping fungi by organic matter incorporation should lead to increased trapping of plant-parasitic nematodes (Wang, 2000).

Soil amended with *C. juncea* to give a 1:100 (w:w) concentration, enhanced parasitic nematode-trapping fungi, nematode egg parasitic fungi, vermiform stage parasites, and bacterivorous nematode population densities more efficiently than soil amended with chopped pineapple tissues or non-amended soil. *Crotalaria juncea* amendment enhanced the population densities of nematode-trapping fungi and the percentage of eggs parasit-ized by the fungi. Enhancement of nematode-trapping fungi was most effective in soils that had not been treated with 1,3-dichloropropene for at least 5 months. Suppression of *R. reniformis* by *C. juncea* amendment was correlated with parasitic nematode-trapping fungi, fungal egg parasites, and bacterivorous nematodes. Nematode-trapping fungi population densities were higher in *C. juncea* planted plots than weed fallow plots. However, four months after removal of *C. juncea*, and replacement with pineapple plants, the population densities of nematode-trapping fungi greatly decreased (Wang, 2000).

Suppressive cropping systems rely on the use of precisely defined sequences of crops to increase populations and activities of naturally occurring antagonistic microorganisms in soil. Some crops such as velvetbean (*Mucuna deerengiana*) produce compounds which are directly toxic to nematodes and stimulate microbial antagonism to plant parasitic nematodes. These 'active' crops when included in cropping systems can increase suppressiveness of the system against nematodes. There are a number of active crops throughout the world which can be used in a practical manner to enhance naturally occurring biological control of plant parasitic nematodes (Wang, 2000)

Rich and Rahi (1995) conducted two greenhouse trials to determine the influence of ground seed of castor (*Ricinus communis*), crotalaria (*Crotalaria spectabilis*), hairy indigo (*Indigofera hirsuta*), and wheat (*Triticum aestivum*) on tomato (*Lycopersicon esculentum*) growth and egg mass production of *Meloidogyne javanica* (test 1) or *M. incognita* (test 2). Ground seed from each plant species was individually mixed with an air-dried, fine sandy soil at rates of 0, 0.5, 1.0, and 2.0% (w/w). The mixtures were placed in one-liter plastic pots, and water was added to bring soil to field capacity. After ten days, 0 or 10 000 *M. javanica* or *M. incognita* eggs and juveniles were added to each pot. A single 'Homestead' tomato seedling was transplanted into each pot and allowed to grow for 70 days in test 1 and 75 days in test 2. Compared to the non-amended control, egg mass production was significantly reduced by all treatments except the 0.5% levels of wheat and castor and the 1.0% castor treatment. The 2.0% levels of ground seed of *Crotalaria* and hairy indigo almost completely suppresses egg mass production of both *M. javanica* or *M incognita.* With the exception of the 1% *Crotalaria* treatment in test 2, total plant weight did not differ between treatments and the control.

Morris and Walker (2002) mixed dried ground plant tissues from 20 leguminous species with *Meloidogyne incognita*-infested soil at 1, 2 or 2.5, and 5% (w/w) and incubated for 1 week at room temperature (21 to 27°C). Tomato ('Rutgers') seedlings were transplanted into infested soil to determine nematode viability. Most tissues reduced gall numbers below the non-amended controls. The tissue amendments that were most effective include: *Canavalia ensiformis*, *Crotalaria retusa*, *Indigofera hirsuta*, *I. nummularifolia*, *I. spicata*, *I. suffruticosa*, *I. tinctoria*, and *Tephrosia adunca*. Although certain tissues reduced the tomato dry weights, particularly at the higher amendment rates (5%), some tissues resulted in greater dry weights. These non-traditional legumes, known to contain bioactive phytochemicals, may offer considerable promise as soil amendments for control of plant-parasitic nematodes. Not only do these legumes reduce root-knot nematodes but some of them also enhance plant height and dry weight.

Nematode management is rarely successful in the long term with unitactic approaches. It is important to integrate multiple-tactics into a strategy. *Crotalaria* offers the potential to be one of the tactics. Some *Crotalaria* species are potential cover crops for managing several important plant-parasitic nematodes including *Meloidogyne* spp. and *R. reniformis*. Unfortunately, the residual effects are short term (a few months). *Crotalaria*, a poor host, generally helps reduce nematode population densities, but the number of nematodes will resurge on subsequent host crops. The damage threshold level, especially on longer-term crops, will often be reached or exceeded. This scenario strongly suggests that integrating the *Crotalaria* rotation system with other nematode management strategies is necessary. Among the possibilities for integration are crop resistance, enhanced crop tolerance, selection for fast growing crop varieties, soil solarization, and biological control. Chemical nematicides should be avoided in a cropping system if the objective is to enhance nematode-antagonistic microorganisms in the cropping system. Several studies have demonstrated the destructive effect of fumigation treatments to nematode antagonistic microorganisms. *Crotalaria juncea* amendments failed to enhance nematode-trapping fungi populations in soils that were recently treated with 1,3-dichloropropane. Wang et al. (2002) concluded that the major impediment to using *Crotalaria* is its short-term effect in agricultural production systems, and suggested that integrating other pest management strategies with *Crotalaria* could offer promising nematode management approaches.

7. CONCLUSIONS AND FINAL REMARKS

In this chapter the different roles that allelopathy can play as a bioregulator tool in agriculture are discussed. A wide spectrum of studies are given on allelopathic plants and other organisms, the chemistry involved in these studies, the mechanisms of action of some allelochemicals, and the use of allelopathy to control weeds, pests (nematodes) and diseases.

Many arguments can be given in favor of organic and sustainable agricultural practices as new forms of resources management such as multiple cropping, cover crops, organic compost, and biological controls for pests. Allelopathy is an emerging tool for a more biorational management of natural resources. However, allelopathy is not a simple panacea for the solution of ecological problems in agroecosystems or in natural ecosystems. It has not been considered as a universal ecological phenomenon;

allelopathy is a challenging and exigent matter of study. At present we have proof that secondary metabolites are involved with biotic interactions, and that allelopathic effects may restrict or enhance, alone or in relation with other environmental factors (light, temperature, humidity and nutrients), the distribution, health and growth of species in natural, artificial or managed communities. In the search for application of allelopathy knowledge is crucial to understand other biotic interactions (competition, defense against herbivory) and also the actual and full significance of a mixture of secondary metabolites all together acting in the environment (Anaya, 1999).

Allelopathy typically operates through the release, modification, and joint action of a number of allelochemicals in a particular situation, and transitions through the soil add to the complications for explaining the phenomenon. The frontiers in research on allelopathy include isolation of additional compounds that may be involved, and determining more precisely how allelochemicals production is regulated and how the compounds function to inhibit growth. Such information may allow modification of crop plants so they have enhanced capability for weed suppression. Alternatively, new herbicides, pesticides, and growth regulators may be developed from some of plant and microorganisms compounds (Einhellig, 1989).

In the study of biological interactions mediated by secondary metabolites it is very important to perform multidisciplinary investigations in a long term approach in order to understand these interactions from an holistic point of view and make use of them for beneficial purposes in the management of natural resources in agroecosystems.

8. REFERENCE

Abrahim, D., Braguini, W.L., Kelmer-Bracht, A.M., Ishii-Iwamoto, E.L. Effects of four monoterpenes on germination, primary root growth, and mitochondrial respiration of maize. J Chem Ecol 2000; 26:611-624.

Ahonsi, M.O., Berner, D.K. Emechebe, A.M., Lagoke, S.T. Selection of rhizobacterial strains for suppression of germination of *Striga hermonthica* (Del.) Benth. Seeds. Biol Con 2002; 24:143-152.

Anaya, A.L., S. del Amo. Allelopathic potential of *Ambrosia cumanensis* H.B.K. (Compositae) in a tropical zone of México. J Chem Ecol 1978; 4:289-304.

Anaya, A.L., Ramos, L., Cruz, R., Hernández, J., Nava, V. Allelopathy in Mexican Traditional Agroecosystems: A case study in Tlaxcala. J Chem Ecol 1987; 13:2083-2101.

Anaya, A.L., Hernández-Bautista, B.E., Torres-Barragán, A., León-Cantero, J. Jiménez-Estrada, M. Phytotoxicity of cacalol and some derivatives obtained from the roots of *Psacalium decompositum* (A. Gray) H. Rob & Brettell (Asteraceae), "matarique" or "maturín". J Chem Ecol 1996; 22:393-403.

Anaya, A.L., Pelayo-Benavides, H.R. Allelopathic potential of *Mirabilis jalapa* L. (Nyctaginaceae): Effects on germination, growth and cell division of some plants. Allelopathy J 1997; 4:57-68.

Anaya, A.L. Allelopathy as a Tool in the Management of Biotic Resources in Agroecosystems. Crit Rev Plant Sci 1999; 18: 697-739.

Anaya, A.L., Mata, R., Sims, J., González-Coloma, A., Cruz-Ortega, R., Guadaño, A., Hernández-Bautista B.E., Ríos, G., Gómez-Pompa, A. Allelochemical potential of *Callicarpa acuminata* (Verbenaceae). J Chem Ecol 2003; 29:2725-2740.

Bais, H.P., Walker, T.S., Stermitz, F.R., Hufbauer, R.A., Vivanco, J.M. 2002. Enantiomeric-dependent phytotoxic and antimicrobial activity of (±)-catechin. A rhizosecreted racemic mixture from spotted knapweed. Plant Physiol 2003; 128: 1173-1179.

Barazani, O. Friedman J. Allelopathic bacteria and their impact on higher plants. Crit Rev Microbiol 2001; 27:41-55.

Barberi, P. Weed management in organic agriculture: are we addressing the right issues? Weed Res 2002; 42:177-193.

Batish, D.R., Tung, P., Singh, H., Kohli, R.K. Phytotoxicity of sunflower residues against some summer season crops. J Agron Crop Sc 2002; 188:19-24.

Bhowmik, P.C., Inderjit. Challenges and opportunities in implementing allelopathy for natural weed management. Crop Prot 2003; 22:661-671.

Blum U., King, L.D., Gerig, T.M., Lehman, M.E., Worsham, A.D. Effects of clover and small grain cover crops and tillage techniques on seedling emergence of some dicotyledonous weed species. Amer J Altern Agricul 1997; 4:146–161.

Blum, U. Effects of microbial utilization of phenolic acids and their phenolic acid breakdown products on allelopathic interactions. J Chem Ecol 1998; 24:685-708.

Blum, U., Staman, K.L., Flint, L.J., Shafer, S. Induction and/or selection of phenolic acid-utilizing bulk-soil and rhizosphere bacteria and their influence on phenolic acid phytotoxicity. J Chem Ecol, 2000; 26:2059-2078.

Blum, U., King, L.D., Brownie, C. Effects of wheat residues on dicotyledonous weed emergence in a simulated no-till system. Allelopathy J 2002; 9:159-176.

Borneman, J., Olatinwo, R., Yin, B., Becker J.O. An experimental approach for identifying microorganisms involved in specified functions: utilisation for understanding a nematode suppressive soil. Australasian Plant Pathol 2004; 33:151-155.

Burgos, N.R., Talbert, R.E. Weed control and sweet corn (*Zea mays* var *rugosa*) response in a no-till system with cover crops. Weed Sci 1996; 44:355-361.

Butler, L.G. Chemical communication between the parasitic weed *Striga* and its crop host. A new dimension of allelochemistry. In: *Allelopathy: Organism, Processes and Application.* Inderjit, Dakshini K.M.M., Einhellig, F.A. eds. ACS Symposium Series Washington, DC 1995; 582:158-168.

Bouwmeester, H.J., Matusova, R., Zhongkui, S., Beale, M.H. Secondary metabolite signalling in host-parasitic plant interactions.Curr Opin Plant Biol 2003; 6:358-364.

Calera, M.R., Mata, R., Anaya A.L, Lotina, B. 5-*O*-b-D-galactopyranosyl-7-methoxy-3', 4'-dihydro-4-phenylcoumarin, an inhibitor of photophosphorylation in spinach chloroplasts. Photosynth Res 1995a; 45:105-110.

Calera, M.R., Soto, F., Sanchez, P., Bye, R., Hernandez-Bautista, B. Anaya, A.L., Lotina-Hennsen, Mata, R. Biochemically active sesquiterpene lactones from *Ratibida mexicana.* Phytochemistry 1995b; 40:419-425.

Calera, M.R. Anaya, A.L., and Gavilanes-Ruiz, M. Effect of a phytotoxic resin glycoside on the activity of the H^+-ATPase from plasma membrane. J Chem Ecol 1995c; 21:289-297.

Calera, M., Mata, R., Lotina-Hennsen, B., Anaya, A.L. Uncoupling behavior of the 4-phenylcoumarins in spinach chloroplasts- Structure-activity relationships. J Agr Food Chem 1996; 44:2966-2969.

Calvet C, Pinochet J, Camprubi A, Estaun V, Rodriguez-Kabana R. Evaluation of natural chemical compounds against root-lesion and root-knot nematodes and side-effects on the infectivity of arbuscular mycorrhizal fungi. J Plant Pathol 2001; 107:601-605.

Chavarria-Carvajal JA, Rodriguez-Kabana R, Kloepper JW, Morgan Jones G. Changes in populations of microorganisms associated with organic amendments and benzaldehyde to control plant-parasitic nematodes. Nematropica 2001; 31:165-180.

Chavarria-Carvajal, J.A., Rodriguez-Kabana, R. Changes in soil enzymatic activity and control of Meloidogyne incognita using four organic amendments. Nematropica 1998; 28:7-18.

Chittapur, B.M., Hunshal, C.S., Shenoy H. Allelopathy in parasitic weed management: Role of catch and trap crops. Allelopathy J 2001; 8:147-159.

Chou, C.H. Allelopathy and sustainable agriculture. ACS Symp. Ser. 1995; 582:211-223.

Conklin, A.E., Erich, M.S., Liebman, M., Lambert, D., Gallandt, E.R., Halteman, W.A. Effects of red clover (*Trifolium pratense*) green manure and compost soil amendments on wild mustard (*Brassica kaber*) growth and incidence of disease. Plant Soil 2002; 238:245-256.

Cruz-Ortega, R., Anaya, A.L., and Ramos, L. Effects of allelopathic compounds of maize pollen on respiration and cellular division of watermelon. J Chem Ecol 1988; 14:71-86.

Cruz-Ortega, R., Anaya, A.L., Hernández, B.E., Laguna G. Effects of allelochemical stress produced by *Sicyos deppei* on seedling root ultrastructure of *Phaseolus vulgaris* and *Cucurbita peppo.* J Chem Ecol 1998; 24: 2039-2057.

Del Amo, S., Anaya, A.L. Effect of some sesquiterpenic lactones on the growth of certain secondary tropical species. J Chem Ecol 1978; 4:305-313.

Doi, F., Ohara, T., Ogamino, T., Sugai, T., Higashinakasu, K., Yamada, K., Shigemori, H., Hasegawa, K., Nishiyama, S. Plant-growth inhibitory activity of heliannuol derivatives. Phytochemistry 2004; 65:1405-1411.

Duke, S.O., Plant terpenoids as pesticides. In Keeler, R.F. and A.T. Tu (Eds.). Handbook of Natural Toxins. Vol. 6. Marcel Dekker, New York, 1991.

Duke, S.O., Romagni, J.G., Dayan, F.E. Natural products as sources for new mechanisms of herbicidal action. Crop Prot 2000; 19:583-589.

Duke, S.O., Dayan, F.E., Rimando, A.M., Schrader, K.K., Aliotta, G., Oliva, A., Romagni, J.G. Chemicals from nature for weed management. Weed Sci 2002a; 50:138-151.

Duke, S.O., Rimando, A,M., Baerson, S.R., Scheffler, B.E., Ota, E. Belz, R.G. Strategies for the use of natural products for weed management. J Pesticide Sci 2002b; 27:298-306.

Duke, S.O., Scheffler, B.E., Dayan, F.E., Dyer, W.E. Genetic engineering crops for improved weed management traits. Crop Biotechnol. ACS Symp Ser 2002c ; 829: 52-66.

Duke, S.O., Baerson, S.R., Dayan, F.E., Rimando, A.M., Scheffler, B.E., Tellez, M.R., Wedge, D.E., Schrader, K.K., Akey, D.H., Arthur, F.H., De Lucca, A.J., Gibson, D.M., Harrison, H.F., Peterson, J.K., Gealy, D.R., Tworkoski, T., Wilson, C.L., Morris, J.B. Research on natural products for pest management. United States Department of Agriculture, Agricultural Research Service. Pest Manag Sci 2003; 59:708-717.

Einhellig, F.A. Interaction among allelochemicals and other stress factors of the plant environment. In: *Allelochemicals, Role in Agriculture and Forestry,* GR Waller ed., ACS Symposium Series, Vol. 330. American Chemical Society, Washington DC, USA 1987; pp. 343–357.

Einhellig, F.A. The Physiology of Allelochemical Action: Clues and Views. In: *Allelopathy*. From Molecules to Ecosystems. Science Publishers, Reigosa, M.J. and Pedrol, N. eds. Inc. Enfield, NH, USA, 2002.

Elakovich, S.D. Terpenoids as models for new agrochemicals. ACS 380. American Chemical Society, Washington, D.C. 1988; 250-261.

Evenari, M. Germination inhibitors. Bot Rev 1949; 15:153-194.

Fischer, N.H. The function of mono and sesquiterpenes as plant germination and growth regulators. In *The Science of Allelopathy.* Putnam, A.R. and Ch-Sh. Tang Eds. John Wiley & Sons. 1986; pp. 203-218.

Fischer, N.H. Plant terpenoids as allelopathic agents. In: *Ecological Chemistry and Biochemistry of Plant Terpenoids.* Harborne, J.B. and F.A. Tomas-Barberan (Eds.). Proceedings of the Phytochemical Society of Europe. Clarendon Press-Oxford.: 1991; pp. 377-398.

Fischer, N.H., Weidenhamer, J.D., Bradow, J.M. Dihydroparthenolide and other sesquiterpene lactones stimulate witchweed germination. Phytochemistry 1989; 28: 2315-2317.

Fischer, N.H., Weidenhamer, J.D., Riopel, J.L., Quijano, L., Menelaou, M.A. Stimulation of witchweed germination by sesquiterpene lactones: a structure-activity study. Phytochemistry 1990; 29:2479-2483.

Fischer, N.H., Williamson, G.B., Weidenhamer, J.D., Richardson, D.R. In search of allelopathy in the Florida scrub: The role of terpenoids. J Chem Ecol 1994; 20:1355-1380.

Galindo, J.C.G., Hernández, A, Dayan F.E., Macías, F.A., Duke, S.O. Dehydrozaluzanin C, a natural sesquiterpenolide, causes rapid plasma membrane leakage. Phytochemistry 1999; 52:805-813.

Galindo, J.C.G., de Luque, A.P., Jorrin, J., Macias, F.A. SAR studies of sesquiterpene lactones as *Orobanche cumana* seed germination stimulants. J Agric Food Chem 2002; 50:1911-1917.

Gómez-Rodríguez, O, Zavaleta-Mejía, E., and González-Hernández, V.A. Livera-Muñoz M., and Cárdenas-Soriano, E. Allelopathy and microclimatic modification of intercropping with marigold on tomato early blight disease development. Field Crops Res 2003; 83:27-34.

Haig, T. Application of hyphenated chromatography-mass spectrometry techniques to plant allelopathy research. J Chem Ecol 2001; 27:2363-2396.

Hallmann, J. and Sikora, R.A.. Toxicity of fungal endophyte secondary metabolites to plant parasitic nematodes and soil-borne plant pathogenic fungi. Eurp J Plant Pathol 1996; 102:155-162.

Hallmann, J., Quadt-Hallmann, A., Rodriguez-Kabana, R., and Kloepper, J.W. Interactions between *Meloidogyne incognita* and endophytic bacteria in cotton and cucumber. Soil Biol Biochem 1998; 30:925-937.

Harper, J.L. Population Biology of Plants. Academic Press, New York, 1977.

Harrison, H.F., Peterson, J.K., Jackson, D.M., and Snook, M.E. Periderm resin glycoside contents of sweetpotato, *Ipomoea batatas* (L.) Lam. clones and their biological activity. Allelopathy J 2003; 12: 53-60.

Hoffmann-Hergarten, S., Gulati, M.K., Sikora R.A.. Yield response and biological control of *Meloidogyne incognita* on lettuce and tomato with rhizobacteria. J Plant Diseas Protect 1998; 105:349-358.

Hyder, P.W., Fredrickson, E.L., Estell, R.E., Tellez, M., Gibbens, R.P. Distribution and concentration of total phenolics, condensed tannins, and nordihydroguaiaretic acid (NDGA) in creosotebush (*Larrea tridentata*). Biochem Systemat Ecol 2002; 30: 905-912.

Inderjit. Plant phenolics in allelopathy. Bot Rev 1996; 62:186-202.

Inderjit and Dakshini, K.M.M. Allelopathic potential of the phenolics from the roots of *Pluchea lanceolata*. Physiol Plant 1994; 92:571-576.

Inderjit and del Moral, R. Is separating resource competition from allelopathy realistic? Bot Rev 1997; 63:221-230.

Inderjit and Mallik, A.U. Effect of phenolic compounds on selected soil properties. Forest Ecol Manag 1997; 92:11-18.

Inderjit, Muramatsu, M., Nishimura, H. On the allelopathic potential of certain terpenoids, phenolics, and their mixtures, and their recovery from soil. Can J Bot 1997; 75: 888-891.

Inderjit, Keating, K.I. Allelopathy: Principles, procedures, processes, and promises for biological control. Adv Agron 1999; 67:141-232.

Inderjit and Duke, S.O. Ecophysiological aspects of allelopathy. Planta 2003; 217:529-539.

Jackson, D.M. and Peterson, J.K. Sublethal effects of resin glycosides from the periderm of sweetpotato storage roots on *Plutella xylostella* (Lepidoptera: Plutellidae). J Econom Entomol 2000; 93:388-393.

Jimenez-Osornio J.J., S.R. Gliessman. Allelopathic interference in a wild mustard (*Brassica campestris* L.) and broccoli (*Brassica oleracea* L. var. *italica*) intercrop agroecosystem. In: *Allelochemicals, Role in Agriculture and Forestry*, Waller G.R. ed. ACS Symp. Ser. 330. American Chemical Society, Washington DC, USA, 1987; pp. 262–274.

John, J. and Narwal, S.S. Allelopathic plants. 9. *Leucaena leucocephala* (Lam.) de Wit. Allelopathy J 2003; 12:13-36.

Keyes, W.J., Taylor, J.V., Apkarian, R.P., Lynn, D.G. Dancing together. Social controls in parasitic plant development. Plant Physiol 2001; 127:1508-1512.

Kloepper, J.W., Rodriguez-Kabana, R., Zehnder, G.W., Murphy, J.F., Sikora, E., Fernandez. C. Plant root-bacterial interactions in biological control of soilborne diseases and potential extension to systemic and foliar diseases. Australas. Plant Pathol 1999; 28:21-26.

Langenheim, J. Higher plant terpenoids: A phytocentric overview of their ecological roles. J Chem Ecol 1994, 20:1223-1280.

Lehman, M.E., Blum, U. Evaluation of ferulic acid uptake as a measurement of allelochemical dose: Effective concentration. J Chem Ecol 1999; 25:2585-2600.

Macias, F.A., J.C.G. Galindo, G.M. Massanet. Potential allelopathic activity of several sesquiterpene lactone models. Phytochemistry 1992; 31:1969-1977.

Macias, F.A., Galindo, J.C.G., Castellano, D. Velasco, R.F. Natural products as allelochemicals. 11. Sesquiterpene lactones with potential use as natural herbicide models (I): trans,trans-germacranolides. J Agric Food Chem 1999; 47:4407-4414.

Macias, F.A., Galindo, J.C.G., Castellano, D., Velasco, R. Sesquiterpene lactones with potential use as natural herbicide models. 2. Guaianolides. J Agric Food Chem 2002a; 48:5288-5296.

Macias, F.A., Torres, A., Galindo, J.L.G., Varela, R., Alvarez, J,A., Molinillo, J.M.G. Bioactive terpenoids from sunflower leaves cv. Peredovick (R). Phytochem 2002b; 61:687-692.

Mankau, R. Reduction of root-knot disease with organic amendments under semifield conditions. Plant Disease Reporter 1968; 52:315-319.

Mazzola, M. Mechanisms of natural soil suppressiveness to soilborne diseases. Antonie Van Leeuwenhoek Internat. J Gen Molec Microbiol 2002; 81:557-564.

Morris, J.B., Walker, J.T. Non-traditional legumes as potential soil amendments for nematode control. J Nematol 2002; 34:358-361.

Muscolo, A., Panuccio, M.R., Sidari, M. The effect of phenols on respiratory enzymes in seed germination - Respiratory enzyme activities during germination of *Pinus laricio* seeds treated with phenols extracted from different forest soils. Plant Growth Regul. 2001; 35:31-35.

Nagabhushana, G.G., Worsham, A.D., Yenish, J.P. Allelopathic cover crops to reduce herbicide use in sustainable agricultural systems. Allelopathy J 2001; 8:133-146.

Nava-Rodriguez, V. Hernandez-Bautista, B.E., Cruz-Ortega, R., Anaya, A.L. Allelopathic potential of beans (*Phaseolus* spp.), other crops, and weeds from Mexico. Allelopathy J

Neori, A., Reddy, K.R., Ciskova-Koncalova, H., Agami, M. Bioactive chemicals and biological-biochemical activities and their functions in rhizospheres of wetland plants. Bot Rev 2000; 66:350-378.

Ngouajio, M., McGiffen, M.E. Going organic changes weed population dynamics. Hort Technol 2002; 12:590-596.

Nilsson, M.C., Gallet, C., Wallstedt, A. Temporal variability of phenolics and batatasin-III in *Empetrum hermaphroditum* leaves over an eight-year period: interpretations of ecological functions. Oikos 1998; 81:6-16.

Ohno, T., Doolan, K., Zibilske, L.M., Liebman, M., Gallandt, E., Berube, C. Phytotoxic effects of red clover amended soils on wild mustard seedling growth. Agricult Ecosyst Environm 2000; 78:187-192.

Pereda-Miranda, R., R. Mata, A.L. Anaya, J.M. Pezzuto, D.B.M. Wickramaratne, A.D. Kinghorn. Tricolorin A, major phytogrowth-inhibitor from *Ipomoea tricolor*. J Natural Products 1993; 56:571-582.

Pérez de Luque, A., Galindo, J.C.G., Macías, F.A., Jorrin, J. Sunflower sesquiterpene lactone models induce *Orobanche cumana* seed germination. Phytochemistry 2000, 2000; 53:45-50.

Persmark, L., H.B. Jansson. Nematophagous fungi in the rhizosphere of agricultural crops. Federation of European Microbiological Societies. Microbiology Ecology 1997; 22:303-312.

Persmark, L., B. Nordbring-Hertz. Conidial trap formation of nematode-trapping fungi in soil and soil extracts. Federation of European Microbiological Societies. Microbiology Ecology 1997; 22:313-323.

Peterson, J.K., Harrison, H.F. Sweet-potato allelopathic substance inhibits growth of purple nutsedge (*Cyperus-rotundus*). Weed Technol 1995; 9:277-280.

Picman, A.K. Biological activities of sesquiterpene lactones. Biochem Syst Ecol 1986; 4: 255-281.

Picman, J., Picman, A.K. Autotoxicity in *Parthenium hysterophorus* and its possible role in control of germination. Biochem Syst Ecol 1984; 12:287-292.

Politycka, B. Phenolics and the activities of phenylalanine ammonia-lyase, phenol-beta-glucosyltransferase and beta-glucosidase in cucumber roots as affected by phenolic allelochemicals. Acta Physiol Plant 1998; 20:405-410.

Politycka, B. Ethylene-dependent activity of phenylalanine ammonia-lyase and lignin formation in cucumber roots exposed to phenolic allelochemicals. Acta Soc Bot Pol 1999; 68:123-127.

Postma J., Montanari M., Van den Boogert P.H.J.F. Microbial enrichment to enhance the disease suppressive activity of compost. European J Soil Biol 2003; 39:157-163.

Putnam, A.R. Allelochemicals from plants as herbicides. Weed Technol 1988; 2:510–518.

Pyrowolakis, A., Westphal, A., Sikora, R.A,. Becker, J.O. Identification of root-knot nematode suppressive soils. Appl Soil Ecol 2002; 19:51-56.

Qiu, J., Hallmann, J., Kokalis-Burelle, N., Weaver, D.B., Rodriguez-Kabana, R., Tuzan, S. Activity and Differential Induction of Chitinase Isozymes in Soybean Cultivars Resistant or Susceptible to Root-knot Nematodes. J Nematology 1997; 29:523-530.

Randhir, R., Lin, Y.T., Shetty, K. Stimulation of phenolics, antioxidant and antimicrobial activities in dark germinated mung bean sprouts in response to peptide and phytochemical elicitors. Process Biochem 2004; 39:637-646.

Reigosa, M.J., Souto, X.C., Gonzalez, L. Effect of phenolic compounds on the germination of six weeds species. Plant Growth Regul 1999; 28:83-88.

Reitz M., Rudolph K., Schroder I., Hoffmann-Hergarten S., Hallmann J., Sikora R.A. Lipopolysaccharides of *Rhizobium etli* strain G12 act in potato roots as an inducing agent of systemic resistance to infection by the cyst nematode *Globodera pallida*. Appl Environ Microbiol 2000; 66(8):3515-3518.

Rice, E.L. Allelopathy. 2nd Ed., Academic Press, Inc. London 1984.

Rich, J.R., Rahi, G.S. Suppression of *Meloidogyne javanica* and *M. incognita* on tomato with ground seed of castor, *Crotalaria*, hairy indigo, and wheat. Nematropica 1995; 25:159-164.

Ridenour, W.M., Callaway, R.M. The relative importance of allelopathy in interference: the effect of an invasive weed on a native bunchgrass. Oecologia 2001; 126:444-450.

Rodríguez-Kábana, R. Organic and inorganic nitrogen amendments to soil as nematode suppressants. J Nematology 1986; 18:129-134.

Rodríguez-Kábana, R., Kloepper, J.W. Cropping systems and the enhancement of microbial activities antagonistic to nematodes. Nematropica 1998; 28:144.

Romagni, J.G., Meazza, G., Nanayakkara, N.P.D., Dayan, F.E. The phytotoxic lichen metabolite, usnic acid, is a potent inhibitor of plant p-hydroxyphenylpyruvate dioxygenase. FEBS Letters 2000; 480:301-305.

Romero-Romero Teresa, Anaya, A.L., Cruz-Ortega, R. Screening the effects of allelochemical stress produced by some plants on cytoplasmic protein synthesis pattern of crop plants. J Chem Ecol 2002; 28:617-629.

Roshchina, V.V. Molecular-cellular mechanisms in pollen allelopathy. Allelopathy J 2001; 8:11-28.

Sashida, Y., Nakata, H. Shimomura, H. and Kagaya, M.. Sesquiterpene lactones from Pyrethrum flowers. Phytochemistry 1983; 22:1219-1222.

Seigler, D.S. Chemistry and mechanisms of allelopathic interactions. Agronomy J 1996; 88:876–885.

Singh, H.P., Batish, D.R., Kaur, S., Kohli, R.K. Phytotoxic interference of *Ageratum conyzoides* with wheat (*Triticum aestivum*). J Agron Crop Sci 2003a; 189:341-346.

Singh, H.P., D.R. Batish, R.K. Kohli. Allelopathic interactions and allelochemicals: New possibilities for sustainable weed management. Crit Rev Plant Sc 2003b; 22:239-311.

Singh, H.P., Batish, D.R., Pandher, J.K., Kohli, R.K. Assessment of allelopathic properties of *Parthenium hysterophorus* residues. Agric Ecosyst Environ 2003c; 95:537-541.

Soler-Serratosa, A., Kokalis-Burelle, N., Rodriguez-Kabana, R., Weaver, C.F., King P.S. Allelochemicals for control of plant parasitic nematodes. Nematropica 1996; 26:57-71.

Souto, C., Pellissier, F., Chiapusio, G. Allelopathic effects of humus phenolics on growth and respiration of mycorrhizal fungi. J Chem Ecol 2000; 26:2015-2023.

Spencer, G.F., R.B. Wolf, D. Weisleder. Germination and growth inhibitory sesquiterpenes from Iva Axillaris seeds. J Nat Prod 1984; 47:730-732.

Staman, K, Blum, U., Louws, F., Robertson, D. Can simultaneous inhibition of seedling growth and stimulation of rhizosphere bacterial populations provide evidence for phytotoxin transfer from plant residues in the bulk soil to the rhizosphere of sensitive species? J Chem Ecol 2001; 27:807-829.

Stanislaus, M.A., Cheng, C.L. Genetically engineered self-destruction: an alternative to herbicides for cover crop systems. Weed Sci 2002; 50:794-801.

Stevens, K.L., G.B. Merrill. Sesquiterpene lactones and allelochemicals from *Centaurea* species. In: *The Chemistry of Allelopathy: Biochemical Interactions Among Plants*. Thompson, A.C. ed. ACS Symp. Ser. 268. American Chemical Society. Washington, D.C., 1985.

Sturz, A.V., Christie, B.R., Nowak, J. Bacterial endophytes: Potential role in developing sustainable systems of crop production. Crit Rev Plant Sci 2000; 19:1-30.

Sturz, A.V., Kimpinski, J. Endoroot bacteria derived from marigolds (*Tagetes* spp.) can decrease soil population densities of root-lesion nematodes in the potato root zone. Plant Soil 2004; 262:241–249.

Toxicity of Pesticides. Pesticide Education Program in Penn State's College of Agricultural Sciences. The Pennsylvania State University. http://pubs.cas.psu.edu/FreePubs/pdfs/uo222.pdf 2004

Upadhyay, R.K., Mukerji, K.G., Rajak, R.L. Integrated Pest Management. In: *IPM System in Agriculture. Vol. I. Principles and Perspectives*. Upadhyaya R.K., Mukerji K.G., Rajak R.L., eds. Aditya Books Pvt. Ltd., New Delhi, 1996; pp. 1-23.

Van Den Boogert, P.H.J.F., Velvis, H., Ettema, C.H., Bouwman, L.A. The role of organic matter in the population dynamics of the endoparasitic nematophagus fungus *Drechmeria coniospora* in microcosms. Nematologica 1994; 40:249-257.

Vargas-Ayala, R., Rodriguez-Kabana, R., Morgan-Jones, G., McInroy, J.A., Kloepper J.W. Shifts in soil microflora induced by velvetbean (*Mucuna deeringiana*) in cropping systems to control root-knot nematodes. Biol Control 2000; 17:11-22.

Vargas-Ayala, R., Rodriguez-Kabana, R. Bioremediative management of soybean nematode population densities in crop rotations with velvetbean, cowpea, and winter crops. Nematropica 2001; 31:37-46.

Vokou, D., Douvli, P., Blionis, G.J., Halley, J.M. Effects of monoterpenoids, acting alone or in pairs, on seed germination and subsequent seedling growth. J Chem Ecol 2003; 29: 2281-2301.

Waller G.R., C.H. Chou. Phytochemical Ecology: Allelochemicals, Mycotoxins, and Insect Pheromones and Allomones. Institute of Botany. Academia Sinica. Taipei, Taiwan 1989.

Waller, G.R. Introduction. In: Recent Advances in Allelopathy. Macías, F.A., Galindo, J.C.G., Molinillo, J.M.G., and Cutler, H.G. eds. Vol. I A science for the Future. International Allelopathy Society and Servicio de Publicaciones de la Universidad de Cádiz, España 1999

Waller, G.R., M.C. Feng, Y. Fujii. Biochemical analysis of allelopathic compounds: plants, microorganisms, and soil secondary metabolites. In: *Principles and Practices in Plant Ecology : Allelochemical interactions,* Inderjit, K.M.M., Dakshini and Foy C.L. eds. CRC Press, Boca Raton, FL, USA. 1999, pp. 75-98.

Wang, K.H., Sipes, B.S., Schmitt, D.P. *Crotalaria* as a cover crop for nematode management: A review. Nematropica 2002; 32 (1):35-57.

Weidenhamer J.D. Distinguishing resource competition and chemical interference: overcoming the methodological impasse. Agron J 1996; 88:866–875.

Weller, S.C., Bressan, R.A., Goldsbrough, P.B., Fredenburg, T.B., Hasegawa, P.M. The effect of genomics on weed management in the 21st century. Weed Sci 2001; 49: 282-289.

Weller, D.M., J.M. Raaijmakers, B.B.M. Gardener, L.S. Thomashow. Microbial populations responsible for specific soil suppressiveness to plant pathogens. Ann Rev Phytopath 2002; 40:309-348.

Weston L.A. Utilization of allelopathy for weed management in agroecosystems. Agron J 1996; 88:860–866.

Weston, L.A., Duke, S.O. Weed and crop allelopathy. Crit Rev Plant Sci 2003; 22:367-389.

Wigchert, S.C.M., Zwanenburg, B. A critical account on the inception of *Striga* seed germination. J Agric Food Chem 1999; 47:1320-1325.

Yenish, J.P., Worsham, A.D., York, A.C. Cover crops for herbicide replacement in no-tillage corn (Zea mays).Weed Technol 1996; 10:815-821.

Yin, B; Scupham, AJ; Menge, JA; Borneman, J. Identifying microorganisms which fill a niche similar to that of the pathogen: a new investigative approach for discovering biological control organisms. Plant Soil 2004; 259:19-27.

Yu, J.Q., Matsui, Y. Effects of root exudates of cucumber (*Cucumis sativus*) and allelochemicals on ion uptake by cucumber seedlings. J Chem Ecol 1997; 23: 817-827.

SCOTT W. MATTNER

THE IMPACT OF PATHOGENS ON PLANT INTERFERENCE AND ALLELOPATHY

Department of Primary Industries, Knoxfield Centre, Victoria, Private Bag 15, Ferntree Gully Delivery Centre, 3156, VIC, Australia.
E-mail : scott.mattner@dpi.vic.gov.au

Abstract. Pathogenesis can have both detrimental and beneficial impacts on plant fitness. As such, pathogens are important forces that influence the structure and dynamics in natural and manipulated plant ecosystems. Plant production and numbers within a community are constrained by environmental limitations, which are often mediated through plant interference. Competition for resources and allelopathy (chemical interactions) are the two most important ways that plants interfere with each other. This chapter reviews the effects of pathogens on the competitiveness and allelopathic ability of their hosts. In most cases, pathogens reduce the competitive ability of their host, making the host prone to displacement by neighbouring, resistant plants. However, pathogens may simultaneously increase the allelopathic ability of their hosts, thereby offsetting their loss in competitiveness to varying degrees. Evidence for enhanced allelopathy by infected plants comes in two forms: (i) pathogens stimulate the production of secondary metabolites by plants, many of which are implicated in allelopathy (eg phenolics), and (ii) field, glasshouse and bioassay studies showing that infected plants may suppress their neighbours more than healthy plants, under conditions of low competition. By conferring the benefit of increased allelopathy on their hosts, pathogens may maintain a self-advantage through increasing the survival chances of their hosts and ultimately themselves. The enhanced allelopathy of infected plants supports the 'new function' hypothesis, which suggests that pathogens evolve toward a mutualistic relationship with their host through the appearance of strains with beneficial effects on the host in addition to their detrimental effects.

1. INTRODUCTION

Plant pathogens are disease agents that live in or on their hosts and there obtain nutriment to the overall detriment of the plant. Fungal pathogens, which cause about 70% of all major crop diseases (Deacon, 1997), are often characterised on the basis of two extremes in trophism, as necrotrophs or biotrophs (Lutrell, 1974; Parbery 1996). Necrotrophic organisms obtain nutriment from necrotic host tissues, which they kill prior to colonisation. Consequently, these pathogens although destructive to the host have little effect on the physiology of the rest of the plant. Biotrophic pathogens draw nutriment directly from living host tissue and can have a critical effect on host physiology. Many thorough reviews concerning the impact of infection on host physiology already occur in the literature (Goodman et al., 1986; Burdon, 1987; Ayres, 1991; Sutic and Sinclair, 1991). By example though, pathogens can alter the partitioning of assimilates and dry matter; reduce net photosynthesis and increase respiration; increase water loss through transpiration and vulnerability to drought;

Inderjit and K.G. Mukerji (eds.),
Allelochemicals: Biological Control of Plant Pathogens and Diseases, 79–101.

and reduce the overall growth, reproductive capacity and yield of their hosts. Consequently, pathogens are important factors that influence the composition and dynamics of natural and manipulated plant ecosystems.

Despite pathogens having an overall detrimental effect on their host, some aspects of infection are beneficial (Burdon, 1991; Parbery, 1996). For example, infection may promote vegetative growth (Catherall, 1966; Wennström and Ericson, 1991), deter or limit grazing by herbivores (Morgan and Parbery, 1980), or increase the host's scavenging ability for nutrients and water (Ahmad et al., 1982; Paul and Ayres, 1988). By conferring some benefit on their host, these pathogens maintain a self-advantage through increasing the chances of survival of their host and ultimately themselves. This is particularly important for obligate biotrophs (e.g. rusts, Uredinales; powdery mildews, Erysiphaceae) that require the host for self-preservation.

In contrast to previous chapters that consider mechanisms for exploiting allelopathy for pathogen control, this chapter addresses the reverse situation – the impact of pathogens on plant interference and allelopathy. It reviews evidence that pathogens decrease the competitiveness and simultaneously increase the allelopathic ability of their hosts. The ability of pathogens to increase their host's allelopathic ability may be an important means by which pathogens confer a benefit on their host. In so doing, these pathogens encourage the continuation of their host's genotype into proceeding generations. Furthermore, the ability of pathogens to enhance allelopathy in their hosts may be one way that obligate biotrophic pathogens are evolving toward multualism.

2. PLANT INTERFERENCE AND ALLELOPATHY

2.1. *Components of Interference*

Harper (1961) introduced the term 'interference' to encompass all effects placed on an organism by the proximity of neighbours. Plant interference is 'the response of an individual plant to its total environment as this is modified by the presence and/or growth of other individuals' (Begon et al., 1986). Interference may have a positive or negative effect on plant growth and may occur between plants of different species, between individuals within the same species or even between individual organs on a single plant. Some of the important ways plants modify the environment in each other's presence are through: competition, allelopathy, protection (e.g. when an unpalatable plant protects a neighbouring palatable plant from grazing), direct stimulation (e.g. when nitrogen fixed by a legume becomes available to a non-legume), direct contact (e.g. when a thorny plant mechanically abrades neighbouring plants in the wind), and non-competitive inhibition (e.g. when a tree provides a rubbing post for grazers and so encourages local trampling). Of these mechanisms, competition is the most dominant in shaping plant populations (Jolliffe, 1988; Tilman, 1988), but research is increasingly demonstrating the importance of allelopathy (Siegler, 1996; Wardle et al., 1998). Comprehensive reviews of competition (Milthorpe, 1961; Tilman,

1988; Grace and Tilman, 1990; Casper and Jackson, 1997) and allelopathy (Putnam and Duke, 1978; Rice, 1984; Inderjit et al., 1995; Anaya, 1999) already occur in the literature.

The view of competition in the present chapter is 'the interaction between individuals brought about by the shared requirements for a resource in limited supply, and leading to a reduction in the survivorship, growth and/or reproduction of the individuals concerned' (Begon et al., 1986). Plants mostly compete for the resources of light, water and nutrients (Donald, 1963), and less frequently for carbon dioxide, oxygen and space (Trenbath, 1974). In the exact sense of the definition, plants do not compete so long as these resources are in excess of the needs of both. When plants do compete, however, the outcome always reduces the growth of at least one of the competitors. The outcome of competition between two plants depends on the ability of each species to secure and utilise resources, i.e. their competitiveness. A highly competitive plant may be one that has a high rate of uptake of a particular resource or a low requirement for that resource (Grace and Tilman, 1990). Plants may differ in their ability to compete for individual resources, while environmental differences may also influence their overall competitive ability. Differences in competitive ability, in turn, help to structure the composition of mixed plant communities.

2.2. *Allelopathy*

Molisch first advanced the term allelopathy in 1937. It derives from the Greek words 'allelon', meaning of each other, and 'pathos', meaning to suffer (Mandava, 1985). Despite the origin of its root words, Molisch used the term to refer to the chemical interactions between all plants (higher plants and microorganisms), including stimulatory as well as inhibitory effects. Some authors have considered that the concept covers only inhibitory effects (Rice, 1974; Putnam, 1985; Boyette and Abbas, 1995), yet others exclude microorganisms from the definition (Putnam and Duke, 1978; Putnam, 1985; Pratley, 1996). Many inhibitory chemicals produced by plants, however, stimulate growth at low concentrations (Liu and Lovett, 1993; Pratley, 1996), and microorganisms can mediate allelopathy (Rice, 1992; Bremner and McCarty, 1993). For these reasons, the definition of allelopathy used in this chapter closely follows that of Molisch's, 'the beneficial and detrimental chemical interaction among [plant] organisms including microorganisms' (Rice, 1984). Although some authors have confused allelopathy with competition, the distinguishing feature is that allelopathy involves an addition of a chemical to the environment, whereas competition involves the shared utilisation of some limited factor required for growth (Muller, 1969; Rice 1984).

Any chemical produced by a plant (donor) that stimulates or inhibits the growth of a neighbour (receiver or receptor) is broadly termed an allelochemical. Typically, allelochemicals are secondary metabolites (Whittaker and Feeney 1971; Rice, 1984; Rizvi et al., 1992), produced as by-products of the acetate and shikimic acid pathways. They may also form as degradation products from the action of microbial enzymes

(Putnam, 1985). Initially, the reason why plants devote resources to the production of these compounds was not understood as they were regarded as functionless waste products (Mothes, 1955). It is now increasingly accepted, however, that these compounds function as defensive agents against pathogens, insects and neighbouring plants (allelopathy). There is an enormous diversity of allelochemicals produced by plants (Bansal, 1994), with classification based on chemical structure (Whittaker and Feeney, 1971; Mandava, 1985; Putnam 1985) or on origin and chemical properties (Rice, 1984). For example, Rice (1984) recognised 14 main categories of allelochemicals. Of these groups, however, the phenolics are considered by many as the most important (Putnam and Duke, 1979; Mandava, 1985; Inderjit, 1996).

Putative allelochemicals have been isolated from a variety of different plant organs, including shoots, roots, flowers, rhizomes, fruit and seed (Rice, 1984). They are stored in cell vacuoles so as not to interfere with the donor plant itself (Chou, 1989). Furthermore, secondary metabolites may be bound to sugars as glycosides or occur as polymers and crystals rendering them ineffective against the donor (Whittaker and Feeney, 1971). The release of allelochemicals into the environment may occur through volatilisation, root exudation, leaching or plant residue decomposition (Rice, 1984).

Allelochemicals induce a wide range of symptoms in receiver plants, ranging from sudden wilting and death (e.g. tomato (*Lycopersicon esculentum*) grown in the vicinity of black walnut (*Juglans nigra*), Hale and Orcutt, 1987), to the more common subtle changes in growth. Determination of this reaction depends on the mode of action, the concentration, and the susceptibility of the receiver plant to the allelochemical. Allelopathic effects may be direct, such as affecting plant metabolism and growth, or indirect, such as altering of soil properties and nutrient status (Inderjit and Weiner, 2001). Rizvi et al. (1992) explained 12 plant functions that allelochemicals may affect, including membrane permeability, stomata function and photosynthesis, and cytology and ultrastructure.

3. THE INFLUENCE OF PATHOGENS ON INTERFERENCE AND ALLELOPATHY

Natural plant populations increase in production and number of individuals until constrained by environmental limitations (Burdon, 1987). The constraint of plant growth by the environment is often mediated through plant interference. Therefore, the ability of plants to interfere with their neighbours is important in determining their abundance in a community. Plant pathogens generally reduce the development, production and longevity of their hosts. In plant communities, this 'burden of a parasite' may result in a partly unoccupied niche that resistant plants re-inhabit. Overall, pathogens play a significant role in the ecology of plant communities by maintaining (Peters and Shaw, 1996) or reducing (Burdon, 1991) species diversity, driving succession (Van der Putten et al., 1993), ensuring plants do not establish under their parents (Augsberger, 1984) and help determine the long-term composition of plant communities (Dobson and Crawley, 1994).

3.1. The Impact of Pathogens on Competition

Several pioneering publications on plant competition suggest that pathogens might alter the balance of competition in mixed communities in favour of the resistant components (de Wit, 1960; Harper, 1977). Since then, there have been numerous reviews and theoretical interpretations of the impact of pathogens in natural communities and on competition (Chilvers and Brittain 1971; Burdon, 1982; 1987; 1991; Dinoor and Eshed, 1984; Gates et al., 1986; Alexander, 1990; Ayres and Paul, 1990; Clay, 1990; Paul, 1990; Dobson and Crawley, 1994; Jarosz and Davelos, 1995; Alexander and Holt, 1998; Mattner 1998), but empirical experimentation is less extensive. Empirical studies have usually involved binary mixtures of a host and a non-host, grown under optimal conditions of nutrient and water availability, and inoculated with a single, copious and homogenous dose of inoculum. With few exceptions, these studies show that the influence of a host-specific pathogen reduces the vigour of the host, rendering it less able to compete with a neighbouring non-host species (Burdon, 1987; Ayres and Paul, 1990; Mattner, 1998). As such, the combination of the 'burden of a parasite' and competition can have a devastating effect on a plant. For example, Groves and Williams (1975) demonstrated that the combined effects of competition from subterranean clover (*Trifolium subterraneum*) and rust infection (*Puccinia chondrillina*) reduced the yield of skeleton weed (*Chondrilla juncea*) by 94%. This combined effect was more marked than the action of either the competitor (yield was reduced by 70%) or the rust (yield was reduced by 51%) alone. Similarly, Friess and Maillet (1996) found that infection by cucumber mosaic virus reduced the vegetative yield of its host purslane (*Portulaca oleracea*), with this effect intensifying when infected plants were in competition with healthy plants. In mixtures of common lambsquarters (*Chenopodium album*) and corn (*Zea mays*) or beetroot (*Beta vulgaris*), foliar infection by *Ascochyta caulina* reduced the competitiveness of its host (lambsquarters) and increased the yield of non-host crops. In corn, infection negated the effects of competition from lambsquarters altogether, highlighting its potential as a biological control agent (Kempenaar et al., 1996). Despite pathogens reducing the competitiveness of their hosts, the effect of infection is often greater on a neighbouring non-host than on the host itself. For example, in mixtures of groundsel (*Senecio vulgaris*) and lettuce (*Lactuca sativa*), rust (*Puccinia lagenophorae*) increased the biomass of lettuce (the non-host) markedly, while only reducing that of its host groundsel marginally (Paul and Ayres, 1987a).

Competitive stress for limited resources intensifies as plant density increases (Shinozaki and Kira, 1956). Consequently, the effect of a pathogen on the competitiveness of its host is most devastating at high densities. For example, rust reduced the competitive ability of groundsel in mixtures with healthy groundsel (Paul and Ayres, 1986) or with lettuce (Paul and Ayres, 1987a) more so at high densities than at low densities. Similarly, Ditommaso and Watson (1995) demonstrated that anthracnose (*Colletotrichum coccodes*) was more detrimental to the growth of velvetleaf (*Abutilon theophrasti*) in mixtures with soybean (*Glycine max*) at high plant densities. Conversely, the performance of the non-host, relative to the infected host, generally increases as density increases. For example, the ability of lettuce to compete with rusted groundsel was greater at high densities than at low densities (Paul and Ayres, 1987a).

Considering that competition occurs for shared resources that are in limited supply, it is not surprising that resource availability influences the interaction between competition and infection by pathogens. Paul and Ayres (1987b) examined the effect of rust on competition between infected and healthy groundsel under conditions of drought. They found that rust reduced the competitiveness of groundsel, however, this was more so in droughted plots than in well-watered plots. They concluded that water stress is important in determining the impact of rust in mixed populations in the field. Owing to the effects pathogens can have on nutrient uptake, Paul and Ayres (1990) also studied the effect of nutrient supply on the interaction between rust and competition. Under high nutrition, groundsel was more competitive than shepherd's purse (*Capsella bursa-pastoris*). This superiority was lost, however, following the inoculation of groundsel with rust. In contrast, shepherd's purse had a greater competitive ability than groundsel under nutrient-poor conditions. Despite this, it did not increase its advantage when groundsel was rusted.

While most studies show that pathogens decrease the competitiveness of their hosts and increase that of neighbouring non-hosts, there are several exceptions. For example, Catherall (1966) found that for most of the year, barley yellow dwarf virus reduced the competitive ability of its host, perennial ryegrass (*Lolium perenne*), when grown with white clover (*Trifolium repens*). During spring, however, the virus stimulated tillering in ryegrass and increased its competitiveness. In another example, the rust *Puccinia pulsatillae* sterilises its host, *Pulsatilla pratensis*, by inhibiting flowering. Yet, infected plants are more vigorous and produce more leaves than healthy plants. In a natural community, Wennström and Ericson (1991) found that diseased plants had a greater survival rate than healthy plants, which they postulated was due to their greater competitiveness. The effects of pathogens on allelopathy may also mediate their influence on competition.

3.2. The Impact of Pathogens on Allelopathy

The term allelopathy is seldom used in plant pathology (Rice, 1984) even though, by definition, allelopathy includes chemical interactions involving microorganisms. Furthermore, the same or similar secondary metabolites implicated in allelopathy between plants are also important in plant pathology in: enhancing the germination of fungal spores (Odunfa, 1978); antibiosis between microorganisms (Di Pietro, 1995); regulating fungal growth and development (Calvo et al., 2002); pathogen/host recognition (Nicol et al., 2003); the development of disease symptoms (Daly and Deverall, 1983); the promotion of infection through the suppression of the host (Toussoun and Patrick, 1963); the breaking of fungistasis (Mol, 1995); and host resistance to pathogens. Similarly, some authors do not consider chemical interactions by microorganisms as part of allelopathy (Putnam and Duke, 1978; Putnam and Tang, 1986; Pratley, 1996). Rice (1984) explained that this reluctance is due to the chemicals involved not always escaping to the environment. This rationale seems invalid, however, because the chemicals involved do enter the environment of the pathogen or the plant. Moreover, microorganisms can mediate allelopathy between plants (Rice, 1992; Bremner and McCartey, 1993). Such difficulties may have hindered the study of the impact of pathogens on plant allelopathy in the past.

Both Rice (1984) and Einhellig (1995) hypothesised that pathogens enhance their host's allelopathic ability, but few studies have observed such a relationship (Tang et al., 1995). This is despite several investigations on the effects of pathogens on plant competition (Burdon, 1987; Ayres and Paul, 1990; Section 3.1), all of which could potentially include allelopathic interactions. However, competition may have obscured allelopathy in these experiments (Trenbath, 1974), since competition is usually the dominant process of interference (Joliffe, 1988; Tilman, 1988). This is particularly so considering that many experiments studying the effect of pathogens on competition have utilised high plant densities, where competitive effects are most intense. Evidence supporting the hypothesis that pathogens can increase allelopathy between plants occurs in at least two forms: (i) pathogens can stimulate secondary metabolite production in their hosts, and (ii) field, glasshouse and bioassay experiments signifying an increased allelopathic ability of infected plants.

3.2.1. Effect of Pathogens on Secondary Metabolite Production

Numerous studies document the ability of plant pathogens to stimulate the metabolic activity (Daly, 1976) and increase the production of secondary metabolites (Stoessl, 1982, 1983; Goodman et al., 1986; Kuc, 1997) by their hosts. Müller and Börger (1941) first proposed that plants produce defensive substances called phytoalexins in response to infection, which are important to host resistance. Phytoalexins are 'low molecular weight; antimicrobial compounds that are both synthesised by and accumulated in plant cells after exposure to microorganisms' (Paxton, 1981). As such, phytoalexins fit the definition of allelochemicals. Like allelochemicals that act against plants, phytoalexins are secondary compounds that belong to a wide range of different chemical classes (Stoessl, 1982), with more than 300 distinct phytoalexins already characterised (Smith, 1996). They are mostly synthesised via the acetate and shikimic acid pathways (Bailey, 1982; Bennett and Wallsgrove, 1994), as are allelochemicals that act against plants. Notwithstanding their similarities, allelopathy between plants and phytoalexin research has developed almost independently.

Many compounds with phytoalexin activity are also implicated in allelopathy between plants (Rice, 1984). For example, isoflavonoids are important phytoalexins (Ingham, 1982; Paxton, 1981; Dakora and Phillips, 1996) and allelochemicals (Tamura et al. 1967; 1969) from the Leguminosae. Parbery et al. (1984) found that the isoflavonoids biochanin A, formononetin and genistein increased in subterranean clover by 62%, 123% and 75% respectively following infection by pepper spot (*Leptosphaerulina trifolii*). In comparison, Tamura et al. (1967; 1969) isolated a succession of isoflavonoids (including biochanin A, formononetin and genistein) from the shoots of red clover (*Trifolium pratense*) that inhibited its own germination by 50% at concentrations of 50 ppm.

As expected, most studies concerned with the toxicology of phytoalexins concentrate on their effects on microorganisms. Increasingly, however, studies show that phytoalexins are also toxic to plants. Despite this, they are seldom referred to as allelochemicals. Pisatin and phaseollin, pterocarpans produced by pea (*Pisum sativum*) and bean (*Phaseolus vulgaris*) respectively, were the first phytoalexins characterised (Perrin and Bottomly, 1962; Cruickshank and Perrin 1963). Skipp et al. (1977) noted that phaseollin inhibited the respiration and growth of cell cultures of bean and tobacco (*Nicotiana tabacum*), eventually causing cell death. Similarly, pisatin reduced the growth of callus cultures of pea (Bailey, 1970) and inhibited the root growth of wheat (Cruickshank and Perrin, 1961). The phytoalexin rishitin (a sesquiterpene) accumulates in potato cells challenged by incompatible isolates of *Phytophthora infestans* (Tomiyama et al., 1968). Studies show that the compound also inhibits pollen germination in three *Solanum* species (Hodgkin and Lyon, 1979); causes lysis of potato and tomato protoplasts (Lyon and Mayo, 1978); and cell death in tubers and epidermal strips of potato (Ishiguri, et al., 1978; Lyon, 1980). Other studies show that phytoaxelins may inhibit seed germination (Chang et al., 1969), growth (Glazener and VanEtten, 1978) and cellular metabolism and function (Lyon, 1980; Boydston et al., 1983; Kurosaki et al., 1984; Giannini et al., 1990; Spessard et al., 1994) of plants. The ability of pathogens to stimulate secondary metabolite production in their hosts and for these to affect plant growth provides strong circumstantial evidence for the hypothesis that pathogens can increase plant allelopathy.

Many studies have established that pathogens can increase the production of phenolic acids in their hosts, which are perhaps the most important group of plant allelochemicals. For example, when challenged by a range of different pathogens, concentrations of phenolics often associated in allelopathic interactions have increased in a variety plants, including carrot (*Daucus carota*) (Phan et al., 1991), chickpea (*Cicer arietinum*) (Singh et al., 2002); date palm (*Phoenix dactylifera*) (Daayf et al., 2003); potato (*Solanum tuberosum*) (Kuc et al., 1956); sorghum (*Sorghum bicolor*) (Woodhead, 1981); sugar cane (*Saccharum officinarum*) (Legaz et al., 1998); sunflower (*Helianthus anuus*) (Spring et al., 1991); and wheat (*Triticum aestivum*) (Siranidou et al., 2002), amongst many others. This increase in phenolics post-infection is one of the most important methods of disease resistance in plants. For example, inoculation with crown rust (*Puccinia coronata* f.sp. *avenae*) induced the production of three phenolic compounds (avenalumins I, II and III) by resistant oats (*Avena sativa*). Purified preparations of these chemicals inhibited the germination and germ tube growth of crown and stem rust (*Puccinia graminis*) at concentrations as low as 200 μg/mL (Mayama, 1981; 1982). Similarly, Mandavia et al. (2000) found that varieties of cumin (*Cuminum cyminum*) tolerant of Fusarium wilt (*Fusarium oxysporum* f.sp. *cumini*) contained higher concentrations of salicylic acid, hydroquinone and umbelliferone in their root, stem and leaf tissues than susceptible varieties. These phenolics inhibited fungal spore germination and mycelial growth of *F. oxysporum*. However, to this author's knowledge, the effect of the increased phenolic concentrations in infected plants on plant allelopathy has not been investigated empirically.

The production of glucosinolates by plants in the Capareles and their subsequent hydrolysis to toxic isothiocyanates, thiocyanates, and nitriles has been one of the most intensively studied systems in allelopathy, due partly to the similarity of these breakdown products to some synthetically produced soil fumigants (and therefore termed biofumigation; Kirkegaard et al., 2000). Glucosinolates are important in plant defence against insects, pathogens and nematodes. Jay et al. (1999) showed that infection of *Brassica napus* with beet western yellows virus increased glucosinolate concentration in tissues by 14%. Similarly, Li *et al.* (1999) found that infection of *B. napus* by *Sclerotinia sclerotiorum* increased glucosinolate content in resistant, but not in susceptible varieties. Exposure to the pathogenic bacterium, *Erwinia carotovora*, triggered the production of glucosinolates in *Arabidopsis thaliana* (Brader et al., 2001). The hydrolysis products from these glucosinolates inhibited the growth of *E. carotovora* in culture. Furthermore, Tierens et al. (2001) demonstrated that a range of pathogens were more aggressive in infecting an *A. thaliana* mutant that did not produce glucosinolates than the wild type, suggesting the importance of glucosinolates in protecting against infection. However, the role of glucosinolates in disease resistance may be species specific, since Andreasson et al. (2001) found that infection of *B. napus* by *Leptosphaeria maculans* had no effect on glucosinolate concentration in the resistant or susceptible host. Despite the ability for at least some pathogens to increase glucosinolate production in the Capareles, no one has investigated whether this directly translates to an increased allelopathic effect against neighbouring plants.

3.2.2. The Effect of Rust on Ryegrass Allelopathy

Perennial ryegrass (*Lolium perenne*) and white clover (*Trifolium repens*) are important components of improved pastures grown in temperate regions worldwide. The ability of ryegrass to become dominant in pastures has led to numerous investigations that demonstrate its potential to interfere with companion plants through allelopathy (Naqvi, 1972; Naqvi and Muller, 1975; Newman and Rovira, 1975, Newman and Miller, 1977; Gussin and Lynch, 1980; Buta et al., 1987; Quigley et al., 1990; Sutherland and Hoglund, 1990; Wardle et al., 1991; Prestidge et al., 1992; Chung and Miller, 1995; Mattner, 1998; Mattner and Parbery, 2001). For example, in a study investigating the allelopathic ability of nine pasture species, Takahashi et al. (1988) found that leachate from soil surrounding ryegrass was the most inhibitory to the growth of the target species, including clover. In a subsequent experiment, they circulated nutrient solution between the roots of ryegrass and clover to eliminate competition effects. They found that the growth of clover declined in the system, particularly when the proportion of ryegrass was high. When they incorporated XAD-4 resin into the system (which selectively traps organic hydrophobic compounds) clover grew normally. Moreover, ryegrass became yellow and stunted when grown alone in the system, but this was prevented by the presence of the resin or clover (Takahashi et al., 1991). In a further experiment, root exudate from ryegrass not only inhibited the growth of clover, but also lettuce seedlings. The phytotoxic fraction of the extract contained *p*-methoxybenzoic, lauric, myristic, pentadecanoic, palmitoleic, palmitic, oleic and stearic acids. Sodium salts of myristic, palmitic, oleic and stearic acids suppressed the growth of clover at concentrations as low as 5 ppm (Takahashi et al., 1993). Similarly, in a study examining the allelopathic effects of a number of crop and pasture species, Halsall et al. (1995) found that aqueous extract from the dried shoots of perennial ryegrass suppressed the germination, radicle elongation, nodulation and seedling root elongation of subterranean and white clover. The magnitude of this inhibition increased as the concentration of the extract increased.

Crown rust, caused by *Puccinia coronata* f.sp. *lolii*, is the most devastating fungal disease of ryegrass, with epidemics regularly occurring between spring and autumn in temperate regions worldwide (Mattner and Parbery, 2001). Severe epidemics reduce ryegrass tillering by 20-38% (Lancashire and Latch, 1966; Mattner, 1998), leaf emergence by 60%, leaf area by 62%, root growth by 75% (Mattner, 1998), and increase the rate of leaf senescence by up to 184% (Lancashire and Latch, 1966; Trorey, 1979; Plummer et al., 1990; Mattner, 1998). Losses of herbage yield in ryegrass from rust have been as great as 94% (Critchett, 1991), with seed yield losses ranging from 12-36% (Hampton, 1986; Mattner 1998). Furthermore, rust infection reduces forage quality (Isawa et al., 1974; Trorey, 1979; Potter, 1987) and palatability to grazers (Cruickshank, 1957; Heard and Roberts, 1975).

As would be expected by the devastating effect that rust has on ryegrass growth, most studies show that rust reduces the competitiveness of ryegrass with non-host plants such as clover. For example, in mixed swards of ryegrass and clover, Lancashire and Latch (1970) found that rust reduced ryegrass yield by 84% and increased the yield of clover by 87%. Furthermore, the proportion of clover in the rusted sward increased from 24% at the beginning to 80% at the termination of their experiment. Thus, their study pointed to a lowered competitiveness of rusted ryegrass. In mixtures of rust resistant and susceptible ryegrass, Potter (1987) found that rust reduced the yield of the susceptible cultivar and increased that of the resistant one, concluding that rust reduced ryegrass competitiveness. Similarly, crown rust infection in swards of ryegrass and cocksfoot (*Dactylis glomerata*) reduced ryegrass composition from 30% to 15%, and was more marked in rust susceptible than resistant cultivars (Trorey, 1979). However, in a series of experiments, Mattner (1998) reported an anomaly to the results of this previous research.

In pot studies consisting of 50:50 mixtures of ryegrass and clover grown over a range of plant densities, Mattner (1998) found that rust reduced the yield of ryegrass by an average of 41%. However, interference from rusted ryegrass suppressed clover biomass by up to 47% compared with interference from the more productive, healthy ryegrass. The onset of the suppression of clover by rusted ryegrass was rapid, occurring as early as 6-13 days after inoculation, which according to some growth parameters was earlier than the effects of rust on ryegrass itself. The suppression of clover by rusted ryegrass was greatest at low plant densities and diminished or disappeared as density increased. In a separate trial, rusted ryegrass again suppressed clover growth, even after the removal of infected tissue by cutting and after the death of the ryegrass. In this instance, ryegrass killed by infection, with a competitive ability of virtually zero, inhibited the growth of clover more than living plants of healthy ryegrass. In further trials, high rust severity and high soil moisture contents increased the suppression of clover by rusted ryegrass.

The ability of rust to directly inhibit or infect clover could not explain the results from these studies, since clover is a non-host of *P. coronata* and inoculation with this rust did not reduce the yield of clover monocultures. Under some conditions, infection by rusts can increase the scavenging ability of some hosts for water and nutrient resources (Ahmad et al., 1982; Paul and Ayres, 1988), potentially increasing their competitive ability. For this to explain the results from Mattner's studies, however, the suppression of clover by rusted ryegrass should have been greatest at high plant densities where competition for resources was most intense. Instead, the suppression of clover by rusted ryegrass was greatest at low densities where resources were plentiful and competition was low. Furthermore, the ability of rusted ryegrass to suppress clover continued even beyond the death of the plant, when it had no capacity to scavenge resources. For these reasons, Mattner (1998) posed the hypothesis that rust reduces the competitiveness of ryegrass, while simultaneously increasing its allelopathic ability. In this way, it was expected that the expression of allelopathy by rusted ryegrass was greatest at low densities, where there was little competition and the effects of rust in reducing ryegrass competitiveness did not obscure its effects on increasing allelopathy. Furthermore, high plant densities may detoxify or dilute the action of allelochemicals (Thijs et al., 1994).

To test the validity of this hypothesis and to separate the effects of competition and allelopathy, four bioassays for allelopathy were conducted (Mattner, 1998; Mattner and Parbery, 2001). Each bioassay highlighted the potential for extracts, leachate, or residues from ryegrass to inhibit the yield of clover through allelopathy, and for rust to enhance this potential. For example, soil previously growing rusted ryegrass suppressed clover biomass by 36% compared with soil previously growing healthy ryegrass. Similarly, leachate from soil supporting rusted ryegrass suppressed clover biomass by 27% compared with that from healthy ryegrass (Mattner and Parbery, 2001). Although some bioassays have confounding and interpretational problems (Stowe, 1979; Inderjit and Weston, 2000), the conformity of results between these different bioassays provides strong evidence for the hypothesis that rust infection increases the allelopathic ability of ryegrass. Furthermore, in the field, the proportion of prickly lettuce (*Lactuca serriola*) reduced in areas of a depleted pasture dominated by rusted ryegrass, compared with areas dominated by non-rusted ryegrass (Mattner, 1998). Further bioassays provided evidence that this association potentially related to an enhanced allelopathic ability of ryegrass rather than to differences in soil chemistry. Thus, this study highlighted the potential for rust to increase ryegrass allelopathy in the field.

Mattner (1998) suggested two mechanisms by which rust may increase ryegrass allelopathy. Firstly, rust may directly stimulate the production of allelochemicals by ryegrass in a defensive response to infection. Alternatively, or additionally, the increased rate of tissue senescence in ryegrass induced by rust may result in a higher concentration of plant residues reaching the soil. These residues may then form an allelochemical source – either directly or following their decomposition. Most evidence gathered from his studies supported the first hypothesis. For example, when ryegrass residues were incorporated into soil, their ability to suppress clover growing in that soil depended on the residues being rusted and not their overall concentration. Furthermore, the knowledge that pathogens stimulate secondary metabolite formation in their hosts and the rapid effect rusted ryegrass had in suppressing clover, supported rust directly stimulating allelochemical production in ryegrass.

Mattner's findings appear to contradict those of Lancashire and Latch (1970) who studied the identical biological system in the field, and found that the proportion of clover in the sward more than doubled following infection of the ryegrass component by rust. However, Lancashire and Latch (1970) conducted their study under conditions that favoured the expression of the reduced competitiveness rusted ryegrass, i.e. at higher plant densities (2150 plants/m^2) than those of Mattner (1998) (57 plants/m^2). More importantly, however, their results only occurred in the highly susceptible ryegrass cultivar, Ruanui. In a more resistant cultivar, Ariki, rust reduced ryegrass yield by 18%, but clover was unable to take advantage of this reduction, producing the same dry weight when grown with rusted ryegrass as when grown with healthy ryegrass. Furthermore, rusted Ariki ryegrass actually suppressed the growth of clover at some harvests. Perhaps rust infection increased the production of defensive chemicals by Ariki, thereby increasing its resistance to rust and its allelopathic ability against clover. Lancashire and Latch (1970) disregarded these results because, overall, the yield of Ariki was abnormally poor in their experiment compared with several previous studies. Nonetheless, the poor yield of Ariki ryegrass occurred in both rusted and non-rusted treatments and does not explain the inability of clover to compensate for the reduced yield of rusted ryegrass, or the suppression of clover by rusted ryegrass. Rather, the hypothesis that rust increases the allelopathic response of ryegrass while also reducing its competitiveness fits their results.

Although there is strong circumstantial evidence from field, glasshouse and bioassay studies that rust increases the allelopathic ability of ryegrass, many questions remain. An important next step is to identify the allelochemicals concerned. Also, do rusted plants simply produce allelochemicals in higher concentrations or do they produce entirely different allelochemicals to healthy plants? Also, what is the allelochemical source in the plant and the mechanism of release to the environment? This information will provide a clearer understanding of the influence of pathogens on allelopathy.

3.2.3. The Effect of Neotyphodium lolii on Ryegrass Allelopathy

The endophytic fungus *Neotyphodium lolii* commonly infects perennial ryegrass forming a mutualistic relationship with its host. Apart from obtaining nutriment, the host provides the endophyte with a relatively exclusive niche and a vehicle for its transmittance through infected seed (Clay, 1987). The host benefits in several ways, including: (i) increased seed germination, dry matter production and tillering compared with non-infected ryegrass (Latch et al., 1985; Quigley, 2000); (ii) resistance to plant diseases caused by nematodes (Stewart et al., 1993), fungi (Latch, 1993) and viruses (Lewis and Day, 1993); (iii) increased tolerance of drought stress (Ravel et al., 1997), (iv) increased nitrogen use efficiency (Arachevaleta et al., 1989); and (v) increased competitiveness (Sutherland and Hoglund, 1989). Additionally, the fungus produces various alkaloids (e.g. peramine, ergovaline, and lolitrem) that protect the host against herbivory (Clay, 1996).

There is much speculation as to the role of the endophyte on the allelopathic ability of its host. This speculation originated with the observation that endophyte-infected ryegrass suppressed the growth of companion clovers to a greater extent than endophyte-free ryegrass (Stevens and Hickey, 1990). In order to explore several hypotheses on how this relationship might arise, Sutherland and Hoglund (1989) grew swards of endophyte-infected and endophyte-free perennial ryegrass in plots with white clover. Endophyte-infected ryegrass produced 16% more dry matter and suppressed the yield of clover by 72% compared with endophyte-free ryegrass. Neither mowing nor grazing by sheep affected the relationship. This demonstrated that selective grazing pressure on clover following a decline in the palatability of endophyte-infected ryegrass was not responsible for the reduction in clover yield. Rather, they suggested that the suppression was due to an increased competitiveness and allelopathic ability of endophyte-infected ryegrass. They implicated allelopathy in this effect because endophyte-infected ryegrass of a comparable yield to endophyte-free ryegrass, suppressed the yield of clover to a greater extent than endophyte-free ryegrass.

In a subsequent experiment, Sutherland and Hoglund (1990) surrounded individual plants of white clover with zero to seven endophyte-infected or endophyte-free perennial ryegrass plants. This experiment failed to demonstrate that endophyte infected ryegrass suppressed clover more than endophyte-free ryegrass. They explained that this might be due to the cutting regime imposed on ryegrass and the non-return of these clippings to the soil, which possibly contained allelochemicals. A bioassay experiment, which showed that aqueous extracts from the shoots of endophyte-infected ryegrass inhibited the shoot growth of clover seedlings by 8%, supported their explanation. In a similar experiment, Quigley et al. (1990) found that aqueous extracts from endophyte-infected ryegrass depressed the root length of four germinating legumes (including white clover), by an average of 10% compared with extract taken from endophyte-free ryegrass. In contrast, after conducting several bioassays for allelopathy, Prestidge et al. (1992) found little evidence to suggest that the presence of endophyte in ryegrass enhanced its allelopathic effect. Field trials also failed to show that clover yield declined when grown with endophyte-infected ryegrass. Similarly, Watson et al. (1993) and Mattner (1998) failed to substantiate an increased allelopathic effect of endophyte-infected ryegrass. Furthermore, there was no apparent interaction between endophyte and rust infection in increasing the allelopathic ability of ryegrass (Mattner, 1998).

Applebee et al. (1999) found that the toxicity of tall fescue (*Festuca arundinacea*) infected with *Neotyphodium coenophialum* increased at elevated concentrations of CO_2 in the atmosphere. In a bioassay study conducted in sterile sand, Sutherland et al. (1999) applied aqueous extracts from ryegrass to potted clover seedlings. Extracts from three ryegrass cultivars infected with three different strains of endophyte, all inhibited the growth of clover, by up to 27% compared with extracts from endophyte-free ryegrass. Both ryegrass cultivar and endophyte strain influenced the degree that ryegrass extracts inhibited clover, but this did not relate to the type of alkaloids produced by the different endophyte strains. These studies suggest that both environmental and genetic influences may moderate the triggers for enhanced allelopathy by endophyte infected grasses, and this may explain the discrepancy in results between individual studies. Nonetheless, the majority of evidence suggests that endophyte infection has the capacity to increase ryegrass allelopathy, which is a further added benefit conferred by this mutualist to its host. Although the endophyte is a non-pathogenic organism, its ability to stimulate plant allelopathy adds further weight to the hypothesis that infection increases the allelopahtic ability of host plants.

3.2.4. Other Systems

The ability of rusts to stimulate allelopathy in their hosts may not be limited to ryegrass, as the rusts *Puccinia hordei* and *Uromyces troflii-repentis* increased the suppression of white clover by barley grass (*Hordeum leporinum*) and subterranean clover (*Trifolium subterraneum*), respectively (Mattner, 1998). Yet, the effect was not universal since *Puccinia coronata* in wild oat (*Avena fatua*), *Puccinia graminis* in cocksfoot (*Dactylis glomerata*) and *Puccinia recondita* in soft brome (*Bromus mollis*) all failed to increase their host's allelopathic ability.

Kong et al. (2002) studied the allelopathic potential of goatweed (*Ageratum conyzoides*) under different environmental stresses, including infection by powdery mildew (*Erysiphe cichoracearum*). Infection stimulated the production of 17 of the 24 volatile chemicals produced by goatweed that they investigated, with total volatile production increasing by 50%. Exposure to the volatiles released by infected goatweed stimulated the growth of peanut (*Arachis hypogaea*), redroot amaranth (*Amaranthus retroflexus*), Italian ryegrass (*Lolium multiflorum*) and cucumber (*Cucumis sativus*) compared with volatiles from healthy goatweed. In contrast, volatiles from infected plants inhibited the growth of three fungal pathogens (*Rhizoctonia solani*, *Botrytis cinerea*, and *Sclerotinia sclerotiorum*). For this reason, they postulated that the allelochemicals stimulated by fungal infection in goatweed are more important in the defence of the plant against infection, rather than against competition from neighbouring plants. Nonetheless, this study currently provides the clearest demonstration of the ability of pathogens to influence the allelopathic ability of plants.

4. IMPLICATIONS FOR PATHOGEN AGGRESSIVENESS AND EVOLUTION

Plant/pathogen interactions in natural populations are often explained in terms of co-evolution, which is 'the joint evolution of two (or more) taxa that have close ecological relationships but do not exchange genes, and in which reciprocal selective pressures operate to make the evolution of either taxon partially dependent upon the evolution of the other' (Pianka, 1978). Burdon (1987) explained the co-evolution of plants and their pathogens in terms of the gene-for-gene concept of resistance and virulence, using the following model:

(a) A uniform host population possessing a single gene for resistance is challenged by a uniform pathogen population with the complementary virulence gene, resulting in pathogenesis.
(b) Under the above conditions, chance mutation favours the appearance of a novel resistance gene preventing pathogenesis through enhanced resistance. These resistant individuals have a greater fitness than their susceptible neighbours and, consequently, increase in frequency within the population.
(c) As the frequency of the resistant genotype increases, selective pressure favours the appearance of a pathogen race with a novel virulence gene capable of attacking the resistant host.
(d) As resistance is broken down, the advantage of the previously resistant host genotype in terms of plant fitness is lost, and its frequency within the population falls.
(e) Under these conditions the appearance of yet another resistant gene is favoured and the cycle continues.

Although Burdon's model provides a good example of the concept of co-evolution, it is important to note that it does not account for the cost of resistance and virulence to the host and pathogen (Parker and Gilbert, 2004), host tolerance to disease (Roy and Kirchner, 2000), or the specificity of a pathogen to its host (Kirchner and Roy, 2002).

The importance of virulence (the degree or measure of pathogenicity) in determining the ability of a pathogen to infect a particular host is central to the concept of co-evolution of plants and pathogens. Despite this, its importance has often been emphasised to the exclusion of another component of pathogen fitness – aggressiveness (Burdon, 1987). In plant pathology, the term aggressiveness has been defined inconsistently (Jarosz and Davelos, 1995; Kirchner and Roy, 2002) or considered synonymous with virulence (Parker and Gilbert, 2004), but here is considered the negative effect of infection on plant fitness (Jarosz and Davelos, 1995). The role of aggressiveness in interactions between plants and biotrophic pathogens is represented by two seemingly opposing views. Harper (1977) suggested that biotrophic pathogens, which need living tissue for their survival, evolve towards minimising their aggressiveness, with evolutionary equilibrium occurring when a pathogen attains commensalistic relationship with its host. On the other hand, less aggressive pathogens are prone to displacement by more aggressive strains (Jarosz and Davelos, 1995), implying that pathogens should evolve towards increased aggressiveness. Despite these arguments portraying the role of aggressiveness differently, they need not be mutually exclusive.

As an adjunct to the concept of co-evolution, the New Function Hypothesis proposed by Clay (1988) suggests that pathogens may evolve towards a mutualistic relationship with their hosts through the appearance of pathogen strains with beneficial effects on the host in addition to their detrimental effects. In this way pathogens minimise their aggressiveness through the acquisition of 'new functions' that increase host fitness and, at the same time, are less prone to displacement by more aggressive strains by maintaining their original capacity for disease. A 'new function' suggested in this chapter is the potential for pathogens to increase the allelopathic ability of their host, which to varying degrees offsets the infected hosts's loss in competitiveness. By conferring some benefit on its host, the pathogen maintains a self-advantage through increasing the survival chances of its host and ultimately itself.

Since biotrophic parasites, such as the rusts, are heavily dependent on the continuity of their host's genotype into succeeding generations, the evolution of interactions that enhance the chances of host survival are important. It is well established that infections of fodder species by biotrophic pathogens can create conditions that either limit the grazing of their hosts, limit the number of grazing animals, or both. Morgan and Parbery (1980) found that infection by *Pseudopeziza medicagnis* lowered protein content, digestibility and palatability of lucerne as well as increasing its oestrogenic activity. Similarly, rust reduces the digestibility and quality of ryegrass (Isawa et al., 1974; Trorey, 1979; Potter, 1987). For this reason ruminants preferentially graze healthy ryegrass rather than rusted ryegrass (Cruickshank, 1957; Heard and Roberts, 1975), indirectly benefiting rusted ryegrass. Furthermore, evidence presented in this chapter supports the ability of rust to add a further benefit to ryegrass, that of increased allelopathy with neighbouring plants. In a similar manner to crown rust, amongst other benefits, the mutualistic endophyte *N. lolli* reduces ryegrass palatability to ruminants (Fletcher and Sutherland, 1993) and increases its allelopathic ability (Sutherland and Hoglund, 1990; Quigley et al., 1990; Sutherland et al., 1999). The parallels between these two systems suggest that the pathogenic relationship between crown rust and ryegrass is evolving toward mutualism.

5. CONCLUSIONS

Pathogens are important forces that influence the structure and dynamics of plant communities. Although there are numerous interpretations and studies on the impact of pathogens on plant competition, few studies have considered their effect on allelopathy. Currently, however, most evidence suggests that pathogens may simultaneously decrease the competitiveness and increase the allelopathic ability of their hosts. By conferring the benefit of increased allelopathy on their hosts, pathogens may offset their host's loss in competitiveness. In so doing, these pathogens make their hosts less prone to displacement by resistant components of the plant ecosystem and encourage the continuation of their host's genotype into proceeding generations, and ultimately their own. Finally, the similarities in the ability of the ryegrass pathogen, *P. coronata*, and the ryegrass mutualist, *N. lolii*, to increase their host's allelopathic capacity suggests that this 'new function' may be one way that the rust pathogen is evolving toward mutualism.

7. REFERENCES

Ahmad, I., Owera, S.A.P., Farrar, J.F., Whitbread, R. The distribution of five major nutrients in barley plants infected with brown rust. Physiol Plant Pathol 1982; 21:335-346.

Alexander, H.M., Holt, R.D. The interaction between plant competition and disease. Perspect Plant Ecol Evol System 1998; 1:206-220.

Alexander, H.M. Dynamics of plant-pathogen interactions in natural plant communities. In: *Pests, Pathogens and Plant Communities*. Burdon J.J. and Leather S.R. eds., Oxford, UK: Blackwell Scientific Publications 1990.

Anaya, A.L. Allelopathy as a tool in the management of biotic resources in agroecosystems. Crit Rev Plant Sci 1999; 18:697-739.

Andreasson, E., Wretblad, S., Graner, G., Wu, X.M., Zhang, J.M, Dixelius, C., Rask, L., Meijer, J. The myrosinase-glucosinolate system in the interaction between *Leptosphaeria maculans* and *Brassica napus*. Mol Plant Pathol 2001; 21:281-286.

Applebee, T.A., Gibson, D.J., Newman, J.A. Elevated atmospheric carbon dioxide alters the effects of allelochemicals produced by tall fescue on alfalfa seedlings. Trans Illinois State Acad Sci 1999; 92:23-31.

Arachevaleta, M., Bacon, C.W., Hoveland, C.S., Radcliffe, D.E. Effect of tall fescue endophyte on plant response to environmental stress. Agron J 1989; 81:83-90.

Augsberger, C.K. Seedling survival of tropical tree species: interactions of dispersal distance, light-gaps, and pathogens. Ecology 1984; 65:1705-1712.

Ayres, P.G. Growth responses induced by pathogens and other stresses. In: *Responses of Plants to Multiple Stresses*. Mooney, H.A., Winner, W.E., Pell, E.J. eds. San Diego, USA: Academic Press 1991

Ayres, P.G., Paul, N.D. The effects of disease on interspecific plant competition. Aspects of Appl Biol 1990; 24:155-162.

Bailey, J.A. Pisatin production by tissue cultures of *Pisum sativum* L. J Gen Microbiol 1970; 61:409-415.

Bailey, J.A. Mechanisms of phytoalexin accumulation. In: *Phytoalexins*. Bailey, J.A., Mansfield, J.W., eds., New York, USA: John Wiley and Sons. 1982.

Bansal, G.L. The science of allelopathy: problems and prospects. In: *Allelopathy in Agriculture and Forestry*. Narwal S.S., Tauro, P. eds. Jodhpur, India: Scientific Publishers, 1994.

Begon, M., Harper, J.L., Townsend, C.R. Ecology: Individuals, Populations and Communities. (1st ed.). Oxford, UK. Blackwell Scientific Publications 1986.

Bennett, R.N., Wallsgrove, R.M. Tansley Review No. 72: Secondary metabolites in plant defence mechanisms. New Phytol 1994; 127:617-633.

Boydston, R., Paxton, J.D., Koeppe, D.E. Glyceollin: a site-specific inhibitor of electron transport in isolated soyabean mitochondria. Plant Physiol 1983; 72:151-155.

Boyette, C.D., Abbas, H.K. Weed control with mycoherbicides and phytotoxins. In: *Allelopathy: Organisms, Processes and Applications*. Inderjit, K.M.M. Dakshini, Einhellig, F.A. eds., Washington DC, USA: American Chemical Society 1995.

Brader, G., Tas, E., Palva, E.T. Jasmonate-dependent induction of indole glucosinolates in Arabidopsis by culture filtrates of the non-specific pathogen *Erwinia caratovora*. Plant Physiol 2001; 126:849-860.

Bremner, J.M., McCarty, G.W. Inhibition of nitrification in soil by allelochemicals derived from plants and plant residues. In: *Soil biochemistry*. Bollag J. Stotzky, G. eds., Vol. 8. New York, USA: Marcel Dekker. 1993.

Burdon, J.J. The effect of fungal pathogens on plant communities. In: *The Plant Community as a Working Mechanism*. Newman E.I. ed., Oxford, UK: Blackwell Scientific Publications, 1982.

Burdon, J.J. Diseases and plant population biology. Cambridge, UK: Cambridge University Press, 1987.

Burdon, J.J. Fungal pathogens as selective forces in plant populations and communities. Aust J Ecol 1991; 16:423-432.

Buta, J.G., Spaulding, D.W., Reed, A.N. Differential growth responses of fractionated turfgrass seed leachates. Hort Sci 1987; 22:1317-1319.

Calvo, A.M., Wilson, R.A., Bok, J.W., Keller, N.P. Relationship between secondary metabolism and fungal development. Microbiol Mol Biol Rev 2002; 66:447-459.

Casper, B.B., Jackson, R.B. Plant competition underground. Ann Rev Ecol System 1997; 28:545-570.

Catherall, P.L. Effects of barley yellow dwarf virus on the growth and yield of single plants and simulated swards of perennial rye-grass. Ann Appl Biol 1966; 57:155-162.

Chang, C.F., Suzuki, A., Kumai, S., Tamura, S. Chemical studies on 'clover sickness.' II Biological functions of isoflavonoids and their related compounds. Agri Biol Chem 1969; 33:398-408.

Chilvers, G.A., Brittain, E.G. Plant competition mediated by host-specific parasites – a simple model. Aus J Biol Sci 1972; 25:749-756.

Chou, C.H. The role of allelopathy in phytochemical ecology. In: *Phytochemical Ecology: Allelochemicals, Mycotoxins and Insect Pheromones and Allomones*. Chou C.H. and Waller, G.R. eds. Taipei, Taiwan: Institute Botanica Academia Senica, 1989.

Chung, I.M., Miller, D.A. Allelopathic influence of nine forage grass extracts on germination and seedling growth of alfalfa. Agron J 1995; 87:767-772.

Clay, K. Effects of fungal endophytes on the seed and seedling biology of *Lolium perenne* and *Festuca arundinacea*. Oecologia 1987; 73:358-362.

Clay, K. Clavicipitaceous fungal endophytes of grasses: coevolution and the change from parasitism to mutualism. In: *Coevolution of Fungi with Plants and Animals*. Pirozynski K.A. and Hawksworth, D.L. eds., London, UK: Academic Press. 1988.

Clay, K. The impact of parasitic and mutualistic fungi on competitive interactions among plants. In *Perspectives on Plant Competition*. Grace J.B., Tilman, D. eds. San Diego, USA: Academic Press. 1990.

Clay, K. Interactions among fungal endophytes, grasses and herbivores. Res Plant Ecol 1996; 38:191-201.

Critchett, C.I. Studies on ryegrass rusts in Victoria. PhD Thesis. Melbourne, Australia: La Trobe University, 1991.

Cruickshank, I.A.M. Crown rust of ryegrass. New Zealand J Sci Technol 1957; 38A:539-543.

Cruickshank, I.A.M., Perrin, D.R. Studies on phytoalexins. III The isolation, assay, and general properties of a phytoalexin from *Pisum sativum* L. Aus J Biol Sci 1961; 14:336:348.

Cruickshank, I.A.M., Perrin, D.R. Phytoalexins of the Leguminosae. Phaseollin from *Phaseolus vulgaris* L. Life Sci 1963; 2:680-682.

Daayf, F., El-Bellaj, M., El-Hassni, M., J'Aiti, F., El-Hadrami, I. Elicitation of soluble phenolics in date palm (Phoenix dactylifera) callus by *Fusarium oxysporum* f.sp. *albedinis* culture medium. Environ Exp Bot 2003; 49:41-47.

Dakora, F.D., Phillips, D.A. Diverse functions of isoflavonoids in legumes transcend anti-microbial definitions of phytoalexins. Physiol Mol Plant Pathol 1996; 49:1-20.

Daly, J.M., Deverall, B.T. Toxins and plant pathogenesis. Sydney, Australia: Academic Press. 1983.

Deacon, J.W. Modern Mycology. (3rd ed.). Oxford, UK: Blackwell Scientific Publications. 1997.

Dinoor, A., Eshed, N. The role and importance of pathogens in natural plant communities. Ann Rev Phytopathol 1984; 22:443-466.

Di Pietro, A. Fungal antibiosis in biocontrol of plant disease. In: *Allelopathy: Organisms, processes, and applications*. Inderjit, K.M.M. Dakshini, Einhellig, F.A. eds. Washington DC, USA: American Chemical Society, 1995.

Ditommaso, A., Watson, A.K. Impact of a fungal pathogen, *Colletotrichum coccodes* on growth and competitive ability of *Abutilon theophrasti*. New Phytol 1995; 131:51-60.

Dobson, A., Crawley, M. Pathogens and the structure of plant communities. Trends Ecol Evol 1994; 9:393-398.

Donald, C.M. Competition among crop and pasture plants. Adv Agron 1963; 15:1-118.

Einhellig, F.A. Allelopathy: current status and future goals. In: *Allelopathy: Organisms, processes, and applications*. Inderjit, K.M.M. Dakshini, Einhellig, F.A. eds. Washington DC, USA: American Chemical Society 1995.

Friess, N., Maillet, J. Influence of cucumber mosaic virus infection on the competitive ability and reproduction of chickweed (*Stellaria media*). New Phytol 1997; 135:667-674.

Fletcher, L.R., Sutherland, B.L. Liveweight changes I lambs grazing perennial ryegrass with different endophyutes. In: Proceedings of the 2nd International Symposium on *Acremonium*/Grass Interactions 1993.

Gates, D.J., Westcott, M., Burdon, J.J., Alexander, H.M. Competition and stability in plant mixtures in the presence of disease. Oecologia 1986; 68:559-566.

Giannini, J.L., Holt, J.S., Briskin, D.P. The effect of glyceollin on proton leakage in Phytophthora megasperma f.sp. glycinea plasma membrane and red beet tonoplast vesicles. Plant Sci 1990; 68: 39-45.

Glazener, J.A., VanEtten, H.D. Phytotoxicity of phaseollin to, and alteration of phaseollin by, cell suspension cultures of *Phaseolus vulgaris*. Phytopathology 1978; 68:111-117.

Goodman, R.N., Király, Z., Wood, K.R. The Biochemistry and Physiology of Plant Disease. Columbia, USA: University of Missouri Press 1986.

Grace, J.B., Tilman, D. Perspectives on Plant Competition. San Diego, USA: Academic Press 1990.

Groves, R.H., Williams, J.D. Growth of skeleton weed (*Chondrilla juncea* L.) as affected by growth of subterranean clover (*Trifolium subterraneum* L.) and infection by Puccinia chondrillina Bubak and Syd. Aust J Agri Res 1975; 36:975-983.

Gussin, E.J, Lynch, J.M. Microbial fermentation of grass residues to organic acids as a factor in the establishment of new grass swards. New Phytol 1981; 89:449-457.

Halsall, D.M., Leigh, J.H., Gollasch, S.E., Holgate, M. The role of allelopathy in legume decline in pastures. II Comparative effects of pasture, crop and weed residues on germination, nodulation and root growth. Aus J Agri Res 1995; 46:189-207.

Hale, M.G., Orcutt, D.M. Physiology of Plants Under Stress. New York, USA: Wiley Interscience 1987.

Hampton, J.G. Fungicidal effects on stem rust, green leaf area, and seed yield in 'Grasslands Nui' perennial ryegrass. New Zealand J Exp Agri 1986; 14:7-12.

Harper, J.L. Population Biology of Plants. London, UK: Academic Press. 1977.

Heard, A.J., Roberts, E.T. Disorders of temporary ryegrass swards in south-east England. Ann Appl Biol 1975; 81:235-279.

Hodgkin, T., Lyon, G.D. Inhibition of Solanum pollen germination in vitro by the phytoalexin rishitin. Ann Bot 1979; 44:253-255.

Inderjit. Plant phenolics in allelopathy. Bot Rev 1996; 62:186-202.

Inderjit, Dakshini, K.M.M., Einhellig, F.A. Allelopathy: Organisms, Processes, and Applications. Washington DC, USA: American Chemical Society 1995.

Inderjit, Weiner, J. Plant allelochemical interference or soil chemical ecology? Persp Plant Ecol Evol System 2001; 4:3-12.

Inderjit, Weston, L.A. Are laboratory bioassays for allelopathy suitable for prediction of field responses? J Chem Ecol 2000; 26:2111-2118.

Ingham, J.L. Phytoalexins from the Leguminosae. In: *Phytoalexins*. Bailey, J.A., Mansfield, J.W. eds., New York, USA: John Wiley and Sons. 1982.

Isawa, K., Abe, A., Nishihara, N. Influence of diseases on the chemical composition and nutritive value of forage crops. I. Chemical composition of Italian ryegrass infected with crown rust. Ann Phytopathol Soc Japan 1974; 40:86-92.

Ishiguri, Y., Tomiyama, A.K., Murai, A., Katsui, N., Masamune, N. Toxicity of rishitin, rishitin-M-1, and rishitin-M-2 to *Phytophthora infestans* and potato tissue. Ann Phytopathol Soc Japan 1978; 44:52-56.

Jay, C.N., Rossall, S., Smith, H.G. Effects of beet western yellows virus on growth and yield of oilseed rape (*Brassica napus*). J Agri Sci 1999; 133:131-139.

Jarosz, A.M., Davelos, A.L. Effects of disease in wild plant populations and the evolution of pathogens aggressiveness. New Phytol 1995; 129:371-387.

Joliffe, P.A. Evaluating the effects of competitive interference on plant performance. J Theoret Biol 1988; 130:447-459.

Kempenaar, C., Horsten, P.J.F.M., Scheepens, P.C. Growth and competitiveness of common lambsquarters (*Chenopodium album*) after foliar application of *Ascochyta caulina* as a mycoherbicide. Weed Sci 1996; 44:609-614.

Kirchner, J.W., Roy, B.A. Evolutionary implications of host-pathogen specificity: fitness consequences of pathogen virulence traits. Evolut Ecol Res 2002; 4:27-48.

Kirkegaard, J.A., Sarwar, M., Wong, P.T.W., Mead, A., Howe, G., Newell, M. Field studies on the biofumigation of take-all by Brassica break crops. Aus J Agri Res 2000; 51:445-456.

Kong, C., Hu, F., Xu, X. Allelopathic potential and chemical constituents of volatiles from *Ageratum conyzoides* under stress. J Chem Ecol 2002; 28:1173-1182.

Kuc, J. Molecular aspects of plant responses to pathogens. Acta Physiologiae Plant 1997; 19:551-559.

Kuc, J., Henze, R.E., Ullstrup, A.J., Quackenbush, F.W. Chlorogenic and caffeic acids as fungistatic agents produced by potatoes in response to inoculation with Helminthosporium carbonum. J Amer Chem Soc1956; 78:3123.

Kurosaki, F., Matsui, K., Nishi, A. Production and metabolism of 6-methoxymellein in cultured carrot cells. Physiol Plant Pathol 1984; 25:313-321.

Lancashire, J.A., Latch, G.C.M. Some effects of crown rust (*Puccinia coronata* Corda) on the growth of two ryegrass varieties in New Zealand. New Zealand J Agri Res 1966; 9:628-640.

Lancashire, J.A., Latch, G.C.M. The effect of crown rust (*Puccinia coronata* Corda) on the yield and botanical composition of two ryegrass/white clover pastures. New Zealand J Agri Res 1970; 13:279-286.

Latch, G.C.M. Physiological interactions of endophytic fungi and their hosts. Biotic stress tolerance imparted to grasses by endophytes. Agri Eco Sys Environ 1993; 44:143-156.

Latch, G.C.M., Hunt, W.F., Musgrave, D.R. Endophytic fungi affect growth of perennial ryegrass. New Zealand J Agri Res 1985; 28:165-168.

Legaz, M.E., de Armas, R., Piñón, D., Vicente, C. Relationships between phenolics, conjugated polyamines and sensitivity of sugar cane to smut (*Ustilago scitaminea*). J Exp Bot 1998; 49:1723-1728.

Lewis, G.C., Day, R.C. Growth of a perennial ryegrass genotype with and without infection by ryegrass endophyte and virus diseases. Ann Appl Biol 1993; 122:144-145.

Li, Y.C., Kiddle, G., Bennet, R.N., Wallsgrove, R.M. Local and systemic changes in glucosinolates in Chinese and European cultivars of oilseed rape (*Brassica napus* L.) after inoculation with *Sclerotinia sclerotiorum* (stem rot). Ann Appl Biol 1999; 134:45-58.

Liu, D.L., Lovett, J.V. Biologically active secondary metabolites of barley. I. Developing techniques and assessing allelopathy in barley. J Chem Ecol 1993; 19:2217-2230.

Lutrell, E.S. Parasitism of fungi on vascular plants. Mycologia 1974; 66:1-15.

Lyon, G.D. Evidence that the toxic effect of rishitin may be due to membrane damage. J Exper Bot 1980; 31:957-966.

Lyon, G.D., Mayo, M.A. The phytoalexin rishitin affects the viability of isolated plant protoplasts. Phytopathol Z 1978; 92:294-304.

Mandava, N.V. Chemistry and biology of allelopathic agents. In: *The Chemistry of Allelopathy*. A.C. Thompson ed., Washington DC, USA: American Chemical Society, 1985.

Mandavia, M.K., Khan, N.A., Gajera, H.P., Andharia, J.H. Parameswaran, M. Inhibitory effects of phenolic compounds on fungal metabolism in host-pathogen interactions in Fusarium wilt of cumin. Allelopathy J 2000; 7:85-92.

Mattner, S.W. The impact of crown rust on interference between perennial ryegrass and white clover. PhD Thesis. Melbourne, Australia: The University of Melbourne 1998.

Mattner, S.W., Parbery, D.G. Rust-enhanced allelopathy of perennial ryegrass against white clover. Agron J 2001; 93:54-59.

Mayama, S., Matsuura, Y., Iida, H., Tani, T. The role of avenalumin in the resistance of oat to crown rust, *Puccinia coronata* f.sp. *avenae*. Physiol Plant Pathol 1982; 20:189-199.

Mayama, S., Tani, T., Matsuura, Y., Ueno, T., Fukami, H. The production of phytoalexins by oat in response to crown rust, *Puccinia coronata* f.sp. *avenae*. Physiol Plant Pathol 1981; 19:217-226.

Milthorpe, F.L. Mechanisms in Biological Competition. Cambridge, UK: Cambridge University Press, 1961.

Mol, L. Effect of plant roots on the germination of microsclerotia of *Verticillium dahliae*. II. Quantitative analysis of the luring effect of crops. Europ J Plant Pathol 1995; 101:679-685.

Mothes, K. Physiology of alkaloids. Ann Rev Plant Physiol 1955; 6:393-432.

Morgan, W.C., Parbery, D.G. Depressed fodder quality and increased oestrogenic activity of lucerne infected with *Pseudopeziza medicagnis*. Aust J Agri Res 1980; 31:1103-1110.

Muller, C.H. Allelopathy as a factor in ecological process. Vegetatio 1969; 18:348-351.

Müller, K.O., Börger, H. Experimentelle Untersuchungen über die Phytophthora-resistenz den Kartoffel. Arbeiten aus der Biologischen Bundensanstalt für Land-und Fortswirtschaft 1941; 23:189-231.

Naqvi, H.H. Preliminary studies of interference exhibited by Italian ryegrass. Biologia (Lahore) 1972; 18: 201-210.

Naqvi, H.H., Muller, C.H. Biochemical inhibition (allelopathy) exhibited by Italian ryegrass (*Lolium multiflorum* L.). Pak J Bot 1975; 7:139-147.

Newman, E.I., Miller, M.H. Allelopathy among some British grasslands species. II Influence of root exudates on phosphorus uptake. J Ecol 1977; 65:399-411.

Newman, E.I., Rovira, A.D. Allelopathy among some British grasslands species. J Ecol 1975; 63:727-737.

Nicol, R.W., Yousef, L., Traquair J.A., Bernards, M.A. Ginsenosides stimulate the growth of soilborne pathogens of American ginseng. Phytochemistry 2003; 64:257-264.

Odunfa, V.S.A. Root exudation in cowpea and sorghum and the effect on spore germination and growth of some soil fusaria. New Phytol 1978; 80:607-612.

Parbery, D.G., Gardner, W.K., Golebiowski, T. Stimulation of isoflavonoid content in subterranean clover by infection with a fungus. J Aus Inst Agri Sci 1984; 50:114-116.

Parbery, D.G. Trophism and the ecology of fungi associated with plants. Biol Rev 1996; 71: 473-527.

Parker, I.M., Gilbert, G.S. The evolutionary ecology of novel plant-pathogen interactions. Ann Rev Ecol Evol Syst 2004; 35:675-700.

Paul, N.D. Modification of the effects of plant pathogens by other components of natural ecosystems. In: *Pests, Pathogens and Plant Communities*. Burdon J.J. and Leather S.R. eds., Oxford, UK: Blackwell Scientific Publications 1990.

Paul, N.D., Ayres, P.G. Interference between healthy and rusted groundsel (*Senecio vulgaris* L.) within mixed populations of different densities and proportions. New Phytol 1986; 104:257-269.

Paul, N.D., Ayres, P.G. Effects of rust infection of *Senecio vulgaris* on competition with lettuce. Weed Res 1987a ; 27:431-441.

Paul, N.D., Ayres, P.G. Water stress modifies intraspecific interference between rust (*Puccinia lagenophorae* Cooke)–infected and healthy groundsel (*Senecio vulgaris* L.). New Phytol 1987b ; 106:555-566.

Paul, N.D., Ayres, P.G. Nutrient relations of groundsel (*Senecio vulgaris* L.) infected by rust (*Puccinia lagenophorae*) at a range of nutrient concentrations. I. Concentrations, content, and distribution of N, P and K. Ann Bot 1988; 61:489-498.

Paul, N.D., Ayres, P.G. Effects of interactions between nutrient supply and rust infection of Senecio vulgaris L. on competition with *Capsella bursa-pastoris* (L.) Medic. New Phytol 1990; 114:667-674.

Paxton, J.D. Phytoalexins – a working redefinition. Phytopathol Z 1981; 101:106-109.

Peters, J.C., Shaw, M.W. Effect of artificial exclusion and augmentation of fungal plant pathogens on a regenerating grassland. New Phytol 1996; 134:295-307.

Perrin, D.R., Bottomley, W. Studies on phytoalexins. V The structure of pisatin from *Pisum sativum* L. J Amer Chem Soc 1962; 84:1919-1922.

Phan, C.T. Phenolics and polyketides in carrots. In: *Mycotoxins and Phytoalexins*. Sharma R.P., Salunkhe D.K., eds., CRC Press, Boca Raton, USA 1991.

Pianka, E.R. Evolutionary Ecology. New York, USA: Harper and Row. 1978.

Plummer, R.M., Hall, R.L., Watt, T.A. The influence of crown rust (*Puccinia coronata*) on tiller production and survival on perennial ryegrass (*Lolium perenne*) plants in simulated swards. Grass Forage Sci 1990; 45:9-16.

Potter, L.R., Cagas, B., Paul, V.H., Birckenstaedt, E. Pathogenicity of some European collections of crown rust (*Puccinia coronata* Corda) on cultivars of perennial ryegrass. J Phytopathol 1990; 130:119-126.

Pratley, J.E. Allelopathy in annual grasses. Plant Protection Quart 1996; 11:213-214.

Prestidge, R.A., Thom, E.R., Marshall, S.L., Taylor, M.J., Willoughby, B., Wildermoth, D.D. Influence of *Acremonium lolii* infection in perennial ryegrass on germination, emergence, survival and growth of white clover. New Zealand J Agri Res 1992; 35:225-234.

Putnam, A.R. Allelopathic research in agriculture, past highlights and potential. In: *The Chemistry of Allelopathy.* Thompson, A.C. ed., Washington DC, USA: American Chemical Society. 1985.

Putnam, A.R., and Duke, W.B. Allelopathy in agroecosystems. Ann Rev Phytopathol 1978; 16:431-451.

Putnam, A.R., Tang, C. Allelopathy: state of the science. In: *The Science of Allelopathy*. Putnam A.R. Tang, C. eds., New York, USA: John Wiley and Sons. 1986.

Quigley, P.E., Snell, F.J., Cunningham, P.J., Frost, W. The effects of endophyte infected ryegrass on the establishment, persistence and production of mixed pastures in Australia. In: Proceedings of the International Symposium on *Acremonium*/Grass Interactions: 1990; 49-51.

Quigley, P.E. Effects of Neotyphodium lolii infection and sowing rate of perennial ryegrass (*Lolium perenne*) on the dynamics of ryegrass/subterranean clover (*Trifolium subterraneum*) swards. Aust J Agri Res 2000; 51:47-56.

Ravel, C, Courty, C., Coudret, A., Charmet, C. Beneficial effects of *Neotyphodium lolii* on the growth and the water status in perennial ryegrass cultivated under nitrogen deficiency or drought stress. Agronomie 1997; 17:173-181.

Rice, E.L. Allelopathy. New York, USA: Academic Press, 1974.

Rice, E.L. Allelopathy. (2nd ed.). Orlando, USA: Academic Press, 1984.

Rice, E.L. Allelopathic effects on nitrogen cycling. In: *Allelopathy: Basic and applied aspects*. Rizvi S.J.H., Rizvi, V. eds., London, UK: Chapman and Hall, 1992.

Rizvi, S.J.H., Haque, H., Singh, V.K., Rizvi, V. A discipline called Allelopathy. In: *Allelopathy: Basic and Applied Aspects*. Rizvi S.J.H., Rizvi, V. eds., London, UK: Chapman and Hall 1992.

Roy, B.A., Kirchner, J.W. Evolutionary dynamics of pathogen resistance and tolerance. Evolution 2002; 54:51-63.

Seigler, D.S. Chemistry and mechanisms of allelopathic interactions. Agron J 1996; 88:876-885.

Skipp, R.A., Selby, C., Bailey, J.A. Toxic effects of phaseollin on plant cells. Physiol Plant Pathol 1977; 10:221-227.

Shinozaki, K., Kira, T. Intraspecific competition among higher plants. VII Logistic theory of the C-D effect. Journal of the Institute of Polytechnics, Osaka City University 1956; 7:35-72.

Singh, U.P., Sarma, B.K., Singh, D.P., Bahadur, A. Studies on exudates-deplete sclerotial development in Sclerotium rolfsii and the effect of oxalic acid, sclerotial exudates, and culture filtrate on phenolic acid induction in chickpea (Cicer arietinum). Can J Microbiol 2002; 48:443-448.

Siranidou, E., Kang, Z., Buchenauer, H. Studies on symptom development, phenolic compounds and morphological defence responses in wheat cultivars differing in resistance to Fusarium head blight. J Phytopathol 2002; 150:200-208.

Smith, C.J. Accumulation of phytoalexins: defence mechanism and stimulus response system. New Phytol 1996; 132:1-45.

Spessard, G.O., Hanson, C., Halvorson, J.S., Giannini, J.L. Effects of phaseollin on membrane leakage in red beet vacuoles and tonoplast vesicles. Phytochemistry 1994; 35:43-47.

Spring, O., Benz, A., Faust, V. Impact of downy mildew (*Plasmopara halstedii*) infection on the development and metabolism of sunflower. Zeitschrift für Planzenkrankheiten und Planzenschutz 1991; 98:597-604.

Stevens, D.R., Hickey, M.J. Effects of endophytic ryegrass on white clover. In: Proceedings of the International Symposium on Acremonium/Grass Interactions, 1990.

Stewart, T.M., Mercer, C.F., Grant, J.L. Development of *Meloidogyne naasi* on endophyte-infected and endophyte-free perennial ryegrass. Aus Plant Pathol 1993; 22:40-41.

Stoessl, A. Biosynthesis of phytoalexins. In: *Phytoalexins*. Bailey, J.A., Mansfield J.W., eds., New York, USA: John Wiley and Sons. 1982.

Stoessl, A. Secondary plant metabolites in preinfectional and postinfectional resistance. In: *The Dynamics of Host Defence*. Bailey, J.A., Deverall B.J. eds., Sydney, Australia: Academic Press 1983.

Stowe, L.G. Allelopathy and its influence on the distribution of plants in an Illinois old-field. J Ecol 1979; 67:1065-1085.

Sutherland, B.L., Hoglund, J.H. Effect of ryegrass containing the endophyte (Acremonium lolii), on the performance of associated white clover and subsequent crops. Proc New Zealand Grassland Assoc 1989; 50:265-269.

Sutherland, B.L., Hoglund, J.H. Effect of ryegrass containing the endophyte *Acremonium lolii* on associated white clover. In: Proceedings of the International Symposium on *Acremonium*/Grass Interactions 1990.

Sutherland, B.L., Hume, D.E., Tapper, B.A. Allelopathic effects of endophyte-infected perennial ryegrass extracts on white clover seedlings. New Zealand J Agri Res 1999; 42:19-26.

Sutic, D.D., Sinclair, J.B. Anatomy and Physiology of Diseased Plants. Boca Raton, USA: CRC Press 1991.

Takahashi, Y., Otani, T., Hagino, K. Studies on interactions among some grassland species. V Collection and isolation of allelopathic hydrophobic compounds from the root exudates of *Lolium perenne* L. J Jap Soc Grassland Sci 1993; 39:236-245.

Takahashi, Y., Otani, T., Uozumi, S., Yoden, Y., Igarashi, R. Studies on the allelopathic interactions among some grassland species. I Effects of root exudates from some grass and legume species on the growth of their own species and other species. J Jap Soc Grassland Sci 1988; 33:334-338.

Takahashi, Y., Otani, T., Uozumi, S., Yoden, Y., Igarashi, R. Studies on the allelopathic interactions among some grassland species. II Assessment of the allelopathic interactions between perennial ryegrass (*Lolium perenne* L.) and white clover (*Trifolium repens* L.) using a root exudate recirculating system. J Jap Soc Grassland Sci 1991; 37:274-282.

Tang, C.S., Cai, W.F., Kohl, K., Nishimoto, R.K. Plant stress and allelopathy. In: *Allelopathy: Organisms, processes, and applications.* Inderjit, K.M.M. Dakshini, Einhellig, F.A. eds., Washington DC, USA: American Chemical Society 1995.

Tamura, S., Chang, C., Suzuki, A., Kymai, S. Isolation and structure of a novel isoflavone derivative in red clover. Agri Biol Chem 1967; 31:1108-1109.

Tamura, S., Chang, C., Suzuki, A., Kymai, S. Chemical studies on 'clover sickness.' I Isolation and structural elucidation of two new isoflavonoids in red clover. Agri Biol Chem 1969; 33:391-397.

Thijs, H., Shann, J.R., Weidenhamer, J.D. The effect of phytotoxins on competitive outcome in a model system. Ecology 1994; 75:1959-1964.

Tomiyama, K., Sakuma, T., Ishizaka, N., Sato, N., Katsui, N., Takasugi, M., Masamune, T. A new antifungal substance isolated from resistant potato tuber tissues infected by pathogens. Phytopathology 1968; 58:115-116.

Toussoun, T.A., Patrick, Z.A. Effect of phytotoxic substances from decomposing plant residues on root rot of bean. Phytopathology 1963; 53:265-270.

Tierens, K.F.M.J., Thomma, B.P.H.J., Brouwer, M., Schmidt, J., Kistner, K., Porzel, A., Mauch-Mani, B., Cammue, B.P.A., Broekaert, W.F. Study of the role of antimicrobial glucosinolate-derived isothiocyantes in resistance of *Arabidopsis* to microbial pathogens. Plant Physiol 2001; 4:1688-1699.

Tilman, D. Plant strategies and the dynamics and structure of plant communities. Princeton, USA: Princeton University Press. 1988.

Trenbath, B.R. Biomass productivity of mixtures. Adv Agron 1974; 26:177-210.

Trorey, G.M. The effect of crown rust (*Puccinia coronata lolii*) on the yield and competitive ability of perennial ryegrass. PhD Thesis. Huddersfield, United Kingdon: Huddersfield Polytechnic. 1979.

Van der Putten, W.H., Van Dijk, C., Peters, B.A.M. Plant-specific soil-borne disease contribute to succession in fordune vegetation. Nature 1993; 362:53-56.

Wardle, D.A., Nicholson, K.S., Ahmed, M. Allelopathic influence of nodding thistle (Carduus nutans L.) seeds on germination and radicle growth of pasture plants. New Zealand J Agri Res 1991; 34:185-191.

Wardle, D.A., Nilsson, M.C., Gallet, C., Zackrisson, O. An ecosystem–level perspective of alleloapthy. Biol Rev Cambridge Philosoph Soc 1998; 73:305-319.

Watson, R.N., Prestidge, R.A., Ball, O.J.P. Suppression of white clover by ryegrass infected with Acremonium endophyte. In: Proceedings of the 2nd International Symposium on *Acremonium*/Grass Interactions, 1993.

Wennström, A., Ericson, L. Variation in disease incidence in grazed and ungrazed sites for the system *Pulsatilla pratensis – Puccinia pulsatillae.* Oikos 1991; 60:35-39.

Whittaker, R.H., Feeny, P.P. Allelochemics: chemical interactions between species. Science 1971; 171:757-770.

de Wit, C.T. On competition. Verslagen van Landbouwkundige Onderzoekingen 1960; 66:1-82.

Woodhead, S. Environmental and biotic factors affecting the phenolic content of different cultivars of *Sorghum bicolor.* J Chem Ecol 1981; 7:1035-1047.

RAMANATHAN KATHIRESAN[1], CLIFFORD H.KOGER[2] AND KRISHNA N. REDDY[3]

ALLELOPATHY FOR WEED CONTROL IN AQUATIC AND WETLAND SYSTEMS

[1]*Professor, Department of Agronomy, Annamalai University, Annamalainagar, Tamilnadu 608 002, India.* [2]*Weed Ecologist, Southern Weed Science Research Unit, USDA-ARS, P. O. Box 350, Stoneville, Mississippi 38776, USA.* [3]*Corresponding author. Plant Physiologist, Southern Weed Science Research Unit, USDA-ARS, P. O. Box 350, Stoneville, Mississippi 38776, USA*
E-mail : kreddy@ars.usda.gov

Abstract. Allelochemicals offer ample scope for ecologically safe and effective weed control in aquatic and wetland systems. This could be attributed to the absence of soil interface in aquatic habitats that contributes largely for rapid degradation of allelochemicals. Simpler strategies involving allelopathy especially for small holder farms, low input agriculture and aquatic environments with appreciable results have been reported. Such strategies include use of allelopathic cultivars, organic manures and plant products. Though allelopathic suppression of weeds could not be construed as an alternate to replace synthetic herbicides, it can fit in an integrated weed management program very well as a prime component. Such strategies are reviewed. Further, a specific case study for the use of plant product along with insect agents for controlling water hyacinth in India and different steps involved in selecting allelopathic plant products for aquatic weed control are discussed.

1. INTRODUCTION

Many of the compounds produced by green plants that are not involved in primary plant metabolism are observed to function as chemical warfare agents against competing plants and pests. Many such natural compounds have the potential to be exploited as herbicides or as leads for discovery of new herbicides (Duke, 1986; Hoagland, 2001). Allelopathy simply refers to chemical warfare between different plants, in which the bio-chemicals from one plant impair germination, growth, survival, reproduction, and behavior of other plants. These allelopathic chemicals are produced by a 'donor' and transmitted to a 'receiver' that can either be 'injured' or 'stimulated'. Allelochemicals act through direct interference with physiological functions of 'receiver' such as seed germination, root growth, shoot growth, stem growth, symbiotic effectiveness or act indirectly through additive or synergistic impact along with pathological infections, insect injury and/or environmental stress. Though many of these allelochemicals exhibit inhibitory response on various morpho-physiological functions of receiver plants and such responses being observed to be dose dependant in a linear fashion, their concentrations required for control of weeds on a field scale are impracticably higher. Further, the degree of selectivity is often a factor limiting

Inderjit and K.G. Mukerji (eds.),
Allelochemicals: Biological Control of Plant Pathogens and Diseases, 103–122.

for their widespread commercial use. The soil-interface in agricultural field conditions also affects the effectiveness of these allelochemicals. However, allelochemicals provide ample opportunity for safe, selective and eco-friendly weed control in aquatic and wetland systems as resistance offered by soil-interface is largely circumvented by the water-interface. Under aquatic environment, the transport of allelochemical to the receiver plant is much more rapid, thus reducing the need for higher doses of allelochemicals for effective weed suppression.

2. ALLELOPATHIC PLANT MATERIALS

In spite of the widespread apprehension that applications of allelopathy in agriculture need to undergo an extensive refinement and techno-commercial perfections, simpler means especially for small holder farms with appreciable results have been reported. Strategies that involve the mechanism of allelopathy to an appreciable magnitude include use of allelopathic cultivars, mulches, organic manures, and plant products in wetlands. In aquatic systems, strategies include use of allelopathic plant products and plant species capable of aggressively replacing invasive undesirable weed species.

2.1. Allelopathic Crop Cultivars in Wetlands

Apparent allelopathic activity in rice accessions against the weed duck salad [*Heteranthera limosa* (Sw.) Willd] was first observed at USDA-ARS in Arkansas, during 1985-86. Within the next five years, 412 rice accessions that imparted allelopathic suppression of duck salad with in a radius of 10 cm were identified by USDA-ARS from their germplasm collection comprising 16,000 rice accessions from 99 countries. The allelopathic suppression of the weed red stem (*Ammania coccinea* Rottb) over an area of 10 cm radius from the rice plant was observed with 145 accessions (Dilday et al., 1994). A hybrid between P1 338046 (allelopathic) and Katy (non-allelopathic) was shown to possess superior agronomic traits in green house studies and quantitative inheritance was indicated for the allelopathic activity in hybrid rice (Dilday et al., 1998). In Egypt, more than 30 rice varieties were shown to be allelopathic on barnyardgrass [*Echinochloa crusgalli* (L.) Beau.] and 10 rice varieties were observed to suppress smallflower umbrella sedge (*Cyperus difformis* L.). These allelopathic rice varieties inhibited the root development and emergence of first or second leaf of both weeds (Hassan et al., 1995). Chemical composition of ethyl acetate extracts from different rice cultivars allelopathic to barnyardgrass in South Korea consisted mainly fatty acid esters, unsaturated ketones, polycyclic aromatic compounds, and alkaloids (Kim and Shin, 1998). Gas chromatography/mass spectrometry (GC/MS) analysis of extracts of water from wetland with allelopathic rice plants showed significant levels of 3-hydroxybenzoic acid, 4-hydroxybenzoic acid, 4-hydroxy cinnamic acid and 3, 4-dihydroxy hydrocinnamic acid (Mattice et al., 1998). Many studies have indicated that the active compounds involved in rice allelopathy are phenolic compounds (Rice, 1987; Chou et al., 1991; Inderjit, 1996; Mattice et al., 1998). The GC/MS analysis of Kouketsumochi, a potential allelopathic rice variety identified several compounds such as sterols, benzaldehydes, benzene derivatives, long chain fatty acids, esters, aldehydes, ketones, and amines as to have been biologically active (Kim et al., 2000). Kim and Shin (2003) observed that more than one allelochemical is likely to be involved in rice allelopathy and that rice allelopathy is not due to one specific phytotoxin. Such leads in rice allelopathy research offer scope for incorporation of this self defense mechanism in improved rice varieties, through breeding programs. A very important element needed for achieving this goal is understanding of the structure of the genetic background of the trait.

However, today's knowledge on allelopathy genetics is limited and this restricts the efforts for intentional breeding to genetically improve allelopathic potential of crops. A valuable milestone in this direction has been reached with the contributions of Jensen et al. (2001). Quantitative Trait Loci (QTL) mapping through 142 DNA markers were located in 142 recombinant inbred lines derived from a cross between a *japonica* upland strain with strong allelopathic traits and an *indica* irrigated strain with weak allelopathic traits. Three main loci, independently contributing for 10% of the promotion of allelochemical synthesis were localized to rice chromosomes 2 and 3. The two QTL traits on chromosome 3 were closely linked, offering scope for easy manipulation. Three different approaches have been attempted recently to produce allelopathic crops. Traditional breeding methods are suggested to be a reasonable alternative by Courtois and Olofsdotter (1998). The first approach involves crossing of two parents with contrasting behavior and derivation of recombinant inbred lines (RILs) through single – seed descent (SSD). The second approach involves introduction of allelopathic traits in to an elite restorer line in developing three – line hybrid rice. Simultaneous back crossing and selfing methods of breeding were attempted to develop hybrid rice with allelopathic activity and its counterpart isogenic hybrid rice with a non-allelopathic effect on weeds. Three lines of rice, Kouketsumochi, Rexmont and IR-24 were used as allelopathic donor, non-allelopathic, and restoring genes, respectively (Lin et al., 2000). Analysis and monitoring of allelopotential and heterotic performance in laboratory and green house indicated a positive and significant allelopathy in this hybrid rice. A third approach suggested is molecular that could be achieved by the regulation of gene expression related to allelochemical bio-synthesis and insertion of allelochemical regulating genes into non-allelopathic crops to induce the synthesis of allelochemicals (Duke et al., 2001).

2.2. *Crop Residues and Organic Manures for Allelopathic Suppression of Weeds*

The residues of crops grown during preceding seasons or tree components of the farm are incorporated in the field before raising field crops to serve as manures. However, such crop residues also offer other environmental benefits that include protection from soil erosion, increase in biological diversity including beneficial organisms and suppression of pests and weeds (Sustainable Agriculture Network, 1998). The residues of such crops can suppress weeds by releasing allelochemicals (Teasdale, 2003). Living mulches, intercrops or smother crops may provide physical weed suppression but their effects in part depend on allelopathy (Bond, 2002). Cover crops suppress weeds during early crop season, but they do not provide full-season weed suppression (Reddy, 2001; Reddy et al., 2003; Teasdale, 1996). Selective activity of tree allelochemicals from *Leucaena leucocephala* (Lam.) Dewit and *Eucalyptus* species on different crops and weed species have been reported by Ferguson and Rathinasabapathi (2003). Field studies at the Experimental Farm, Annamalai University, India assessed the impact of different crop residues applied as green leaf manure to both rice nurseries and wetland (transplanted) rice fields (Parthiban and Kathiresan, 2002; Kathiresan, 2004). The green leaves of *Eucalyptus globulus* L., *L. leucocephala* and *Calotropis gigantea* L. were applied at 5 t ha^{-1}. Green leaves were soil incorporated at the time of final land preparation, flooded with water, and allowed to decompose. After 7 days, the land was leveled and used as nursery to raise rice seedlings. Similarly, in wetland rice fields, green leaves were incorporated and rice seedlings were transplanted into puddled soil. The results showed that *Eucalyptus* and *Leucaena* leaves significantly reduced the density of *Echinochloa* spp. and *Cyperus rotundus* L. in the nursery. In wetland rice fields, the densities of *C. rotundus, Cyperus littoralis* Gaud., *C. difformis* and *Sphenoclea zeylanica* Gaertn. were drastically reduced by these two crop residues. The weed suppression was even better than the rice herbicide, butachlor. However, use of these crop residues in the nursery reduced rice populations. Laboratory studies supported this observation and clearly showed a direct inhibitory response of higher magnitude on rice seed and *Echinochloa* seed by the leaf leachates whose concentrations corroborate with field doses. The crop management strategy of applying green leaf manures of *Eucalyptus* and *Leucaena* could compliment weed suppression (comparable to that of a single application of pre-emergence herbicide) in wetland transplanted rice. The risk of inhibition of rice seed germination restricts the application of this technique to rice nursery. However, it is possible to include these green leaves as component of an integrated weed management strategy in wetland rice production as rice seedlings are less sensitive to these crop residues. An array of compounds such as cineol, pinene, caffeic acid, gallic acid, eucalyptin, hyperoside, and rutin, which constitute primary allelochemicals in *Eucalyptus*, might have suppressed the germination of weed seed as well as rice in the nursery as observed in other studies (Rao et al., 1994; Sivagurunathan et al., 1997). Aqueous extracts of leaves, seeds and litter of *Leucaena* were shown to be significantly phytotoxic on many test species in South Korea (Chou, 1990) and the chemicals isolated include mimosine, quercetin and gallic, protocatechuic, *p*-hydroxybenzoic, vanillic, caffeic and p-coumaric acids.

Recuperation of nutrients absorbed by the crops in small holder farms of developing countries has been traditionally taken care of through the incorporation of several kinds of bulky organic manures. Besides crop residues, other farm wastes such as cattle manure and some post harvest byproducts of crops are frequently used for manuring the crops. Filter pressmud, a bi-product of sugar factories, that crystallizes sugar from cane juice is the precipitated impurity from cane juice. Pressmud from cane juice accounts for about 3% and all sugar factories accumulate enormous quantities of pressmud. This bi-product is widely used in South Asian countries as organic manure. This pressmud upon incorporation in the soil, prior to transplanting of rice in wetlands, releases allelopathic metabolites that contribute to weed suppression.

Allelopathic suppression of weeds by the application of pressmud as organic manure in wetland agriculture has been included as a vital weed management component in Integrated Systems of Farming (Arulchezian and Kathiresan, 1990; Kathiresan, 2004a).

2.3. Invasive Weeds with Allelopathic Potential

Invasive species, particularly those non-indigenous to terrestrial, aquatic, wetland, and wildlife habitat, cause extensive environmental and economical damage to native ecosystems. Invasive plant species alone cause losses in excess of $35 billion annually in the U.S.A. (Pimentel et al., 2000). Non-indigenous weeds in the U.S.A. are estimated to invade 700,000 hectares of wildlife habitat per year (Babbitt, 1998). Invasive aquatic weeds are a significant problem in the U.S.A. Aquatic weeds choke waterways, reduce recreational use of lakes and rivers, and alter aquatic animal species (Pimentel et al., 2000). In the U.S.A. alone, an estimated $100 million is spent annually for managing aquatic weed species (OTA, 1993). An estimated $15 million is spent on managing the aquatic weed species hydrilla [*Hyrdilla verticillata* (L.F.) Royle] in the state of Florida, U.S.A. (Center et al., 1997). Invasive weeds in terrestrial ecosystems such as pastures cause estimated losses of $2 billion annually in the U.S.A. (Pimentel, 1991). Examples of invasive weeds in pastures are cogongrass [*Imperata cylindrica* (L.) Beauv.], yellow starthistle (*Centaurea solstitialis* L.), and purple loosestrife (*Lythium salicaria* L.). Specifically, cogongrass is a C_4, rhizomatous, perennial monocot that has become an invasive weed in many gulf-states of the southeastern U.S.A. since its introduction to the U.S.A. in the late 19th and early 20th centuries (Byrd and Bryson, 1999; Dickens, 1974; Dickens and Buchanan, 1971; Elmore, 1986). Cogongrass grows in tropical, subtropical, and some temperate regions of the world (Akobundu and Agyakwa, 1998; Bryson and Carter, 1993), and is found on all continents except Antarctica (Holm et al., 1977). Cogongrass is among the most troublesome weeds worldwide (Falvey, 1981; Holm et al., 1977). It is extremely competitive with crops and neighboring plant communities (Eussen and Wirjahardja, 1973). It has been reported to reduce corn (*Zea mays* L.) grain yield by 80 to 100% (Koch et al., 1990; Udensi et al., 1999). Koch et al. (1990) also reported > 90% yield reduction for intercropped corn and cassava (*Manihot esculenta* Crantz) grown in cogongrass infested fields.

Cogongrass competes with other plant species commonly found in similar ecosystems (roadways, pastures, mining areas, parks, and other natural and recreational areas) via allelopathic-type mechanisms. Residues and liquid exudates of cogongrass foliage and root residues have been shown to reduce germination and shoot and root growth of monocot and dicot weed and crop species (Koger and Bryson 2004; Koger et al., 2004). Germination of bermudagrass [*Cynodon dactylon* (L.) Pers.] and Italian ryegrasses (*Lolium multiflorum* Lam.) was reduced by as much as 97%, and shoot and root growth by as much as 96% at the highest concentrations. However, cogongrass had no allelopathic effect on hemp sesbania [*Sesbania exaltata* (Raf.) Rydb. Ex A. W. Hill] (Koger and Bryson, 2004). Phenolic compounds present in foliage and roots of cogongrass may be responsible for the allelopathic inhibition of germination and seedling development of other species. Inderjit and Dakshini (1991) reported several phenolic compounds extracted from leachates of cogongrass foliage and roots/rhizomes reduced germination and shoot and root length of mustard [*Brassica juncea* (L.) Czern and Coss.] and tomato (*Lycopersicon esculentum* Mill.). Inderjit and Dakshini (1991) also found phenolic compounds in leachates of soil collected near the rhizosphere of cogongrass as well as up to 3 m away that were not present in control soils. These reports suggest that allelopathic effect of cogongrass is species-specific. Thus, allelochemicals from cogongrass may serve as potential leads in discovery of new selective herbicides.

2.4. Plant Products for Allelopathic Control of Weeds

Presence of abundant moisture in wetlands allow faster transport of allelochemicals from the applied plant products to the target weeds. Aquatic systems have the advantage of no soil-interface that restricts the activity of allelochemicals on susceptible flora. As a result, use of plant products for allelopathic control of weeds has more potential in wetlands and aquatic systems than soil containing terrestrial systems. Carrot grass (*Parthenium hysterophorus* L.), an invasive obnoxious weed originated from Mexico has shown to be allelopathic on another introduced invasive aquatic weed water hyacinth [*Eichhornia crassipes* (Mart.) Solms. Laubach] in India. Dried residues of leaves and flower of carrot grass applied in to the water at 0.5% (w/v) killed water hyacinth within one month. Residues of leaves of carrot grass showed the highest biological activity followed by flowers, stem and roots. The flowers and leaves of carrot grass also had higher total phenolic acid levels in the medium which was responsible for the inhibition (Pandey et al., 1993). Dry powder of an epiphytic plant *Cassytha* sp dissolved in water infested with water hyacinth at 1-2% (w/v) resulted in complete desiccation of leaves of water hyacinth with in 15 days and drastic reduction in biomass (Kauraw and Bhan, 1994). Parthenin, a sesquiterpene lactone that is one of the major toxins in the weed, proved lethal to water lettuce (*Pistia stratiotes* Linn.) and duck weed (*Lemna perpusilla* Torr.) at 50 ppm concentration and to water hyacinth at 100 ppm concentration. The mechanism of allelopathic inhibition by parthenin was identified to be associated with decline in water use, root dysfunction, excessive leakage of solutes resulting in massive damage to cellular membranes, loss of dehydrogenase activity in the roots and destruction of chlorophyll in the leaves (Pandey, 1996a).

Among several allelochemicals (*p*-hydroxybenzoic acid, anisic acid, salicylic acid, coumaric acid, fumaric acid, tannic acid, gallic acid, chlorogenic acid, vanillic acid, caffeic acid and ferulic acid), *p*-hydroxybenzoic acid had highest phytotoxicity on aquatic weeds and was lethal at 50 ppm to all weeds tested, including floating aquatic weeds like salvinia (*Salvinia molesta* Mitchell), azolla (*Azolla nilotica* Decne. ex Mett.), spirodella (*Spirodella polyrhiza* (L.) Schieiden) and lemna (*Lemna paucicostaha* Hegelm.), and submerged weeds like hydrilla (*H. verticilata*), ceratophyllum (*Ceratophyllum demersum* L.) and najas (*Najas graminea* Del.). However, *p*-hydroxybenzoic acid was lethal for water hyacinth and water lettuce only at 100 ppm. Water hyacinth (*E. crassipes*), water lettuce (*P. stratiotes*) and water fern (*S. molesta* Baker.) were relatively more tolerant to allelochemicals except for *p*-hydroxybenzoic acid compared to other floating or submerged weeds (Pandey, 1996b).

An Indian medicinal herb *Coleus amboinicus* Lour., commonly used for curing cold, flu and other such ailments upon raw consumption, showed remarkable activity on water hyacinth among different weeds and herbs tried for their allelopathy on water hyacinth. Dried leaves of this medicinal herb *C. amboinicus* were ground to powder and applied to the water system as a suspension (30 g L^{-1}). Death of water hyacinth occurred within 24 h and nearly 100% reduction in biomass was achieved within 9 days. However, spraying *C. amboinicus* over the foliage of water hyacinth at 100 g L^{-1} proved ineffective due to lack of penetration and absorption of allelochemicals from dried powder into the plant system. Applied as a suspension in water, the natural product was active, killing the weed even at lower dosages of 12.5 g L^{-1} within 14 days (Kathiresan and Kannan, 1998). In the Philippines, 57 different compounds, including α humulene, carvacrol, thymol, α-pinene and α-terpine were identified from *C. amboinicus.* Some of these compounds showed a high degree of biological activity, proving lethal to several micro organisms, insects and snails (Vasquez et al., 1999). *C. amboinicus* powder was shown to inhibit algal growth in static-water systems (Kathiresan, 1998).

Graded dosages of the natural products of *C. amboinicus* were evaluated for their inhibitory effect on water hyacinth (Kathiresan, 1999) through specific bio-assay methods, involving whole plants and cut leaves, separately. The study indicated that dry powder of *C. amboinicus* is a candidate plant product for the control of floating aquatic weeds water hyacinth, water lettuce and duck weed with rapid biological activity. The biological activity is due not only to mere water loss but also physiological effects (as compared to sodium chloride at 40 g L^{-1}, the standard desiccant with 25% weed biomass reduction after 9 days). The data also implies a higher magnitude of inhibition on root system (Kathiresan, 2000).

The lowest dosage of 1% concentration (10 g L^{-1}) may seem high, but is similar to that widely used for therapeutic purposes. Earlier studies have shown that water hyacinth was relatively tolerant to allelochemicals, as 100 ppm of *p*-hydroxybenzoic acid was required to cause death, whereas 50 ppm concentration was sufficient for other weeds (Pandey, 1996b). Similarly, fungal oxalates and pure oxalic acid that caused considerable and severe chlorosis in other aquatic weeds induced only slight chlorosis in water hyacinth (Charudattan and Lin, 1974). Under these circumstances, a crude extract of a safe medicinal herb *C. amboinicus,* causing death of water hyacinth within 9 days, might offer an effective control strategy. Further, a stable inhibitory response caused by *C. amboinicus* powder applied to cut leaves of water hyacinth under controlled conditions in static water at dosages ranging from 30 g L^{-1} down to 1.0 g L^{-1} reduced fresh weight from 61 to 49% respectively, suggesting that *C. amboinicus* dry powder could exert adequate allelopathy at low dosages. The lowest dosage of 0.1 g L^{-1} (100 ppm) of *C. amboinicus* caused 24% fresh weight reduction. Apparently, the dosage of *C. amboinicus* powder lethal to water hyacinth could be reduced drastically, if either the natural product or the active principle is formulated to enhance absorption through foliage. In India, when the sources of irrigation water recede during the summer, the smaller volume of water is more accessible for treating with *C. amboinicus* at 10 g L^{-1} (1% w/v). *C. amboinicus* also hold promise for bio-control of water hyacinth on small farms. In another aquatic habitat, *Myriophyllum spicatum* L. has been shown to be allelopathic on submerged macrophytes and the compound that was identified as major active allelochemical Tellimagrandin II (Gross et al., 1996) also interferes with photosystem II and photosynthetic oxygen evolution in *Anabaena* sp. (Leu et al., 2002).

Barnyardgrass is a problem weed in wetland transplanted rice. The weed infestation begins with use of contaminated rice seeds. Morphology of barnyardgrass closely resembles rice during seedling stage. The seedlings are pulled along with rice seedlings from the nursery and transplanted in the main field. Transplanted seedlings of barnyardgrass may escape control with pre-emergence herbicides in the main field. Allelopathy of rice seeds is shown to be stimulatory to the germination of barnyardgrass. Farmers in developing countries normally soak the gunny bags with rice seeds harvested in the previous season, in water canals overnight prior to seeding. The allelochemiclas similar to gibberellin released from the rice seeds help break the dormancy of barnyardgrass and stimulate germination (Kathiresan and Gurusamy, 1995). To overcome this allelopathic stimulation of barnyardgrass, it is recommended that small farmers soak the rice seeds in open containers, skim the floating barnyardgrass seeds (due to differential specific gravity) as a preventive weed control measure. However, in other studies rice hull extracts have been shown to exert allelopathic inhibition of germination and growth in barnyardgrass (Ahn and Chung, 2000).

3. ALLELOPATHIC MICROORGANISMS

Microbes play a vital role in designing and implementing allelopathic control of aquatic weeds and certain weeds of wetland eco-systems. They serve as agents for classical bio-control in several parts of the world with a simple idea of locating highly host specific agents from the native range of the weed and introducing them in new regions requiring control where the weed has established, after rigorous experimental evaluation. The microbial toxins either in their parental form or as metabolites also offer scope to serve as effective herbicides for weed control. Some of the microbial toxins exhibit a new mode of action with novel target site that could stimulate discovery of new herbicidal chemistry.

3.1. Pathogens for Weed Control

Bio-control using pathogens holds promise mostly in non-cropland situations because of the slow pace of control of weeds and the wider window for control as compared with the shorter window of the cropping season and associated disturbances under cropland situations. However, in aquatic systems they appear to be more realistic, due to the absence of such cropping barriers and enhanced mode of dispersal in water. Among the wetland weeds, *Echinochloa crusgalli* and *Cyperus rotundus* are being evaluated for biological control using the fungi *Exserohilum monocerus* and *Dactylaria higginsi* (Luttrell) MB Ellis, respectively (Charudattan and Dinoor, 2000). Fungal pathogens of aquatic weeds have been identified from Neotropical regions since the late 1970s. Alligator weed (*Alternanthera philoxeroides* (Mart.) Griseb) introduced from Brazil, became a serious invader of aquatic systems in Australia, Asia and North America. A preliminary survey of fungal pathogens on alligator weed in Brazil has yielded two species *Nimbya alternantherae* (Holocomb and Antanopoulos) Simmons and Alcorn and *Cercospora alternantherae* Ellis and Langois (Barreto and Torres, 1999). Both caused leaf spots on the alligator weed and their virulence varied with geographic range and altitude. Aquatic ferns such as *Salvinia molesta* and *Salvinia auriculata* JB Aublet., are regarded as invasive aquatic weeds in many of the water habitats in Neotropics, Africa and Asia. Necrotic spots caused by the fungus *Drecshlera* sp. on *S. auriculata* and the inoculum was easily produced and pathogenic (Muchovej and Kushalappa, 1979). Cattail (*Typha domingensis* Pers.,) native to Neotropics has also evolved to be invasive in aquatic systems. The fungal pathogen *Phoma typhae – domingensis* is regarded as a potential candidate for the development of mycoherbicide for cattail (Barreto and Evans, 1996). *Egeria densa* Planch, a worse submerged aquatic weed interrupts hydro electricity generation, damages grids, and causes substantial economic loss to hydroelectric companies. Among eight fungal isolates that have shown promise, *Fusarium graminearum* Schw. was pathogenic to *E. densa* causing chlorosis followed by necrosis and complete tissue disintegration (Barreto et al., 2000). Three different fungi (*Chaetomella raphigera* Swift, *Cercospora* sp. and *Mycosphaerella* sp.) caused severe blight in parrot's feather (*Myriophyllum aquaticum* (Vell.) Verdc), an invasive aquatic weed (Barreto et al., 2000). Water hyacinth, recognized as the world's worst aquatic weed (Holm et al., 1977), has been the prime target weed for bio-control. One of the first pathogens to be patented as a mycoherbicide was *Cercospora piaropi* Tharp. (Charudattan, 1996). Other promising pathogens are *Uredo eichhorniae* Fragosco and Ciferri, *Myrothecium roridum* Tode ex Fr. *Alternaria eichhorniae* NagRaj and Ponnappa (Barreto et al., 2000).

3.2. Microbial Toxins as Leads for Herbicides

The unique relationship between plants and their pathogens suggest that microorganisms may be a better source of future herbicides than allelochemicals produced by higher plants (Duke, 1986). The major constraint with most allelochemicals from higher plants is that their range of selectivity is narrow and they are often autotoxic. Perhaps, this may be one reason that breeding for crop plants to produce higher levels of allelochemicals has not been aggressively pursued. In contrast, many microbial phytotoxins are both selective and efficacious at low rates. Bialaphos [L-2-amino-4-(hydroxyl) (methyl) o-phosphinyl)-butyryl-L-alanyl-L-alanine], a tripeptide extract from *Streptomyces viridichromogens* Schulz. Freiburg. marketed as Herbiace®, tentoxin, a cyclic tetrapeptide produced by *Alternaria alternata* (Fr.) Keissl., rhizobitoxine produced by *Rhizobium japonicum* Kirchner and an analogue of cystathionine are a few examples. Several non-host selective phytotoxins produced by microorganisms have been reported (Duke, 1986; Hoagland, 2001).

4. ALLELOPATHY –LEAD FOR NOVEL HERBICIDAL CHEMISTRY

Allelochemicals can provide potential leads in discovery of new herbicides with novel molecular target sites of action. Hydantocidin from *Streptomyces hygroscopicus* B. Straubinger is an analogue of the allelochemical phosphinothricin and inhibits adenylosuccinate synthetase. This analogue mimics the substrate inositol monophosphate (IMP) and binds the enzyme 1000 fold tighter than IMP, thereby forming a dead-end complex. Hadacidin and alanosine, both microbial products are also inhibitors of this enzyme (Duke et al., 2000). To date, bialaphos and glufosinate derived from microbial compounds are the two most successfully commercialized herbicides (Hoagland, 2001). Glufosinate is the synthetic version of phosphinothricin, a breakdown product of bialaphos (Hoagland, 2001; Lydon and Duke, 1999). Bialaphos sold as herbicide in Japan is derived from *Streptomyces* species is a proherbicide that is metabolically degraded to phosphinohricin by target plant in order to be herbicidally active. Glufosinate is the only commercial herbicide that inhibits glutamine synthetase despite plethora of natural and synthetic products known to inhibit glutamine synthetase (Lydon and Duke, 1999). A gene coding for an enzyme that acetylates the active acid of phosphinothricin rendering it non-phytotoxic, phosphinothricin acetyl transferase (Pat) gene has been isolated from *S. hygroscopius*. This gene has been used in truncated form to transform crops such as maize to impart tolerance to glufosinate (Copping, 2002). Other examples of herbicides derived from natural product chemistry are triketones from Syngenta and cinmethlin from BASF. Another potential herbicide of microbial origin is AAL-toxin, a natural metabolite from *Alternaria alternata* f sp. *lycopersici*. Monocots were generally immune to the effects of AAL-toxin, whereas several broad leaved species are susceptible. Abbas et al. (1995) proposed that the selective weed control could be possible through AAL-toxin.

5. SCREENING OF ALLELOPATHIC PLANT MATERIALS

Allelopathy is essentially a chemical defense mechanism used by plants to keep other plants out of their space. Though allelopathy forms a part of network of chemical communication between plants, it is part of plant interference along with competition for resources. Competition involves the removal or a diminution of shared resources, while allelopathy involves the addition of a chemical compound to the environment through different processes (Rice, 1984; Putnam, 1985). Allelopathic chemicals in general affect seed germination, root growth, shoot growth and/or seedling vigor in the early stages of the receiver's growth and may interfere with metabolic functions like photosynthesis, membrane permeability, biosynthesis of enzymes, lipids, protein, etc. as the receiver progress in growth processes. In aquatic systems and wetlands, screening of allelopathic plant materials for biological efficiency is relatively easier as the allelochemicals are frequently absorbed through the roots of the receiver and transported from the donor directly through water without much of resistance from the soil-interface. However, screening processes have different phases involving either larger size or larger population of target weeds in different environments. Screening is used mainly to confirm the biological efficiency of the substances to a magnitude which would at least serve as component of integrated weed management though not as a holistic weed control measure. Such a rigorous or repeated experimentation with different sets or modes of bio-assay becomes imperative as many of the allelochemicals exhibit inhibitory response on seedling germination and establishment but seldom lethal on large sized receiver plants. Screening techniques for allelopathy in aquatic systems or allelochemicals which are transported mainly through water body should include some critical points raised by Willis (1985), Putnam and Tang (1986), and Cheng (1992). They are:

- Pattern of inhibition of one species by another must be shown using suitable controls, describing the symptoms and quantitative growth reduction;
- The putative aggressor plant must produce a toxin;
- There must be a mode of toxin release from the plant to the environment and thus the target plant;
- The afflicted plant must have some means of toxin uptake, be exposed to the chemical in sufficient quantities and time to cause damage, and to notice similar symptoms;
- The observed pattern of inhibition should not be explained solely by physical factors or other biotic factors, especially competition.

The first phase or initial stage of screening include bio-assays. Those plant materials that are confirmed to possess biological activity through bio-assay need to be further studied for their dose response pattern under controlled environment. Plant materials elucidating appreciable response even at minimal doses could be further evaluated for their allelopathic potential under natural habitats.

5.1. Bio-Assays

Bioassays are an integral part in all studies of allelopathy. They have multiple uses such as evaluating allelopathic potential of different plant material, tracing activity during extraction, purification, and identification of bioactive compounds. The techniques used vary greatly and one has to standardize the procedure independently, and modify to suit the occasion and conditions. According to Rice (1984) and Putnam and Tang (1986), seed germination is used as a test in most bioassays. Though different types of bioassays are used, all of them in general include seed placed on substrate saturated with the test solution. Germination, as indicated by the emergence of the radical 2 mm beyond the seed coat is scored over a period of time. The factors that need to be kept constant are oxygen availability, osmotic potential of the test solution, pH, and temperature. The elongation of the hypocotyls or coleoptiles are often observed along with germination. Dry biomass which is easier to measure could be used as a measure of growth (Bhowmik and Doll, 1984). Sensitivity is normally higher in growth bioassays than in germination bioassays. When the quantity of allelochemicals is limited, agar cultures can be used. Pre-germinated seeds can be exposed to the agar cultures containing test solution.

Bioassay for searching of allelopathy in aquatic weeds is comparatively easier and this could be designed under both laboratory or greenhouse conditions using either part or whole plant of the aquatic weed (Kathiresan, 2000; Kannan, 2002). Whole plants of floating or submerged aquatic weeds targeted for allelopathic control can be grown in suitable containers with water containing standardized nutrients. The powder of candidate allelopathic substances or plant products are added to water either on w/v or v/v basis with appropriate untreated controls. Periodic observations at designated intervals could include reduction in root length and mass, reduction in shoot length and mass, desiccation, scorching and bleaching or mortality score similar to herbicide injury. Based on the screening data, plant products could be classified in to highly allelopathic, moderately allelopathic, and less allelopathic. For example, 55 different plant products were screened for allelopathic inhibition of water hyacinth involving whole plants as well as cut leaves at Annamalai University. In bio-assays involving whole plants, ten of them including *C. amboinicus, P. hysterophorus* and *L. leucocephala* were highly allelopathic based on fresh weight reduction (>30%) of water hyacinth within 48 hr after treatment. Another 12 including *Acalypha indica* Linn., *Trianthema portulacastum* L. and *Sesbania grandiflora* (L.) Pers. showed moderate allelopathy, fresh weight reduction of water hyacinth was 15-30% within 48 hr after treatment. Twelve other plant products including *Croton sparsiflorus* Morong, *Cleome viscosa* L. and *Eclipta alba* L. showed less allelopathy, fresh weight reduction of water hyacinth was less than 15% within 48 hr after treatment (Table 1). The remaining 21 plant species including *Leucas aspera* Spreng. *Curcuma longa* L. and *Euphorbia hirta* L. did not show any allelopathic effect on water hyacinth (Kannan, 2002). To measure dose dependant responses of allelopathic substances more precisely

Table 1 : Percentage reduction in fresh weight of water hyacinth (Eichhornia crassipes.) due to various plant products. (Kannan, 2002)[a].

Treatments	*Days After Treatment*			
plant products @ 30 g L^{-1}	*2*	*4*	*6*	*8*
Coleus amboinicus	56.57 (69.66)	90.00 (100.00)	90.00 (100.00)	90.00 (100.00)
Acalypha indica	29.98 (24.97)	44.24 (48.68)	90.00 (100.00)	90.00 (100.00)
Leucas aspera	-[b]	-	-	-
Croton sparsiflorus	20.53 (12.30)	34.51 (32.11)	45.37 (50.66)	90.00 (100.00)
Curcuma longa	-	-	-	-
Trianthema portulacastum	27.06 (20.70)	45.09 (50.17)	90.00 (100.00)	90.00 (100.00)
Cleome viscosa	21.10 (12.96)	35.77 (34.17)	45.66 (51.16)	90.00 (100.00)
Leucaena leucocephala	34.84 (32.65)	90.00 (100.00)	90.00 (100.00)	90.00 (100.00)
Sesbania grandiflora	27.17 (20.86)	44.51 (49.16)	90.00 (100.00)	90.00 (100.00)
Parthenium hysterophorous	36.92 (36.10)	90.00 (100.00)	90.00 (100.00)	90.00 (100.00)
Euphorbia hirta	-	-	-	-
Eclipta alba	21.27 (13.17)	33.93 (31.17)	44.94 (49.91)	90.00 (100.00)
Control	-	-	-	-
SE_D	2.40	2.34	1.25	-
CD ($P = 0.05$)	4.80	4.69	2.51	-

[a] Figures in parenthesis are original values before arc-sine transformation.
[b] No allelopathic effect.

and to detect the sensitivity of the target weed to minute quantities of allelopathic substances, incised plant parts of aquatic weeds would serve better than whole plants. For screening of different allelopathic substances for inhibition of water hyacinth, a specific bio-assay method was designed with cut leaves of water hyacinth (Kathiresan, 2000; Kannan, 2002). This bioassay involved exposure of cut leaves of water hyacinth to graded doses of plant materials to be tested (dissolved in water). The leaves of water hyacinth plants (with healthy leaves submerged in water) were detached by cutting the petiole with a razor blade with care to retain the incision point below water level. The detached portions of leaf with a part of petiole intact was kept submerged in water for 90 seconds to ensure that no air was trapped internally. Then these leaves were transferred to scintillation vials with water where in different plant products were dissolved and individually compared with an untreated control. The percent fresh weight reduction of the cut leaves was calculated using the formula

$$\frac{\text{Initial weight of the cut leaves} - \text{weight after 24 hr of treatment}}{\text{Initial weight of cut leaves}} \text{ X } 100$$

5.2. Dose Response Studies

Even if a plant product proves appreciably allelopathic on aquatic weeds in bio-assay, tracing the pattern of allelopathic inhibition with graded doses of the plant product becomes vital as aquatic systems requires enormous quantities of plant product due to the quantum of water body (dilution), which in many cases is not practically feasible. Furthermore, this dose response has to be plotted for differing morpho-physiological states of the weed that occurs in common, as the quantity of allelochemical required for a knock down effect is less with a small statured weed compared to that of larger sized weed. For this purpose, before screening of allelopathic plant products on water hyacinth, the latter was classified in to different morphological stages (large, medium and small) prevalent in the state of Tamilnadu, India using discriminant analysis (Kannan and Kathiresan, 1998). Both whole plant and cut leaf bio-assays were done on each of the three stages of water hyacinth. Higher (10 to 30 g L^{-1}) doses were used in whole plants to impart a near lethal effect on the whole plant and lower doses were used in cut leaf bio-assays to cause some allelopathic injury, if not lethal. The doses used for water hyacinth cut leaf bio-assay were 30, 25, 20, 15 10, 5 2.5, 1, 0.5, 0.25, and 0.1 g L^{-1}. The dose response data revealed that cut leaf-bioassay was superior to whole plant assay. For example, *C. amboinicus* at as low as 0.1 g L^{-1} dose caused 24% fresh weight reduction in water hyacinth within a week of exposure (Kathiresan, 2000). Therefore, cut leaf bio-assay could be useful to detect allelopathic potential of plant products which otherwise may have been missed if whole plant assay was used. In contrast, the dose response study data revealed that lethal doses for large, medium, and small plants of water hyacinth were relatively closer. *C. amboinicus* at low (10 g L^{-1}) dose caused death of the water hyacinth after 20 days whereas at high (25 g L^{-1}) dose caused death within one week.

5.3. Field Testing of Allelopathic Substances

In many instances, the materials or plant substances that prove to be allelopathic in laboratory or pot culture experiments may not elucidate similar magnitude of allelopathic response on aquatic weeds in aquatic environments, watersheds, or wetlands. Hence, it is imperative to confirm plant products for their allelopathic potential on weeds in their own natural habitat. A search was made on allelopathic plant products for use in water hyacinth control programs at Department of Agronomy, Annamalai University. Ten of 55 different plant products that showed allelopathic suppression of water hyacinth within 48 h of treatment were selected and tested for their efficacy in natural habitats. The field testing was done in a two tier model. First, the ten plant products were tested in microponds (simulated natural habitat). Second, the plant products that confirmed to be allelopathic in microponds were further evaluated in natural watersheds.

5.4. Preparation of Microponds

Small (microponds) ponds of 1m x 1m x 1m dimension were dug in the field and canals with running water on both sides of the pond were maintained to minimize seepage loss of water from the ponds. Initially, the microponds were conditioned by flooding and maintaining water as needed for 20 days. After 20 days, healthy water hyacinth plants were introduced and allowed to grow for 2-3 days. The plant products to be tested were dissolved in the water of microponds, separately. Allelopathic potential of the plant products was evaluated in comparison with untreated control using data on fresh weight, number of healthy leaves, and chlorophyll content of the water hyacinth.

5.5. Evaluation in Natural Habitats

Top three best performing plant products in the micropond study were selected for realistic confirmation of their allelopathic potential in natural habitats. To this end, three watersheds of 168, 123, and 492 m^2 area were selected at three different locations within a radius of 50 km from Annamalai University. Water hyacinth was allowed to establish in watersheds and the surface area of water was completely covered by weed. Each watershed was divided in to three equal strips using polyethylene sheets stretched between bamboo poles running down the entire depth of water, with the poles anchored to the bottom of the watershed. The plant products were applied to water individually to one of the strips in each watershed. The data were collected on fresh weight, number of healthy leaves, and chlorophyll content at five days interval. Results proved that the plant products showing higher magnitude of allelopathic inhibition on water hyacinth in initial bio-assays like *C. amboinicus* continued to retain and exhibit the same magnitude of inhibition on water hyacinth with out any dissipation in natural watersheds or aquatic environments.

6. ALLELOPATHY AND INTEGRATED WEED MANAGEMENT

Because of limited resources, an average farmer in a developing country can neither afford to take big economic risks nor opt for technologies associated with a lot of external inputs. As a result, research on vegetation management strategies capable of minimizing weed infestation and simultaneously favoring sustainable crop production that are economical and eco-friendly needs attention (Akobundu, 2000). Allelopathy fits in to this approach as one of the integral principles in any such cropping systems involving crop rotation, inter-cropping, cover crop, and off-season land management (such as raising green manures and ploughing *in situ*). Linking similar integrated farming approaches to integrated weed management and integrated pest management helps to address bio-diversity concerns with a simultaneous reduction in agrochemical use especially in low input agriculture and small hold farms. A 3-yr study of weed management in wetland transplanted rice, rice - mung bean cropping sequence with treatments assigned to the same plots every season at Annamalai University revealed that lowland weeds like *C. difformis* was drastically reduced by the introduction of a relay crop of mung bean in the sequence (Kathiresan, 2002). Raising a green manure crop of *Sesbania aculeata* Poir in the off-season (May - July) and ploughing it *in situ* at the age of 45 days, before the cultivation of rice in the first (August - January) as well as second (January - April) season, helped in reducing weed competition in both the rice crops (Gnanavel and Kathiresan, 2002). Off-season land management such as raising green manure crop significantly reduced the weed seed reserves in the soil through allelopathic interference, whereas rotation of an upland crop like mung bean with rice interrupted the weed flora in lowland through mung bean residues.

Integrated weed management assumes significance in managing aquatic systems. Use of herbicides are constrained with drastic reduction in water quality and ultimate ill effect on associated non target organisms. In countries like India, herbicides are yet to get registered for use in aquatic systems. Under these conditions managing infestations of water hyacinth, water fern, and water lettuce is challenging. In one of the recreational lakes with tourist attraction in a hill resort in Ooty, in the state of Tamilnadu, India, the public authority has spent heavily (Indian Rupees 1.25 crores, about US $200,000) for manual clearing of water hyacinth for one time. Similarly, thousands of army personnel were used for clearing water hyacinth in a lake in Bangalore, the capital city of Karnataka State, India. Classical biological control is the only option available and that too is difficult in situations where the water body dries off in the peak summer, leaving the released insects to starve and die due to interrupted host range. Accordingly, integration of short term control measures with classical biocontrol might offer excellent results. Allelopathy reinforced classical bio-control research has been targeted and taken up at Department of Agronomy, Annamalai University through National Agricultural Technology Project funded by Indian Council of Agricultural Research. This project originated from the basic concept of allelopathic inhibition of water hyacinth by *C.amboinicus* as mentioned earlier. However, the requirement of plant product for treating larger watersheds might pose practical difficulties. Previous results also indicated that if absorbed in to plant through foliage, the plant product could be very effective even under very low doses (0.1 g L^{-1}). The only hurdle faced for application of the plant product on the foliage is retention of plant product due to the repulsion by leaf cuticle. Any rupture and/or damage to leaf cuticle could potentially enhance absorption of plant product. To this end, the well established insect bio-control agents in India, *Neochetina bruchii* Hustache / *eichhorniae* Warner were chosen for the study to serve as a component of integrated weed management. These weevils normally scrape on the leaves of water hyacinth.

An attempt was made to integrate both these bio-control tools viz. classical bio-control using *N. bruchii / eichhorniae* and application of the plant product *C. amboinicus*. Integrating both the tools are possible with two different sequences. Treating the water body first with plant product at a lesser dose with the expectation that it will reduce the vigor of the weed, predisposing it for faster and rapid destruction by the insect agents that are to be released later is one possibility. Whereas releasing the insect agents first on the weed host, allowing them to make leaf scrapings that might help foliar uptake of plant product that could be sprayed later is another. Both these sequences were compared in the study. It was observed that treating the water body first with plant product followed by the release of insect agents on the weed showed an antagonistic interaction, as the insects migrated from treated, partially killed plants to healthy plants. The second sequence of releasing the insect agents first followed by spraying of the plant product on the weed foliage produced an additive or synergistic response with rapid and complete weed control with in a single season. The optimum inoculation loads of insect agents, concentration of the spray fluid of plant product required, length of interlude between the release of insect agents and spraying of plant product were standardized for all three different growth stages of the weed, and the success of this integrated approach was demonstrated at three different watershed environments in the state of Tamilnadu, India. The plant product was also shown to be safe for the insect agents with out inducing migratory behavior and without causing any histo-pathological injury on different tissues of the insects like salivary gland, gut, cutin, testis, and brain. Further, the integrated approach also proved safe for non-target organisms and water quality (Kathiresan, 2004b).

7. CONCLUSIONS

Plants can interfere with each other through allelopathy or competition for resources. Allelopathy can be used in weed management in several ways including cover crops, smother crops, green manure crops, breeding for allelopathic crop cultivars, mulching and crop residue management. Allelopathic suppression of weeds will not replace synthetic herbicides which are the dominant method of weed control in many countries nor will it be economically competitive with herbicides. However, allelopathy can fit in an integrated weed management strategy very well as a vital component. This approach could reduce the sole dependence on synthetic herbicides for solving many complex weed problems. The examples discussed herein include an aggressive rice cultivar for complimenting weed control in direct seeded rice and plant product-reinforced classical bio-control (through weevils) of water hyacinth.

Disclaimer

Mention of trade names or commercial products in this publication is solely for the purpose of providing specific information and does not imply recommendation or endorsement by the United States Department of Agriculture.

8. REFERENCES

Abbas H.K., Tanaka T., Duke S.O., Boyette C.D. Susceptibility of various crop and weed species to AAL-toxin, a natural herbicide. Weed Technol 1995; 9:125-130.

Ahn J.K., Chung I.M. Allelopathic potential of rice hulls on germination and seedling growth of barnyardgrass. Agron J 2000; 92:1162-1167.

Akobundu I.O. Getting weed management technologies to farmers in the developing world. Proceedings of III International Weed Science Congress, Brazil, CD-ROM. International Weed Science Society, Oxford, MS, USA 2000; 2 p.

Akobundu I.O., Agyakwa C.W. A Handbook of West African Weeds. 2nd ed. Ibadan, Nigeria: International Institute of Tropical Agriculture 1998; 496 p.

Arulchezian M.P., Kathiresan R.M. Effect of organic manures and herbicides on transplanted rice cv. ADT.37. In Proceedings 3rd Tropical Weed Science Conference, Kuala Lumpur, Malaysia, 1990; pp. 313-317.

Babbitt B. Statement by Secretary of the Interior on invasive alien species. National Weed Symposium. Bureau of Land Management Denver, CO. 1998; pp. 8-10.

Barreto R., Charudattan R., Pomella A., Hanada R. Biological control of neotropical aquatic weeds with fungi Crop Prot 2000; 19:697-703.

Barreto R.W., Evans H.C. Fungal pathogens of some Brazilian aquatic weeds and their potential use in biocontrol. In Proceedings of the IX International Symposium on Biological Control of Weeds. V.C. Moran, J. H. Hoffmann, eds. University of Cape Town, Cape Town, South Africa, 1996; pp. 121-126.

Barreto R.W., Torres A.N.L. Nimbya alternantherae and Cercospora alternantherae: two new records of fungal pathogens on *Alternanthera philoxeroides* (alligator weed) in Brazil. Aust Plant Pathol 1999; 28:103-107.

Bhowmik P.C., Doll J.D. Allelopathic effects of annual weed residues on growth and nutrient uptake of corn and soybeans. Agron J 1984; 76:383-388.

Bond W. Non-chemical weed management. In Weed Management Handbook. E.L.Robert Nayler, ed. BCPC Black Well Publishers, UK. 2002.

Bryson C.T., Carter R. Cogongrass, *Imperata cylindrica*, in the United States. Weed Technol 1993; 7:1005-1009.

Byrd J.D., Bryson C.T. Biology, ecology, and control of cogongrass [*Imperata cylindrica* (L.) Beauv.]. Mississippi Dept. Agric. and Commerce, Bureau of Plant Industry, Mississippi State, Mississippi, USA. Fact Sheet 1999 01. 2 p.

Center T.D., Frank H.J., Dray F.A. Biological control. In Strangers in Paradise. D. Simberloff, D.C. Schmitz, T.C. Brown, eds. Island Press, Washington DC, 1997. pp 245-266.

Charudattan R. Pathogens of Biological control of water hyacinth In Strategies for water hyacinth control. R.E. Charudattan, R. Labrada, T.D. Center, C. Kelly–Begazo. eds. FAO, Rome, 1996; pp. 189-196.

Charudattan R., Dinoor A. Biological control of weeds using plant pathogens: accomplishments and limitations. Crop Prot 2000; 19:691-695.

Charudattan R., Lin Y. Isolates of *Penicillium, Aspergillus* and *Trichoderema* toxic to aquatic plants. Hyacinth Control J 1974; 12:70-73.

Cheng H.H. A conceptual framework for assessing allelochemicals in the soil environment. In *Allelopathy: Basic and Applied Aspects*. S.J.H. Rizvi, V. Rizvi, eds. Chapman and Hall Publishers, 1992; pp. 21-29.

Chou C.H. The role of allelopathy in agro-ecosystem: studies from tropical Taiwan. In: Gliessman, S.R. (ed). Agroecology: Researching the ecological basis for sustainable agriculture. Ecological studies #78. Springer – Verlag. Berlin., 1990; pp. 105-121.

Chou C.H., Chang F.J., Oka H.I. Allelopathic potential of wild rice *Oryza perennis*. Taiwania 1991; 36:201-210.

Copping G.L. Herbicides discovery. In Weed Management Handbook. BCPC. Blackwell Publishers, UK. 2002.

Courtois B., Olofsdotter M. Incorporating the allelopathy trait in upland rice breeding programs. In *Allelopathy in Rice*. M. Olofsdotter, ed. International Rice Research Institute, Manila, Philippines: 1998; pp. 57-68.

Dickens R. Cogongrass in Alabama after sixty years. Weed Sci 1974: 22:177-179.

Dickens R, Buchanan G.A. Old weed in a new home-that's cogongrass. Highland Agri Res 1971; 18:2.

Dilday R.H., Lin J., Yan W. Identification of allelopathy in the USDA-ARS rice germplasm collection. Aust J Exp Agric 1994; 34:907-910.

Dilday R.H., Yan W.G., Moldenhauer K.A.K., Gravois K.A.. Allelopathic activity in rice for controlling major aquatic weeds. In *Allelopathy in Rice*. M. Olofsdotter, ed. International Rice Research Institute, Manila, Philippines, 1998; pp. 7-26.

Duke S.O. Naturally occurring chemical compounds as Herbicides. Reviews of Weeds Science 1986; 2:15-44.

Duke S.O., Romagni J.G., Dayan F.E. Natural products as source of new mechanisms of herbicidal action. Crop Prot 2000; 19:583-589.

Duke S.O., Scheffler B.E., Dayan F.E., Weston L.A., Ota E. Strategies for using transgenes to produce allelopathic crops. Weed Technol 2001; 15:826-834.

Elmore C.D. Weed survey-southern states. Research Report from Southern Weed Science Society 1986; 39:36-158.

Eussen J.H.H, Wirjahardja S. Studies of an alang-alang (*Imperata cylindrica* (L.) Beauv.) vegetation. Biotrop Bull 1973; 6:1-24.

Falvey J.L. *Imperata cylindrica* and animal production in southeastern Asia: A review of Tropical Grasslands 1981; 15:52-56.

Ferguson J.J., Rathinasabapathi B. Allelopathy : How plants suppress other plants. HS944 series Institute of food and agricultural sciences, University of Florida, online publication 2003; http://edis.ifas.ufl.edu

Gnanavel I., Kathiresan R.M. Sustainable weed management in rice-rice cropping system. Indian J Weed Sci 2002; 34:192-196.

Gross E.M., Meyer H., Schilling G. Release and ecological impact of algicidal hydrolysable polyphenols in *Myriophyllum spicatum*. Phytochemistry 1996; 41:133-138.

Hassan S.M., Aidy I.R., Bastawisi A.O. Allelopathic potential of rice varieties against major weeds in Egypt. Paper presented at annual meeting of Weed Science Society of America. Seattle, Washington, USA. 1995.

Hoagland, R.E. Microbial allelochemiclas and pathogens as bioherbicidal agents. Weed Technol 2001; 15:835-857.

Holm L.G., Plucknett D.L., Pancho J.V., Herberger J.P. The World's Worst Weeds – Distribution and Biology. University Press of Hawaii, Honolulu. 1977.

Inderjit. Plant phenolics in allelopathy. Bot Rev 1996; 62:186-202.

Inderjit, Dakshini K.M.M. Investigations on some aspects of chemical ecology of cogongrass. *Imperata cylindrica* (L.) Beauv. J Chem Ecol 1991; 17:343-352.

Jensen L.B., Courtois B., Shen L., Li Z., Olofsdotter M., Mauleon R.P. Location genes controlling rice allelopathic effects against barnyardgrass in upland rice. Agron J 2001; 93:21-26.

Kannan C. Integrated control of water hyacinth (*Eichhornia crassipes* (Mart.) Solms). Ph.D. Dissertation. Annamalai University, Tamilnadu, India 2002.

Kannan C., Kathiresan R.M. Biological control at different growth stages of water hyacinth. Proceedings of the 1st IOBC global working group meeting for the biological and integrated control of water hyacinth. 16-19 November, Harare, Zimbabwe. M.P. Hill, M.H. Julien, T.D. Center, eds. 1998; pp. 87-90.

Kathiresan R.M. Allelopathic control of water hyacinth. In *Annual Report 1997 of Centre of Aquatic Plant Management*, J.R. Newman, ed. UK. 1998; pp. 54-56.

Kathiresan R.M. Allelopathic potential of naïve plants on water hyacinth. XIV International Plant Production Congress, Jerusalem, Israel 1999; Abstract, pp. 146.

Kathiresan R.M. Allelopathic potential of native plants against water hyacinth. Crop Prot 2000; 19:705-708.

Kathiresan, R.M. Weed Management in Rice-Blackgram Cropping System. Indian J Weed Sci 2002; 34:220-226.

Kathiresan R.M. Integration of elements of farming system for sustainable weed and pests management in the tropics. IV International Weed Science Congress, Durban, South Africa. 2004a; Abstract, pp. 6.

Kathiresan, R.M. Integration of botanical herbicide *Coleus amboinicus / aromaticus* with insect biological control of water hyacinth. Completion report on research project funded from National Agricultural Technology Project – Indian Council of Agricultural Research, 2004b.

Kathiresan R.M., Gurusamy. Allelopathic potential of paddy seeds and some plant products on Barnyardgrass. Biennial Conference of Indian Society of Weed Science. Annamalai University, Annamalainagar 1995; Abstract, pp.110.

Kathiresan R.M., Kannan C. Allelopathy for the control of water hyacinth. First Meeting of Global Working Group on Integrated and Biological Control of Water Hyacinth. IOBC, Harare, Zimbabwe 1998; Abstract, pp. 26.

Kauraw L.P., Bhan V.M. Efficacy of cassytha powder to water hyacinth and marigold to *Parthenium* population. Weed News 1994; 1:3-6.

Kim K.U., Shin D.H. Rice allelopathy research in Korea. In Allelopathy in Rice. M. Olofsdotter, ed. International Rice Research Institute, Manila, Philippines, 1998; pp. 39-43.

Kim K.U., Shin D.H. The importance of allelopathy in breeding new cultivars. In Weed Management for Developing Countries. Addendum I. Labrada, ed., FAO Plant Production and Protection Paper 2003; pp.195-210.

Kim K.W., Kim K.U., Shin D.H., Lee I.J., Kim H.Y., Koh J.C., Nam S.H. Searching for allelochemicals from the allelopathic rice cultivar, Kouketsumochi. Korean J Weed Sci 2000; 20:197-207.

Koch W, Grobbmann F, Weber A, Lutzeyer H.J, Akobundu I.O. Weeds as components of maize/cassava cropping systems. In Standortgemaesse landwirtschaft in West Africa, M. von Oppen, ed. Stuttgart, Germany: Universitaet Hohenheim,1990; pp. 219-244

Koger C.H., Bryson C.T. Effect of cogongrass (*Imperata cylindrica*) extracts on germination and seedling growth of selected grass and broadleaf species. Weed Technol 2004; 18:236-242.

Koger C.H., Bryson C.T, Byrd J.D. Response of selected grass and broadleaf species to cogongrass (*Imperata cylindrica*) residues. Weed Technol 2004; 18:353-357.

Leu E., Liszkay A.K., Goussias C., Gross E.M. Polyphenolic allelochemicals from the aquatic angiosperm myriophyllum spicatum inhibit photosystem II. Plant Physiol 2002; 130:2011-2018.

Lin W., Kim K.U., Liang K., Guo Y. Hybrid rice with allelopathy. In Proceedings of the International Workshop in Rice allelopathy, K.U. Kim, D.H. Shin, eds., Kyungpook National University, Taegu, Korea, 17-19 August 2000. Institute of Agricultural Science and Technology, Kyungpook National University, Taegu, 2000; pp. 49-56.

Lydon, J., Duke, S.O. Inhibitors of glutamine biosynthesis. In *Plant Amino Acids: Biochemistry and Biotechnolgy*, B.K. Singh, ed. Marcel Dekker, New York. 1999. pp. 445-464.

Mattice J., Lavy T., Skulman B., Dilday R. Searching for allelochemicals in rice that control duck salad. In Allelopathy in Rice. M. Olofsdotter, ed. International Rice Research Institute, Manila, Philippines, 1998; pp. 81-98.

Muchovej J.J., Kushalappa A.C. Dreschslera leaf spot of *Salvinia auriculata*. Plant Dis Rep 1979; 63:154.

[OTA] Office of Technology Assessment. Harmful non-indigenous species in the United States. Washington DC: Office of Technology Assessment, US Congress 1993.

Pandey D.K. Phytotoxicity of sesquiterpene lactone parthenin on aquatic weeds. J Chem Ecol 1996a; 22:151-160.

Pandey D.K. Relative toxicity on allelochemicals to aquatic weeds. Allelopathy J 1996b; 3:241-246.

Pandey D.K., Kauraw L.P., Bhan V.M. Inhibitory effect of parthenium (*Parthenium hysterophorus* L.) residue on growth of water hyacinth (*Eichhornia crassipes* Mart. Solms.). J Chem Ecol 1993; 19:2663-2670.

Parthiban C., Kathiresan R.M. Use of certain plant materials for weed management in transplanted rice. Indian J Weed Sci 2002; 34:187-191.

Pimentel D., Lach L., Zuniga R., Morrison D. Environmental and economic costs of nonindigenous species in the United States. BioScience 2000. 50:53-65.

Pimentel D. Handbook on pest management in Agriculture. CRC Press, Boca Raton FL 1991.

Putnam A.R. Weed allelopathy. In *Weed Physiology, Volume 1: Reproduction* and *Ecophysiology*, S.O. Duke, ed., CRC Press Boca Raton FL 1985; pp.131-155.

Putnam A.R., Tang C.S. Allelopathy: State of Science. In *The Science of Allelopathy*, A.R. Putnam, C.S. Tang, eds., Wiley, New York, 1986; pp. 1-19.

Rao O.P., Saxena A.K., Singh B.P. Allelopathic effects of certain agroforestry tree species on the germination of wheat, paddy and gram. Ann Forestry. 1994; 2:1.

Reddy K.N. Effects of cereal and legume cover crop residues on weeds, yield, and net return in soybean (*Glycine max*). Weed Technol 2001; 15:660-668.

Reddy K.N., Zablotowicz R.M., Locke M.A., Koger, C.H. Cover crop, tillage, and herbicide effects on weeds, soil properties, microbial populations, and soybean yield. Weed Sci 2003; 51:987-994.

Rice E.L. Allelopathy: An overview. Allelochemical: Role in agriculture and forestry. American Chemical Society Symposium Series 330, 1987; pp. 8-22.

Rice E.L., Allelopathy. 2nd edition. Academic Press. 1984.

Sivagurunathan, M., Sumithradevi G., Ramasamy K. Allelopathic compounds in *Eucalyptus* spp. Allelopathy J 1997; 4:313-320.

Sustainable Agriculture Network. 1998. Managing cover crops profitably. Second edition. Handbook Series Book 3. Beltsville, MD.

Teasdale J.R. Contribution of cover crops to weed management in sustainable agricultural systems. J Prod Agric 1996; 9:475-199.

Teasdale J.R. Principles and practices of using cover crops in weed management systems. In *Weed Management for Developing Countries*. Addendum I. Labrada, ed., FAO Plant Production and Protection Paper 2003; pp. 169-178.

Udensi E.U., Akobundu I.O, Ayeni A.O, Chikoye D. Management of cogongrass (*Imperata cylindrica*) with velvet bean (*Mucuna pruriens* var. *utilis*) and herbicides. Weed Technol 1999; 13:201-208.

Vasquez E.A., Morallo-Rejesus B., Punzalan E.G, Kraus W. Biological studies on the essential oil of *Coleus amboinicus*. IXV International Plant Production Congress, Jerusalem, Israel 1999; Abstract. pp. 29.

Willis R.J. The historical bases of the concept of allelopathy. J Hist of Bio 1985; 18:71-102.

ANTONY V. STURZ

BACTERIAL ROOT ZONE COMMUNITIES, BENEFICIAL ALLELOPATHIES AND PLANT DISEASE CONTROL

Prince Edward Island Department of Agriculture, Fisheries, Aquaculture and Forestry, P.O. Box 1600, Charlottetown, P.E.I., C1A 7N3, Canada.
Tel: 902-368-5664, FAX: 902-368-5661
E-mail:avsturz@gov.pe.ca

Abstract. The release of root exudates from plants encourages the formation of beneficial bacterial communities in the root zone capable of generating secondary metabolites that improve plant health and crop yield. Metabolites with antibiotic or lytic action have been identified, while others are known to induce systemic disease resistance in the host plant, or interfere with the nutritional requirements of phytopathogens. However, despite existing positive relationships between bacterial communities and their plant hosts, man-made attempts at applying bacteria for biocontrol purposes have met with limited success. Inconsistent performance of biocontrol bacteria in the field may be due to the variable expression of genes involved in the biocontrol action, or simply the resistance of established soil communities to a sudden and inundative influx of adventive bacterial species or strains. Regardless of the inherent capacity of 'naturally occurring' soil microbial ecosystems to buffer anthropogenic interference, crop management systems are regularly used to distort agro-ecosystems through, for example, the use of tillage operations, alternate cropping systems, monoculture, crop rotation length, fertilizer and organic amendments, and various crop protection chemistries. The management of soil microbial communities for disease control appears to involve, in part, the creation of short term chaos in the microbial community through the application of such perturbation stresses. While hope remains that bacterial communities with biocontrol activity will one day be used as an adjunct to, or replacement for, agrichemical crop protectants, reliable biological controls that moderate pathogen attack remain elusive. In the interim, disease suppressive soils may be encouraged to form through the use of modest perturbation stresses that promote microflora species' diversity and functionalities underpinning natural bioantagonism.

1. INTRODUCTION

In its broadest sense allelopathy has been defined as "...any process involving secondary metabolites produced by plants, microorganisms, viruses and fungi that influence the growth and development of biological systems..." (IAS, 1998). In the bacterial kingdom, the production of secondary metabolites (allelochemicals) can result in the development of mutualistic, beneficial or antagonistic relationships (Smith and Goodman, 1999; Boller, 1995) amongst bacteria, or between bacteria and other living organisms. As such, bacterial secondary metabolites can act at the trophic level, directly affecting nutrient uptake or metabolism, or at the informational level, being recognized as signals by appropriate chemoperception systems (Dusenbery, 1992; Boller, 1995). Consequently, allelopathy may be viewed as an ecological phenomenon (Romeo, 2000)

Inderjit and K.G. Mukerji (eds.),
Allelochemicals: Biological Control of Plant Pathogens and Diseases, 123–142.

capable of regulating ecosystem health and biodiversity (Wardle et al., 1997; Mallik, 2000).

Many secondary metabolites can act at both the trophic and the informational level, becoming attractants or repellents, toxins or growth stimulants, depending upon the microbial partnerships involved (Dusenbery, 1992). Accordingly, microbially produced allelochemicals have been reported to express themselves as lytic agents or enzymes (Fridlender et al., 1993; Jacobson et al., 1994; Glick et al., 1999), antibiotics (Lynch, 1976; Bender et al., 1999), siderophores (Buysens et al., 1994; Marschener and Crowley, 1997) auxins (Patten and Glick, 1996; Glickman et al., 1998; Glick et al., 1999), volatile compounds (Claydon et al., 1987; Bakker and Schippers, 1987) and phytotoxic substances (Hoagland and Cutler, 2000).

An intimacy is seen to extend between plants and microbes, in as much as plants appear able to influence the composition of the microbial community around their root systems by leaking specific carbohydrates, carboxylic and amino acids (Grayston et al., 1998) into the root zone, as well as through the 'carbon-loading' that occurs as root cell material is sloughed off during root growth (Hawes and Brigham, 1992). Hawes et al., 1998). In turn, rhizobacteria appear able to induce root exudation responses in plants (Bolton et al., 1993; Merharg and Killham, 1995). The result is a circular allelopathic cascade initiated by plant root exudates that trigger a positive microbial 'allelopathic feedback' in which the final receptor organism is also the initiator.

It is this allelochemical interaction amongst soil microbial communities and the way in which their relationships subsequently influence plant health and disease development that will form the emphasis of the present chapter.

2. MICROBE-MICROBE INTERACTIONS

2.1. Commensal Relationships, Protocooperative Assemblages and Plant Pathogens

A large proportion of successful biocontrol events in the infection court can be attributed to the positive outcome of multiple allelopathic episodes governed by the interplay among bacterial populations. In the root zone - that region characterized as occurring outside the host plant, but within the sphere of influence of the root system - bacterial populations often form protocooperative assemblages (loose associations) that are mutually beneficial (though not obligatory). By contrast, commensal relationships where the microorganism (commensal) benefits, while the host (plant) is neither harmed nor helped, are a feature of the closer physical juxtaposition of bacteria and host plant, often marked by the sharing of the same food resource. This latter relationship can be a feature of bacteria inhabiting the root surface (rhizoplane), but can also be extended to include those bacterial colonists found within plants (endophytes)-specifically in the endoroot tissues. However, not withstanding the above, distinctions between commensal or protocooperative assemblages become equivocal in the face of community antagonism to pathogen invasion, and should, perhaps, be viewed independently of any plant health benefits.

While pathogen antagonism has been collectively termed biological control and defined by Baker (1987) as '... a decrease of inoculum or the disease-producing activity of a pathogen accomplished through one or more organisms, including the host plant...', it can also be argued that antagonism, at the communal level, is not necessarily directed at phytopathogens specifically, but at any group of organisms 'invading' the trophic level of an established community. In this sense, interactions amongst autochthonous (established) consortia (functional groupings) of microflora can be classified as beneficial where they promote plant health or inhibit phytopathogen attack (Mukerji et al., 1999).

Accordingly, any exochthonous latecomer (colonist or pathogen) is likely to provoke a negative response from an established community, since its arrival will upset any balance amongst the community members (Atlas, 1986). In this respect, the resilience of a soil microbial community, when expressed in terms of its ability to inhibit invasions (colonization) by 'non-community' species will, in part, define its stability.

As such, exochthonous species, enter the niche environment by chance but cannot maintain themselves in an active condition (Cooke and Rayner 1984). By contrast, indigenous communities may be subdivided into the slow growing autochthonous groups and the fast growing transient zymogenous groups - the former surviving on refractory substrates, and so, for the most part, remaining constantly active, while the latter only become active when a suitable food resource presents itself, and so are otherwise quiescent (Cooke and Rayner 1984).

2.2. 'Self-awareness' in Bacterial Communities

It is believed that population density of a consortium component species mediates population function through 'self-awareness' mechanisms such as 'quorum sensing', that enable bacteria to communicate among and between species in a consortium (Miller and Bassler, 2001)

The degree to which bacterial consortia behave as commensal, protocooperative or pathogenic assemblages is dictated (in part) by the component populations' density, which will vary according to the prevailing abiotic conditions affecting secondary metabolite production - including those with antibiotic properties (Grimwood et al., 1989; Tateda et al., 2001). Key abiotic factors include, among others, pH, temperature, moisture, salinity, oxygen concentration and carbon availability (Duffy and Défago, 1999; Gaballa et al., 1997; Gutterson et al., 1988; Nakata et al., 1999; Shanahan et al., 1992; Slininger and Jackson, 1992; Slininger and Sheawilbur, 1995)

A population's ability to identify 'itself' through the recognition of diffusible signaling molecules (autoinducers) - generally acylated homoserine lactones (acyl-HSLs) for gram-negative bacteria, and oligopeptides for gram-positive bacteria - elicits the modulation of gene expression, that can alter bacterial function in ways that may ultimately destabilize cooperative behaviour (Salmond et al., 1995; Albus et al., 1997; Surette and Bassler, 1998).

Paradoxically, the ability to exchange low molecular weight diffusible signal molecules suggests that certain species specific traits are repressed to allow individual bacteria to form consortia, so enabling them to survive within a specific habitat until such time as it benefits that species population to dissolve the co-operative group. This event is often catalyzed by the build up of sufficient amounts of signal molecule to activate receptor proteins that trigger changes in gene expression.

Quorum sensors in bacteria have been implicated in regulating a range of physiological activities including conjugal transfer (Piper et al., 1993), swarming responses (Givskov et al., 1997), sporulation (Dworkin and Kaiser, 1985; Hoch, 1995), biofilm formation (Davies et al., 1998), and, most importantly, the regulation of virulence genes that initiate extracellular polysaccharides, enzymes, surfactants and antibiotics prior to and during pathogen attack (Beck von Bodman and Farrand, 1995; Fuqua, et al., 2001; Parsek et al., 1999; Whithers et al., 2001).

Thus, for example, the causal agent of potato soft rot, *Erwinia carotovora*, will only initiate exoenzyme secretion - involving cellulases and various pectinases - when cell density levels reach a certain threshold (Pirhonen et al., 1993); presumably to ensure that invading bacteria do not prematurely elicit a host-defence response prior to achieving sufficient bacterial cell numbers to mount a successful infection. Interestingly, *E. carotovora* exoenzymes are produced in tandem with the antibiotic carbapenem, which it is believed inhibits other competitor bacteria in the infection court (Bainton et al., 1992). Consequently, it appears that some pathogen's attack and self-defence measures are co-ordinated during attempts at plant infection.

2.3. Community Self-regulation

The biological basis of community homeostasis is generally believed to result from the dynamic balance which member populations in the community must exert to inhibit the recognition of 'population self' at the expense of the 'community self'. Where complex regulatory cascades control gene expression of colonization and pathogenicity, gene induction is generally modulated by the rate at which a population's signal molecule strength is diluted away. Accordingly, any situation that allows for signal molecule build-up - such as the close proximity of cells, or an environment limited diffusion rate - will result in the greater the likelihood that population self-recognition and gene expression will be triggered.

To-date, compounds reported to regulate population density dependent behaviours include N-acyl-homoserine lactones, among plant associated Proteobacteria (Eberl, 1999; Fuqua et al., 2001), ã-butryolactone in the Streptomyces (Yamada and Nihira, 1998), oligopeptides in a variety of gram-positive species (Dunny and Winans, 1999; Kleerebezem and Quadri, 2001), cyclic dipeptides in some gram-negative species (Holden et al., 1999) and fatty acid and butyrolactone derivatives in the plant pathogenic bacteria *Xanthomonas campestris* and *Ralstonia solanacearum* (Barber et al., 1997).

2.4. *Microbe-microbe Disruption of Quorum Sensing Mechanisms*

It should now be apparent that quorum sensing plays a significant role in the biology and regulation of both plant-microbe and microbe-microbe interactions. And while pathogenic bacteria depend significantly on quorum sensing regulation to coordinate the saprophytic and parasitic phases of their life cycles, plants and their adherent root zone microbial communities have evolved mechanisms by which to disrupt this strategy.

For example, *Variovorax paradoxus*, a relatively common soil organism, is able to utilize (degrade) acyl-HSL signaling molecules as an energy source (Leadbetter and Greenberg, 2000) with the effect that those classes of bacteria relying on acyl-HSLs for cell-to-cell signaling molecules will be kept 'unaware' of their own presence and population density. Pierson et al. (1998) demonstrated that approximately 8% of rhizobacteria recovered at random from the surfaces of wheat roots could specifically stimulate quorum sensing gene regulated expression in adjacent *P. aureofaciens* bacteria. Thus different bacteria appear able to exchange quorum sensing signals, with the possibility of forming functional mixed communities (Bauer and Robinson, 2002).

Several instances have been reported of soil bacteria in possession of enzymes designed to degrade or inactivate acyl-HSLs (Bauer and Robinson, 2002; Dong et al., 2001, 2002; Leadbetter, 2001; Whitehead et al., 2001). A case in point is the lactonase enzyme, AiiA, from *Bacillus cereus*, which, it is believed, opens the lactone ring in acyl-HSLs, thereby reducing signal strength in the order of a 1000 fold (Dong et al., 2000). In circumstances where *B. cereus* and *E. carotovora* co-exist as commensals in field soils, AiiA is able to inactivate the acyl-HSL autoinducer in *E. carotovora* rendering the pathogen avirulent.

Clearly, understanding the mechanism of signal synthesis, and being able to identify signal synthesis inhibitors, has implications for developing quorum sensing-targeted antivirulence molecules, or engineering beneficial communities which utilize acyl-HSL signals as an energy source and so inhibit pathogenic trait expression. In the former case, AHL signal-inactivating molecules (the so-called 'quorum quashing' moieties) - namely AHL-lactonases and AHL-acylases - have already been identified in a *Ralstonia* sp. isolated from a mixed-species biofilm (Lin et al., 2003).

However, while selection pressures upon component bacterial populations in a community will include biological pressures created by microbially mediated moderation of habitats, external abiotic pressures created by environmental perturbations - such as climate or crop management practices - also exist. As a result, even though "... microorganisms are potentially everywhere, [the] environment selects..." (Alexander, 1997).

3. MICROBIAL ANTAGONISM AND DISEASE CONTROL

3.1. *Microbially Induced Biological Control in Soils*

Since every living soil sample will yield organisms with antagonistic activity to some other organism, or group of organisms, it has almost become axiomatic that

"...antagonistic potential resides in every soil microorganism..." (Baker and Cook, 1974). Consequently, it is generally held that most soils possess the biological propensity to inhibit or reduce their soil microflora's tendency toward disease, and so can be considered disease suppressive to some extent (Hornby, 1983). As a result, there exists in the literature a vast array of a terms used to describe soils that are inhospitable to plant pathogens. For example, i) soils where plant pathogens fail to become established have been referred to as resistant (Walker and Snyder, 1933), long-life, immune, intolerant, or antagonistic (Baker and Cook, 1974; Huber and Schneider, 1982), ii) soils where pathogens become established but fail to produce disease have also been termed suppressive (Schroth and Hancock, 1982); while iii) soils where disease incidence diminishes with continued monoculture have been termed decline soils (Shipton, 1975; Hornby, 1979, 1983).

Attempts to simplify the biological basis for disease suppression in agricultural soils have reduced this concept to two broad mechanisms; namely that of i) a "general suppression" based upon the activity of the total microbial biomass that is not transferable between soils, and ii) a "specific suppression" that depends upon the activity of specific groups of microorganisms (Weller et al., 2002).

Whether a bacterial population behaves pathogenically or not will be a function of that component species' genetics, the restraints which other members in the community are able to impose, and the result of any overriding selection pressures dictated by environmental factors governing habitat type and host predisposition to disease.

Environmentally mediated host predisposition to disease have been linked with obligate pathogen performance, and included exposure to cold (Schulz and Bateman, 1969), low light intensity, or short day lengths (Foster and Walker, 1947), salinity stress (MacDonald, 1982), high temperature (Edmunds, 1964), and drought or moisture stress (Boyer, 1995; Duniway, 1977).

In contrast, factors predisposing host plants to attack by rogue members of a commensal community, or protocooperative assemblage, are less well understood, though it appears likely that any dramatic change in the niche environment can provide an ecological advantage that benefits some community members at the expense of others. During the resulting population increase of the favoured community population, cell density dependent pathogenesis is triggered.

3.2. Disease Suppression and Pathogen Evasion

The wide array of nomenclature used to describe disease suppression in agricultural soils, is matched by an equally wide variety of individual microbial mechanisms postulated to explain these phenomena. However, it should be noted that these mechanisms are fairly presumptive, and, if they occur *in vivo*, are likely to operate in parallel with each other (Figure 1).

In general, microbial biocontrol mechanisms have been classified according to effect (Baker, 1968) and have included such actions as parasitism/predation, niche competition, antibiosis and systemic induced resistance - the latter three falling

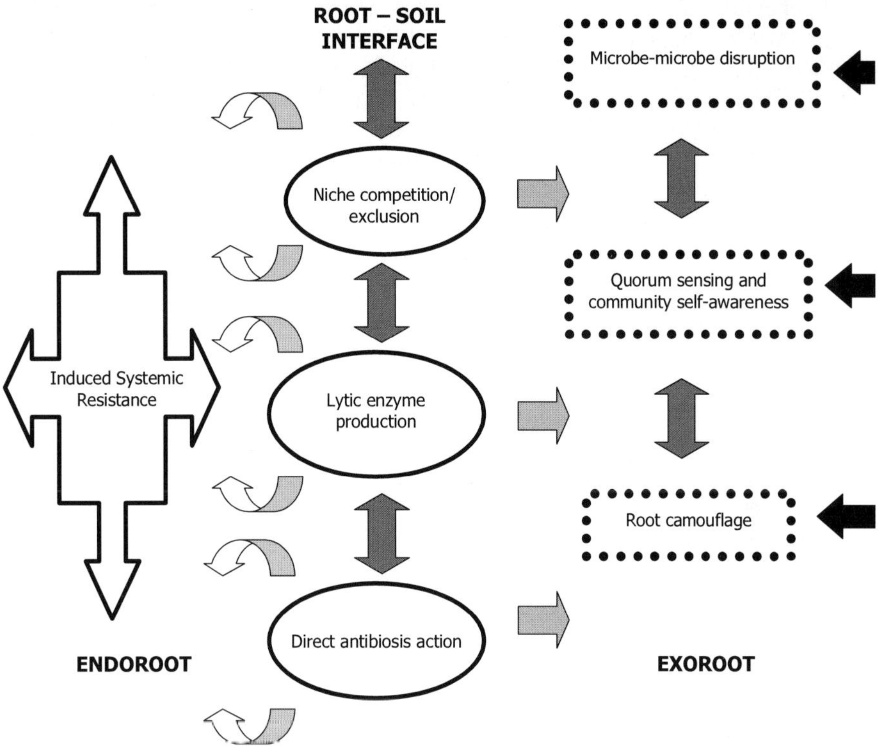

Figure 1. Parallel action of disease suppression mechanisms operating within the host plant (endoroot) and in the surrounding soil (exoroot).

reasonably comfortably within the ambit of allelopathy (see Keel and Défago, 1997; Mukerji et al., 1999; Weller, 1988).

Five common mechanisms are usually cited, namely:

i) *Resource(niche) competition (or exploitation)* (O'Sullivan and O'Gara, 1992; Stephens et al., 1993; Whipps, 2001) - for example, siderophore (chelator)-producing bacteria with high affinities for, and capable of sequestering, specific mineral elements, can inhibit phytopathogens with the same requirement if that mineral is limited in the soil (Schroth and Hancock, 1982; Dowling et al, 1996; Loper and Henkels, 1997, 1999).

ii) *Antibiosis action* - through the production of specific or non-specific microbial metabolites with antibacterial, antifungal and anti-nematode activity (Levy and Carmelli, 1995; Fujimoto et al., 1995; Raaijmakers et al., 2002; Thomashow et al., 1997). To-date, several antibiotic substances have been identified of which those produced by the pseudomonads have been particularly well characterized. Of these, antibiotics identified with biocontrol properties include the phloroglucinols, phenazine derivatives, pyoluteorin, pyrrolnitrin, cyclic lipopeptides and hydrogen cyanide (Haas and Keel, 2003). Among the other antibiotics characterized are agrocin 84 (*Agrobacterium* sp.), herbicolin A (*Erwinia* sp.), iturin A, surfactin, and zwittermicin A (*Bacillus* sp.) and xanthobacin (*Stenotrophomonas* sp.) (Hashidoko et al., 1999; He et al., 1994; Sayre and Starr, 1988; Thomashow et al., 1997; Silo-Suh et al., 1994). A comprehensive account of bacterially produced antibiotics may be found in Raaijmakers et al. (2002).

iii) *Lytic enzyme action* - a feature of several bacteria with proven biocontrol ability, and generally involves the direct degradation of pathogen cell wall material, or the disruption of a particular developmental stage. Thus, for example, chitinase production by *Serratia plymuthica* has been reported to inhibit spore germination and germ-tube elongation in *Botrytis cinerea* (Frankowski et al., 2001), while ß-1,3-glucanase synthesized by *Paenibacillus* sp. and *Streptomyces* sp. can lyse fungal cell walls of *Fusarium oxysporum* f. sp. *cucumerinum* (Singh et al., 1999). Other enzymes produced by bacteria with biocontrol activity include hydrolase (Chernin and Chet, 2002), laminarinase (Lim et al., 1991) and protease (Kamensky et al., 2003).

iv) *Induced systemic resistance* (ISR) in plants (Wei et al., 1991; Tuzun and Kloepper, 1994) - whereby non-pathogenic rhizobacterial stimulation of defence-related genes is elicited through the encoded production of jasmonate (van Wees et al., 1999), peroxidase (Jetiyanon et al., 1997) or enzymes involved in the synthesis of phytoalexins (van Peer et al., 1991). Though no specific ISR-eliciting signal has been identified, thus far, evidence for the involvement of lipopolysaccharides, siderophores and phloroglucinols has been submitted (Hoffland, et al., 1995; Leeman et al., 1995, 1996; Maurhofer et al., 1994; van Wees et al., 1997), and,

v) *Root camouflage* (Gilbert et al., 1994) - proposed as a mechanism to explain the observation that certain rhizobacterial populations in disease resistant cultivars are able to minimize the 'attractive' nature of the host's root system so masking its presence to potential plant pathogens by restricting local population density development. Such microbial systems may operate in tandem with those that desensitize the chemoperception systems of microorganisms in the root zone, through the over production of chemical stimulii (Armitage, 1992; Dusenbery, 1992).

The parallel operation of all these biocontrol mechanisms in a four dimensional soil space makes their action and interaction difficult to follow. Biocontrol strains only occupy a small fraction of the root surface, in microcolonies spread out unevenly along the root surface (Bowen and Rovira, 1976; Normander et al., 1999). Disease suppression, when it occurs through antibiosis, is most likely restricted to local action only, and most probably at sub-inhibitory levels. Even so, antibiotics can cause intense physiological effects upon neighbouring organisms at subinhibitory concentrations. Quinolone and macrolide antibiotics have been reported to block cell- to cell signaling, and the production of virulence factors in *P. aeruginosa* (Grimwood et al., 1989; Tateda et al., 2001). Similarly, subinhibitory concentrations of antibiotics can suppress adherence mechanisms in bacteria (Breines and Burnham, 1994), and the production of extracellular virulence factors in bacteria (Herbert et al., 2001). Accordingly, secondary metabolites can impact soil microbial ecosystems in a variety of ways, and at a variety of levels (Haas and Keel, 2003).

3.3. Partitioning of disease suppressive bacteria in the endo and exoroot

Most research into soil bacterial communities has been restricted to the exoroot - that fraction of the microfloral community found in the rhizoplane, plant rhizosphere or root zone soil (Haas and Keel, 2003). Endoroot bacteria have largely been ignored, despite plant-bacteria interactions extending into the endoroot of all plants (Conn et al., 1997; Bensalim et al., 1998). The frequent recovery of communities of endophytic bacteria, in the absence of any pathological condition (Chanway, 1996) and the finding that bacterial endoplant communities are capable of mediating against phytopathogen invasion (Benhamou et al., 1996) has led to the suggestion that plants may have co-opted bacteria as part of a disease suppressive response to phytopathogen attack.

Several instances have been reported of endophytic bacteria as effective biocontrol agents (van Buren et al., 1993; Brooks et al., 1994). van Peer et al. (1990) found that endo- and exoroot bacteria from the same genera formed discrete sub-populations, each suited to colonizing their respective environmental niches. Tissue-specific relationships can form between communities of bacterial endophytes and their host plant, and endobacteria have been shown to adapt functionally to certain tissue sites and among certain tissue types (Sturz et al., 1999). Unfortunately, the population densities of endophytic bacteria tend to be highly variable among plant tissues and so may be of little practical value in terms of affording plants a comprehensive first line of defence to pathogen attack.

4. 'ENGINEERING' BENEFICIAL MICROBIAL ALLELOPATHIES

4.1. Anthropogenic intrusions

The premise underlying most anthropogenic biocontrol systems is the notion that it is possible to encourage the occurrence and development of beneficial rhizobacterial allelopathies in the root zone, primarily through the direct application of specific biocontrol agents, and or soil conditioning amendments (Sturz and Christie, 2003). It must be acknowledged at the outset that such systems have had varying degrees of success.

4.2. Inundative approaches in biological control

In general, anthropogenic attempts to import biological control agents into the field have been through the inundative application of super quantities of a few key biocontrol or plant growth promoting agents. These have been applied directly as drenches or sprays, or alternatively as a cell suspensions incorporated in mulches or compost material. Carriers such as granular peat formulations, mineral soils (Chao and Alexander, 1984), bacterial encapsulations within polymer gels (Bashan, 1986) or in natural gum or talc mixtures (Kloepper and Schroth, 1981) have also been tried.

Perhaps, in hindsight, it is not surprising that such inundative approaches have been relatively unsuccessful, with most biocontrol agents failing to fulfill their initial promise. Such failures have usually been attributed to poor competence of biocontrol agents in the infection court and the difficulties associated with the instability of biocontrol agents in culture (Schroth et al., 1984; Weller, 1988); not least because in the case of bacterial agents, the accumulation of extracellular signalling molecules within large population densities of individual strains, can modulate a diverse range of metabolic processes, some of which are incompatible with the goals of phytoprotection (see above).

However, the use of single antagonists may itself be an inappropriate strategy, arising from the belief that a given plant disease can be attributed to a single pathogen only (Baker, 1987). Such one-on-one syndrome concepts follow from the belief that successful control of a pathogen is achievable with a single fungicide, or by single-factor resistance, coupled with the observation that single antagonists have often provided effective control in presumptive tests for antagonists in *in vitro* studies, or as biocontrol agents applied to sterilized soil (Baker, 1987)

Needless to say, the very practice of applying massive quantities of a single bacterial species to the infection court will not only alter the putative biocontrol agent's physiology, but also its niche behaviour. Perhaps, more intriguingly, it may also garner a general and antagonistic response from the resident population.

Several attempts at engineering bacterial strains with more reliable biocontrol performance have been tried, whereby biosynthetic genes for various antibiotics have been designed to be constitutively over-expressed. On occasion, such engineered strains have provided improved plant protection in the soil microcosm (Delany et al., 2001; Ligon, et al., 2000; Timms-Wilson et al., 2000). However, in long term field evaluations, engineered derivatives have also lacked consistency; loss of stable performance and lack of superiority to wild-type strains being cited as the principal reasons for failure (Bakker et al., 2002).

Although it would be premature to generalize the findings of such studies, it appears that the engineering of a single trait (antibiotic production) in a single biocontrol strain can not overcome the problem of inconsistent performance in the field, given the multi-factor nature of biocontrol mechanisms and the potential for interaction with wild-type species in the soil microbial community.

While one-on-one antagonism may indeed be the sole operating mechanism involved in microbial disease suppression, an equally valid interpretation might be that the inundative addition of biocontrol agents can stimulate a general antagonistic response from the autochthonous microbial population to the 'invader' (inundating) species. In this circumstance and irrespective of any inconsistencies in the field performance of biocontrol agents attributed to unfavourable edaphic factors - such as temperature, soil moisture, pH, clay content, soil type - biocontrol success or failure may simply be due to the resident community's antagonistic response following the inundative insertion of a non-indigenous species. Consequently, both pathogen and biocontrol agent are inhibited in collateral fashion and to various degrees; a scenario that is congruent with the defensive mutualism theory proposed by Clay (1988).

4.3. Modifying Soil Agro-ecosystems

The extent to which producers can develop beneficial root zone allelopathies amongst microbial communities will depend largely upon the resilience of the soil in question (Szabolcs, 1994) and the type of crop management and tillage systems being practised (Sturz and Christie, 2003). Plant species are known to apply a selective and specific influence on microbial diversity in the rhizosphere through their differential root exudate spectra (Grayston et al., 1998), and the plastic nature of the relationship between resident microbial communities. Thus the level of disease suppressiveness in a soil is eminently amenable to deformation through the use of selected cultural practices. This regardless of the inherent capacity of 'natural' soil microbial ecosystems to buffer anthropogenic interference.

Crop management systems are regularly used to distort agro-ecosystems through, for example, the use of tillage operations, alternate cropping systems, monoculture, crop rotation length, fertilizer and organic amendments, and various crop protection chemistries. The management of soil microbial communities for crop yield maximization appears to involve, in part, the creation of short term chaos in the microbial community through the application of a plethora of perturbation stresses (Odum et al., 1979). Moderate levels of 'input perturbation' are considered to improve ecosystem performance, while higher levels of perturbation stress result in performance loss.

Input perturbations have commonly been used to modify soil microbial agro-ecosystems at the expense of pathogen populations. The subsequent variation in habitat and increase in niche heterogeneity - though on a microscale and at multiple sites along the root - is believed to encourage microbial biodiversity and consequently increase the potential for root zone competition (Smucker, 1993; Andrews and Harris, 2000). Thus, for example, increasing soil acidity (Davis and Callihan, 1974, Sturz et al., 2003), applying irrigation soon after tuber initiation (Lapwood et al., 1973; Oestergaard and Nielsen, 1979) and the addition of soil amendments, green manures and mulches (Tremblay and Beauchamp, 1998) have all been relatively successful in reducing the development common scab on potatoes.

Disease suppression has also been achieved against a wide range of pathogens by incorporating green manures (plough-down crops) (Tu and Findlay, 1986), animal manures (Gorodecki and Hadar, 1990) and composts (including organic solid wastes) (Nelson and Hoitink, 1983; Cohen et al., 1998) into field soils. All these amendments can encourage aggressive competition among microbial communities (Hoitink and Boehm, 1999; Hoitink and Fahy, 1986; Hoitink et al., 1997), with the added effect that manure and compost decomposition can release both volatile and non-volatile toxic compounds that inhibit phytopathogenic nematodes (Sayre et al., 1965; Abawi and Widmer, 2000) and reduce the survival rates of pathogenic microbes (De Brito et al., 1995; Chen et al., 1987 a, b).

Though time consuming, unfashionable and often slow to show effect, traditional crop production practices that involve environmentally sustainable practices, such as conservation tillage (Sturz et al., 1997; Bockhus and Shroyer, 1998), 'creative' fallowing options (Sturz et al., 2001), manuring (Hoitink and Boehm, 1999), long term crop rotations (Peters et al., 2003) and compatible cropping systems (Sturz et al., 2003), can yield plant health and crop yield benefits. Whether such knowledge-based, time-intensive management practices can be made more popular remains the challenge.

5. CONCLUSION

The close relationships formed between plant root systems and their respective rhizobacterial communities can lead to profoundly positive allelopathies that improve plant health and crop yield. The selective release of root exudates and plant leachates activates and sustains specific beneficial rhizobacterial communities in the root zone (endo- and exoroot). In turn these bacterial communities are able to generate a wide array of secondary metabolites which can improve plant health, either directly through biological control mechanisms, or by the modulation of phytopathogen populations in root zone. Simultaneously, certain root zone bacteria can induce systemic disease resistance in their plant host. Certain microbial communities are also able to interact functionally with each other through a variety of sensing mechanisms and gene expression triggers that are prompted by bacterial secondary metabolites with multiple activity.

Despite the natural occurrence of strong positive relationships that can develop between bacteria and plants, anthropocentric attempts at applying bacteria for biocontrol purposes have had limited commercial success, notwithstanding advances in our understanding of the molecular mechanisms governing biocontrol interactions in the rhizosphere. The inconsistent performance of biocontrol bacteria in the field could be due to variable expression of genes involved in biocontrol, or merely the resistance of established soil communities to a sudden and inundative influx of adventive bacterial species or strains.

While the hope remains that bacteria with biocontrol activity will one day be used as an adjunct to, or replacement for, commercial chemical fungicides, it will be necessary to better understand the influence on, within and between those microbial communities resident in the soil agro-ecosystems. In the interim, more traditional 'hit and miss' methods for encouraging serendipitous beneficial allelopathies to arise in root zone communities - through, for example, the application of perturbation stresses - still have a role to play in modern systems of crop management, since they encourage the development of microflora species' diversity and functionalities that underpin the antagonism systems promoting biological control in disease suppressive soils.

6. REFERENCES

Abawi, G.S., Widmer, T.L. Impact of soil health management practices on soilborne pathogens, nematodes and root diseases of vegetable crops. Appl Soil Ecol 2000; 15:47-47.

Albus, A.M., Pesci, E. C., Runyen-Janecky, L. J., West, S. E. H., Iglewski, B.H. Vfr controls quorum sensing in *Pseudomonas aeruginosa*. J Bacteriol 1997; 179: 3928-3935.

Alexander, M. Microbial communities and interactions: a prelude. In *Manual of Environmental Microbiology*. Hurst, C.J., Knudsen, G.R., McInerney, M.J., Stetzenbach, L.D., Walter M.V., eds., ASM Press: Washington, 1997; pp. 5-13.

Andrews, J.H., Harris, R.F. The ecology and biogeography of microorganisms on plant surfaces. Annu Rev Phytopathol 2000; 38:145-180.

Atlas, R.M. Applicability of general ecological principles to microbial ecology. In, *Bacteria in Nature*. Poindexter, J.S., Leadbetter, E.R., eds. Volume 2, Plenum Press: New York, 1986; pp. 339-370.

Armitage, J.P. Behavioral responses in bacteria. Annu Rev Physiol 1992; 54:683-714

Bainton, N.J., Stead, P., Chhabra, S.R., Bycroft, B.W., Salmond, G.P., Stewart G.S.A.B, Williams, P. N–(3-oxohexanoyl)-L-homoserine lactone regulates carbapenem antibiotic production in *Erwinia carotovora*. Biochem J 1992; 288:997-1004.

Baker, R. Mechanisms of biological control of soil-borne pathogens. Annu Rev Phytopathol 1968; 6:263-94.

Baker, K.F., Cook, R.J. Examples of biological control. In, *Biological Control of Plant Pathogens,* Baker K.F, Cook, R.J. Freeman: San Francisco, 1974; pp. 61-106.

Baker, K. Evolving Concepts of Biological Control of Plant Pathogens. Ann Rev Phytopathol 1987; 25: 67-85.

Bakker, P.A.H.M., Glandorf, D.C.M., Viebahn, M., Ouwens, T.W.M., Smit, E., Leeflang, P., Wernars, K., Thomashow, L.S., Laureijs, E., Thomas-Oates, J.E., Bakker, P.A.H.M., Leendert C. van Loon, L.C. Effects of *Pseudomonas putida* modified to produce phenazine-1-carboxylic acid and 2,4-diacetylphloroglucinol on the microflora of field grown wheat. Antonie van Leeuwenhoek Int. J Gen Mol Microbiol 2002; 81:617-24.

Bakker, A.W., Schippers, B. Microbial cyanide production in the rhizosphere in relation to potato yield reduction and *Pseudomonas* spp. - mediated plant growth-stimulation. Soil Biol Biochem 1987; 19: 451-57.

Barber, C.E. Tang, J.L., Feng, J.X., Pan M.Q., Wilson, T.J. et al. A novel regulatory system for the pathogenicity of *Xanthomonas campestris* is mediated by a small diffusible signal molecule. Mol Microbiol 1997; 24:555-566.

Bashan, Y. Field dispersal of *Pseudomonas syringae* pv. *tomato, Xanthomonas campestris* pv. *vesicatoria*, and *Alternaria macrospora* by animals, people, birds, insects, mites, agricultural tools, aircraft, soil particles, and water sources. Can J Bot 1986; 64:76-281.

Bauer, W.D., Robinson, J.B. Disruption of bacterial quorum sensing by other organisms. Curr Opin Biotechol 2002; 13:234-237.

Beck von Bodman, S., Farrand, S.K. Capsular polysaccharide biosynthesis and pathogenicity in Erwinia stewartii require induction by an N-acylhomoserine lactone autoinducer. J Bacteriol 1995; 177: 5000-5008.

Bender, C.L., Rangaswamy, V., Loper J. Polyketide production by plant associated pseudomonads. Annu Rev Phytopathol 1999; 37:175-196.

Benhamou, N., Kloepper, J.W., Quadt-Hallman, A., Tuzun, S. Induction of defence-related ultrastructural modifications in pea root tissues inoculated with endophytic bacteria. Plant Physiol 1996; 112: 919-929.

Bensalim, S., Nowak, J., Asiedu, S. A plant growth promoting rhizobacterium and temperature effects on performance of 18 clones of potato. Am Pot J 1998; 75:145-152.

Bockhus, W.W., Shroyer, J.P. The impact of reduced tillage on soilborne plant pathogens. Annu Rev Phytopathol 1998; 36:485-500.

Boller, T. Chemoperception of Microbial Signals in Plant Cells. Annu. Rev. Plant Physiol. Plant Mol. Biol 1995; 46:189-214.

Bolton, H.J., Frederickson, J.K. and Elliott, L.F. Microbial ecology of the rhizosphere. In *Soil Microbial Ecology,* Metting F.B.J. ed. Marcel Dekker: New York, 1993; pp. 27-63.

Bowen, G.D., Rovira, A.D. Microbial colonization of plant roots. Annu Rev Phytopathol 1976; 14:121-44.

Boyer, J.S. Biochemical and Biophysical Aspects of Water Deficits and the Predisposition to Disease. Annu Rev Phytopathol 1995; 33:251-74.

Breines, D., Burnham, J. Modulation of Escherichia coli type 1 fimbrial expression and adherence to uroepithelial cells following exposure of logarithmic phase cells to quinolones at subinhibitory concentrations. J Antimicrob Chemother 1994; 34:205-21.

Brooks, D.S., Gonzalez, C.F., Appel, D.N., Filer, T.H. Evaluation of endophytic bacteria as potential biological control agents for oak wilt. Biol Cont 1994; 4:373-381.

Buysens, S., Poppe, J., Hofte, M. Role of siderophores in plant growth stimulation and antagonism by Pseudomonas aeruginosa 7NSK2. In, *Improving plant productivity with rhizosphere bacteria*. Ryder, M.H., Stephens P.M., Bowens G.D., eds. CSIRO: Adelaide, 1994; pp. 139-141.

Chanway, C.P. Endophytes: they're not just fungi! Can J Microbiol 1996; 74:321-322.

Chao, W.L., Alexander, M., Mineral soils as carriers for *Rhizobium* inoculants. Appl Environ Microbiol 1984; 47:94-97.

Chen, W., Hoitink, H.A J., Schmittenner, A.F. Factors affecting suppression of *Pythium* damping-off in container media amended with composts. Phytopathology 1987a; 77:755-60.

Chen, W., Hoitink, H.A.J., Tuovinen, O.H. The role of microbial activity in suppression of damping-off caused by *Pythium ultimum*. Phytopathology 1987; 78:314-22.

Chernin, L., Chet I. Microbial enzymes in biocontrol of plant pathogens and pests. In, *Enzymes in the Environment: Activity, Ecology, and Applications*. Burns, R.G., Dick, R.P., eds. Marcel Dekker: New York, 2002; pp. 171-225.

Clay, K. Fungal endophytes of grasses: a defensive mutualism between plants and fungi. Ecology 1988; 69:10-16.

Claydon, N., Allan, M., Hanson, J. R., Avent, A.G., Antifungal alkyl pyrones of *Trichoderma harzianum*. Trans Br Mycol Soc 1987; 88:503-13.

Cohen, R., Chefetz, B., Hadar, Y., Suppression of soil-borne pathogens by composted municipal solid waste. In, *Beneficial co-utilization of agricultural, municipal and industrial bi-products*. Brown, S., Angle, J.S., Jacobs, L., eds. Kluwer Academic Publishers: Dordrecht, Netherlands, 1998, pp. 113-130.

Conn, K.L., Nowak, J., Lazarovits, G. A gnotobiotic bioassay for studying interactions between potatoes and plant growth-promoting rhizobacteria. Can J Microbiol 1997; 43:801-808.

Cooke, R.C., Rayner, A.D.M. Ecology of Saprotrophic Fungi. Longman Group Limited: London, 1984.

Davies, D.G., Parsek, M.R., Pearson, J.P., Iglewski, B.H., Costerton, J.W., Greenberg, E.P. The involvement of cell-to cell signals in the development of a bacterial biofilm. Science 1998; 280:295-298.

Davis, J.G., Callihan, R.H., Effects of gypsum, sulfur, terractor, and terractor super-x for potato scab control. Am Pot J 1974; 51:35-43.

De Brito Alvarez, M.A., Gagne S., Antoun., H. Effect of compost on rhizosphere microflora of the tomato and on the incidence of plant growth-promoting rhizobacteria. App Envir Microbiol 1995; 61(1):194-199.

Delany, I.R., Walsh, U.F., Ross, I., Fenton, A.M,, Corkery, D.M., O'Gara, F. Enhancing the biocontrol efficacy of *Pseudomonas fluorescens* F113 by altering the regulation and production of 2,4-diacetylphloroglucinol. Plant Soil 2001; 232:195-205.

Dong, Y.H, Xu, J.L., Li, X.Z., Zhang, L.H. AiiA, an enzyme that inactivates the acyl-homoserine lactone quorum sensing signal and attenuates the virulence of *Erwinia carotovora*. Proc Natl Sci 2000; 97: 3526-3531.

Dong, Y.H., Wang, L.H., Xu, J.L., Zhang, H.B., Zhang, X.F., Zhang, L.H. Quenching quorum sensing-dependent bacterial infection by an N-acyl homoserine lactonase. Nature 2001; 411:813-817.

Dong, Y.H., Gusti, AR, Zhang, Q., Xu, J.L., Zhang, L.H. Identification of quorum-quenching N-acyl homoserine lactonases from *Bacillus* species. Appl Environ Microbiol 2002; 68:1754-1759.

Dowling, D.N., Sexton, R., Fenton, A., Delany, I., Fedi, S., McHugh, B., Callanan, M., Moënne-Loccoz, Y., O'Gara, F. Iron regulation in plant-associated Pseudomonas fluorescens M114: implications for biological control. In *Molecular Biology of Pseudomonads*. Nakazawa, K., Furukawa, K., Haas D. and Silver S. eds. ASM Press: Washington, 1996; pp. 502-511.

Duffy, BK., Défago, G. Environmental factors modulating antibiotic and siderophore biosynthesis by *Pseudomonas fluorescens* biocontrol strains. Appl Environ Microbiol 1999; 65:2429-2438.

Duniway, J.M. Predisposing effect of water stress on the severity of *Phytophthora* root rot of safflower. Phytopathology 1977; 67:884-89.

Dunny, G.M., Winans, S.C. Bacterial life: neither dull nor boring. In, *Cell-Cell Signaling in Bacteria*. Dunny G.M., Winans S.C., eds., ASM Press: Washington, 1999; pp. 1-5.

Dusenbery, D.B. Sensory Ecology: How Organisms Acquire and Respond to Information. Freeman:New York, 1992.

Dworkin, M., Kaiser, D. Cell interactions in myxobacterial growth and development. Science 1985; 230: 18-24.

Eberl, L. N-acyl homoserine lactone mediated gene regulation in gram-negative bacteria. Syst Appl Microbiol 1999; 22:493-506.

Edmunds, L.K. Combined relation of plant maturity, temperature and soil moisture to charcoal rot development in grain sorghum. Phytopathology 1964; 54:514-17.

Foster, R.E., Walker, J.C. Predisposition of tomato to Fusarium wilt. J Agr Res 1947; 74:165-85.

Fox, R.T., Manners, J.G, Myers, A. Ultrastructure of entry and spread of *Erwinia carotovora* var *atroseptica* into potato tubers. Potato Res 1971; 14:61-73.

Frankowski, J., Lorito, M., Scala, F., Schmidt, R., Berg, G, Bahl, H. Purification and properties of two chitinolytic enzymes of *Serratia plymuthica* HRO-C48. Arch Microbiol 2001; 176:421-426.

Fridlender, M., Inbar, J., Chet, I. Biological control of soilborne plant pathogens by a â-1,3 glucanase-producing *Pseudomonas cepacia*. Soil Biol Biochem 1993; 25:1211-1221.

Fujimoto, D.K., Weller, D.M., Thomashow, L.S. Role of secondary metabolites in root disease suppression. In, *Allelopathy: organisms, processes and applications*. Inderjit, K.M.M. Dakshini, Einhellig F.A., eds., ACS Symposium Series: Washington, D.C. 1995; pp. 330 - 347.

Fuqua, C., Parsek, M.R., Greenberg, E.P.Regulation of gene expression by cell-to-cell communications: acyl-homoserine lactone quorum sensing. Annu Rev Genet 2001; 35:439-468.

Gaballa, A., Abeysinghe, P.D., Urich, G., Matthijs, S., Degreve, H. Trehalose induces antagonism towards *Pythium debaryanum* in *Pseudomonas fluorescens* ATCC 17400. Appl Environ Microbiol 1997; 63:4340-4345

Gilbert, G.S., Handelsman, J., Parke, J.L. Root camouflage and disease control. Phytopathology 1994; 84:222-225.

Givskov, M., Eberl, L., Molin, S. Control of exoenzyme production, motility and cell differentiation in *Serratia liquefaciens*. FEMS Microbiol Lett 1997; 148:115-122.

Glick, B.R., Patten, C.L., Holguin, G., Penrose, D.M. Auxin Production. In, *Biochemical and genetic mechanisms used by plant growth promoting bacteria*. Glick, B.R., Patten, C.L., Holguin, G, Penrose, D.M. eds. Imperial College Press: London, 1999; pp. 86 -133.

Glickman, E., Gardan, L., Jaquet, S., Hussain, S., Elasri, M., Petit, A., Dessaux, Y. Auxin production is a common feature of most pathovars of *Pseudomonas syringae*. Mol Plant-Microbe Interact 1998; 11:156-162.

Grayston, S.J., Wang, S., Campbell, C.D., Edwards, A.C. Selective influence of plant species on microbial diversity in the rhizosphere. Soil Biol Biochem 1998; 30:369-378.

Grimwood, K., To, M., Rabin, H.R., Woods, D.E. Inhibition of *Pseudomonas aeruginosa* exoenzyme expression by subinhibitory antibiotic concentrations. Antimicrob Agents Chemother 1989; 33:41-47.

Gorodecki, B., Hadar, Y. Suppression of *Rhizoctonia solani* and *Sclerotium rolfsii* in container media containing composted separated cattle manure and composted grape marc. Crop Prot 1990; 9:271-274.

Gutterson, N., Ziegle, J.S., Warren, G.J., Layton, T.J. Genetic determinants for catabolite induction of antibiotic biosynthesis in *Pseudomonas fluorescens* HV37a. J Bacteriol 1988; 170:380-385.

Haas, D., Keel, C. Regulation of antibiotic production in root-colonization *Pseudomonas* spp. and relevance for biocontrol of plant disease. Annu Rev Phytopathol 2003; 41:117-153.

Hashidoko, Y., Nakayama, T., Homma, Y., Tahara, S. Structure elucidation of xanthobaccin A, a new antibiotic produced from *Stenotrophomonas* sp. strain SB-K88. Tetrahedron Lett 1999; 40:2957-2960.

Hawes, M.C., Brigham, L.A., Wen, F., Woo, H.H., Zhu, Y. Function of root border cells in plant health: pioneers in the rhizosphere. Annu Rev Phytopathol 1998; 36:311-327.

Hawes, M.C., Brigham, L.A. Impacts of root border cells on microbial populations in the rhizosphere. Adv Plant Pathol 1992; 8:119-148.

He, H., Silo-Suh, L.A., Clardy, J., Handelsman, J. Zwittermycin A, an antifungal and plant protection agent from *Bacillus cereus*. Tetrahedron Lett 1994; 35:2499-502.

Herbert, S., Barry, P., Novick, R.P. Subinhibitory clindamycin differentially inhibits transcription of exoprotein genes in *Staphylococcus aureus*. Infect Immun 2001; 69:2996-3003.

Hinton, D.M., Bacon, C.W. *Enterobacter cloacae* is an endophytic symbiont of corn. Mycopathologia 1995; 129:117-125.

Hoagland, R.E., Cutler, S.J. Plant microbial compounds as herbicides. In, *Allelopathy in Ecological Agriculture and Forestry,* Narwal, S.S., Hoagland, R.E., Dilday, R.H., Reigosa, M.J. eds., (Proceedings of the III

International Congress on Allelopathy in Ecological Agriculture and Forestry, Dharwad, India. 18-21 August 1998). Kluwer Academic Publications: London, 2000; pp.73-99.

Hoch, J.A. Control of cellular development in sporulating bacteria by the phosphorelay two-component signal transduction system. In, *Two component signal transduction.* Hoch, J.A., Silvay, T.J. eds., ASM Press: Washington, D.C. 1995; pp. 129-144.

Hoffland, E., Pieterse, C., Bik, L., van Pelt, J.A. Induced systemic resistance in radish is not associated with accumulation of pathogenesis-related proteins. Physiol Mol Plant Pathol 1995; 46:309-320.

Hoitink, H.A.J, Boehm, M.J. Biocontrol within the context of soil microbial communities: a substrate-dependent phenomenon. Annu Rev Phytopathol 1999; 37:427-446.

Hoitink, H.A.J., Fahy, P.C. Basis for the control of soilborne plant pathogens with composts. Ann Rev Phytopathol 1986; 24:93-114.

Hoitink, H.A.J., Stone, A.G., Han, D.Y. Suppression of plant disease by composts. HortScience 1997; 32:184-197.

Holden, M.T.G., Chhabra, S.R. de Nys, R., Stead, P., Bainton, N.J. , Hill, P.J., Manefield, M., Kumar, N., Labatte, M., England, D., Rice, S., Givskov, M., Salmond, G.P.C., Stewart, G.S.A.B., Bycroft, B.W., Williams, P. Quorum sensing cross-talk: isolation and chemical characterization of cyclic dipeptides from *Pseudomonas aeruginosa* and other gram-negative bacteria. Mol Microbiol 1999; 33:1254-1266.

Hornby, D. Suppressive soils. Ann Rev Phytopathol 1983; 21:65-85.

Hornby, D. Take-all decline: a theorist's paradise. In, Soilborne plant pathogens. Schippers B., Gams W., eds., Academic Press: New York, 1979; pp. 133-156.

Huber, D.M., Schneider, R.W. The description and occurrence of suppressive soils. In, *Suppressive Soils and Plant Disease*. Schneider, R.W. ed. APS: St. Paul, Minnesota, 1982; pp. 1-7.

International Allelopathy Society (IAS). Article 2, IAS Constitution (draft), Second Newsletter 1998, http://www2.uca.es/dept/quimica_organica/ias/newslett2.htm#draft International Allelopathy Society. Department of Biochemistry, Oklahoma State University, Stillwater, Oklahoma, USA, 1998.

Jacobson, C.B., Pasternak, J.J., Glick, B.R. Partial purification and characterization of ACC deaminase from plant-growth promoting rhizobacterium *Pseudomonas putida* GR12-2. Can J Microbiol 1994; 40:1019-1025.

Jetiyanon, K., Tuzun, S., Kloepper J. Lignification, peroxidase and superoxide dismutases as early plant defence reactions associated with PGPR-mediated induced systemic resistance. In, *Plant growth-promoting rhizobacteria: present status and future prospects.* Kobayashi, K., Homma, Y., Kodama, F., Kondo N., Akino, S. eds. OECD: Paris, France, 1997; pp. 265-268.

Kamensky , M., Ovadis, M., Chet, I., and Chernin, L. Soil-borne strain IC14 of Serratia plymuthica with multiple mechanisms of antifungal activity provides biocontrol of *Botrytis cinerea* and *Sclerotinia sclerotiorum* diseases. Soil Biol Biochem 2003; 35:323-331.

Keel, C., Défago, G. Interactions between beneficial soil bacteria and root pathogens: mechanisms and ecological impact. In, *Multitrophic Interactions in Terrestrial Systems*. Gange A.C., Brown, V.K. eds. Blackwell Science Ltd.: London, UK, 1997; pp. 27-46.

Kleerebezem, M., Quadri, L.E. Peptide pheromone-dependent regulation of antimicrobial peptide production in Gram-positive bacteria: a case of multicellular behaviour. Peptides 2001; 22:1579-1596.

Kloepper, J.W., Schroth, M.N. Development of a powder formulation of rhizobacteria for inoculation of potato seed pieces. Phytopathology 1981; 71:590-92.

Lapwood, D.H., Wellings, L.W., Hawkins, J.H. Irrigation as a practical means to control potato common scab (Streptomyces scabies): final experiment and conclusions. Plant Pathol 1973; 22:35-41.

Leadbetter, J.R. Plant microbiology. Quieting the raucous crowd. Nature 2001; 411:748-749.

Leadbetter, J.R., Greenberg, E.P. Metabolism of acyl-homoserine lactone quorum-sensing signals by *Variovorax paradoxus*. J Bacteriol 2000; 182:6921-6926.

Leeman, M., Denouden, E.M., Vanpelt, J.A., Dirkx, F., Steijl, H., Bakker, P.A.H.M., Schippers, B. Iron availability affects induction of systemic resistance to fusarium wilt of radish by *Pseudomonas fluorescens*. Phytopathology 1996; 86:49-55.

Leeman, M., Vanpelt, J.A., Denouden, F.M., Heinsbroek, M., Bakker, P.A.H.M, Schippers, B. Induction of systemic resistance against fusarium wilt of radish by lipopolysaccharides of *Pseudomonas fluorescens*. Phytopathology 1995; 85:1021-27.

Levy, E., Carmelli, S. Biological control of plant pathogens by antibiotic-producing bacteria. In, *Allelopathy: organisms, processes and applications*. Inderjit, K.M.M. Dakshini, Einhellig, F.A. eds., ACS Symposium Series: Washington, D.C., 1995, pp. 300-309.

Ligon, J.M., Hill, D.S., Hammer, P.E., Torkewitz, N.R., Hofmann, D., Kempf, H.J., van Pée, K.H. Natural products with antifungal activity from *Pseudomonas* biocontrol bacteria. Pest Manage Sci 2000; 56: 688-695.

Lim, H.S., Kim, Y.S., Kim, S.D. *Pseudomonas stutzeri* YPL-1 genetic transformation and antifungal mechanism against *Fusarium solani*, an agent of plant root rot. Appl Environ Microbiol 1991; 57:510-516.

Lin, Y.H., Xu, J.L., Hu, J., Wang, L.H., Ong, S.L., Leadbetter, J.R., Zhang, L.H. Acyl- homoserine lactone acylase from Ralstonia strain XJ12B represents a novel and potent class of quorum-quenching enzymes. Mol Microbiol 2003; 47:849-860.

Lynch, J.M. Products of soil microorganisms in relation to plant growth. CRC Crit Rev Microbiol 1976; 5: 67-107.

Loper, J.E., Henkels, M.D. Availability of iron to *Pseudomonas fluorescens* in rhizosphere and bulk soil evaluated with an ice nucleation reporter gene. Appl Environ Microbiol 1997; 63:99-105.

Loper, J.E., Henkels, M.D. Utilisation of heterologous siderophores enhances levels of iron available to *Pseudomonas putida* in the rhizosphere. Appl Environ Microbiol 1999; 65:5357-5363.

MacDonald, J.D. Effect of salinity stress on the development of *Phytophthora* root rot of chrysanthemum. Phytopathology 1982; 72:214-19.

Mallik, A.U. Challenges and opportunities in allelopathy research: a brief overview. J Chem Ecol 2000; 26:2007-2009.

Marschener, P., Crowley, D.E. Iron stress and pyoverdin production by a fluorescent pseudomonad in the rhizosphere of white lupin (*Lupinus albus* L.) and barley (*Hordeum vulgare* L.) Appl Environ Microbiol 1997; 63:277-281.

Maurhofer, M., Hase, C., Meuwley, P., Métraux, J.P., Défago, G. Induction of systemic resistance of tobacco to tobacco necrosis virus by the root-colonizing *Pseudomonas fluorescens* strain CHA0: influence of the gacA gene and of pyoverdine production. Phytopathology 1994; 84:139-46.

Merharg, A.A., Killham, K. Loss of exudates from roots of perennial rye-grass inoculated with a range of microorganisms. Plant Soil 1995; 170:345-349.

Miller, M.B., Bassler, B.L. Quorum sensing in bacteria. Annu Rev Microbiol 2001; 55:165-199.

Misaghi, I.J., Donndelinger, C.R. Endophytic bacteria in symptom-free cotton plants. Phytopathology 1990; 80:808-811

Mukerji, K.G., Chamola, B.P., Upadhyay, R.K. eds. *Biotechnolocal approaches in Biocontrol of Plant Pathogens*. Kluwer Academic/Plenum Publishers, New York, Dordrecht, London, 1999; 255 p.

Nakata, K., Yoshimoto, A., Yamada, Y. Promotion of antibiotic production by high ethanol, high NaCl concentration, or heat shock in *Pseudomonas fluorescens* S272. Biosci Biotechnol Biochem 1999; 63: 293-297.

Nelson, E.B., Hoitink, H.A.J. The role of microorganisms in the suppression of *Rhizoctonia solani* in container media amended with composted hardwood bark. Phytopathology 1983, 73.274-78.

Normander, B., Hendriksen, N.B., Nybroe, O. Green fluorescent protein-marked *Pseudomonas fluorescens*: localization, viability, and activity in the natural barley rhizosphere. Appl Environ Microbiol 1999; 65:4646-51.

O'Sullivan, D.J., O'Gara, F. Traits of fluorescent *Pseudomonas* spp. involved in suppression of plant root pathogens. Microbiol Rev 1992; 56:662-676.

Odum, E.P., Finn, J.T., Franz, E.H. Perturbation theory and subsidy-stress gradient. BioScience 1979; 29:349-352.

Oestergaard, S.P., Nielsen, S. Control of potato scab (*Streptomyces scabies*) by irrigation. Tidsskrift for Planteavl 1979; 83:201-204.

Parmar, N. Dardarwal, K.R. Stimulation of nitrogen fixation and induction of flavonoid like compounds by rhizobacteria. J Appl Microbiol 1999; 86:36-44.

Parsek, M.R., Val, D.L., Hanzelka, B.L., Cronan, J.E. Jr., Greenberg E.P. Acyl homoserine-lactone quorum-sensing signal generation. Proc Natl Acad Sci USA. 1999; 13:4360-4365.

Patten, C.L., Glick, B.R. Bacterial biosynthesis of indole-3-acetic acid. Can J Microbiol 1996; 42:207-220.

Peters, R.D., Sturz, A.V., Carter, M.R., Sanderson, J.B. Developing disease-suppressive soils through crop rotation and tillage management practices. Soil Till Res 2003; 72:139-152.

Petersen, C.A., Emanuel, M.E., Humphreys, G.B. Pathway of movement of apoplastic fluorescent dye tracers through the endodermis at the site of secondary root formation in corn (*Zea mays*) and broad bean (*Vicia faba*). Can J Bot 1981; 59:618-625.

Pierson, E.A., Wood, D.W., Cannon, J.A., Blachere, F.M. III. LSP: interpopulation signaling via homoserine lactones among bacteria in the wheat rhizosphere. Mol Plant Interact 1998; 11:1078-1084.

Piper, K.R., Beck von Bodman, S., Farrand S.K. Conjugation factor of *Agrobacterium tumefaciens* regulates Ti plasmid transfer by autoinduction. Nature 1993; 362:448-450.

Pirhonen, M., Flego, D., Heikinheimo, R., and Palva, E.T. A small diffusible signal molecule is responsible for the global control of virulence and exoenzyme production in the plant pathogen *Erwinia carotovora.* EMBO J 1993; 12:2467-2476.

Raaijmakers, J.M., Vlami, M., de Souza, J.T. Antibiotic production by bacterial biocontrol agents. Antonie van Leeuwenhoek Int Genet Mol Microbiol 2002; 81:537-547.

Romeo, J.T. Raising the beam: moving beyond phytotoxicity. J Chem Ecol 2000; 26:2011-2014.

Roos, I.M.M., Hattingh, M.J. Scanning electron microscopy *Pseudomonas syringae* pv. *morsprunorum* on sweet cherry leaves. Phytopathol Z 1983; 108:18-25.

Salmond, G.P.C., Bycroft, B.W., Stewart, G.S.A.B., Williams, P., The bacterial 'enigma': cracking the code of cell-cell communication. Mol Microbiol 1995; 16:615-624.

Sayre, R.M., Patrick, Z.A., Thorpe. H.J. Identification of a selective nematicidal component in extracts of plant residues decomposing in soil. Nematologica 1965; 11:263-68.

Sayre, R.M., and Starr, M.P. Bacterial diseases and antagonisms of nematodes. In *Diseases of nematodes.* Poinar, G.O., Jansson H.B. eds. CRC Press: Florida, U.S.A, 1988; pp. 69-101.

Schroth, M.N., Loper, J.E., Hildebrand, D.C. Bacteria as agents of plant edisease. In *Current Perspectives in Microbial Ecology*. Klug, M.J., Reddy, C.C. eds. Am Soc Microbiol: Washington, D.C., U.S.A. 1984; pp. 362-369.

Schroth, M.N., Hancock, J.G. Disease-suppressive soil and root-colonizing bacteria. Science 1982; 216:1376-81.

Schulz, F.A., Bateman, D.F. Temperature response of seeds during the early stages of germination and its relation to injury by *Rhizoctonia solani*. Phytopathology 1969; 59:352-55.

Scott, R.I., Chard, J.M., Hocart, M.J., Lennard, J.H., Graham, D.C. Penetration of potato tuber lenticels by bacteria in relation to biological control of blackleg disease. Potato Res 1996; 39:333-344.

Shanahan, P., O'Sullivan, D.J., Simpson, P., Glennon, J.D., O'Gara, F. Isolation of 2,4-diacetylphloroglucinol from a fluorescent pseudomonad and investigation of physiological parameters influencing its production. Appl Environ Microbiol 1992; 58:353-358.

Shipton, P.J. Take-all decline during cereal monoculture. In *Biology and Control of Plant Pathogens,* Bruehl G.W. ed., St. Paul, MN: Am. Phytopathol. Soc. 1975; pp. 137-144.

Siciliano, S.D., Theoret, C.M., de Freitas, J.R., Hucl, P.J., Germida, J.J. Differences in the microbial communities associated with the roots of different cultivars of canola and wheat. Can J Microbiol 1998; 44:844-851.

Silo-Suh, L.A., Lethbridge, B.J., Raffel, S.J., He, H., Clardy, J., Handelsman, J. Biological activities of two fungistatic antibiotics produced by *Bacillus cereus* UW85. Appl Environ Microbiol 1994; 60:2023-30.

Singh, P.P., Shin, Y.C., Park, C.S., Chung, Y.R. Biological control of Fusarium wilt of cucumber by chitinolytic bacteria. Phytopathology 1999; 89:92-99.

Slininger, P.J., Jackson, M.A. Nutritional factors regulating growth and accumulation of phenazine 1-carboxylic acid by *Pseudomonas fluorescens* 2-79. Appl Microbiol Biotechnol 1992; 37:388-392.

Slininger, P.J., Sheawilbur, M.A. Liquid culture pH, temperature, and carbon (not nitrogen) source regulate phenazine productivity of the take-all biocontrol agent *Pseudomonas fluorescens* 2-79. Appl Microbiol Biotechnol 1995; 43:794-800.

Smith, K.P., Goodman, R.M. Host-variation for interaction with beneficial plant associated microbes. Annu Rev Phytopathol 1999; 37:473-491.

Smucker, A.J.M. Soil Environmental Modifications of Root Dynamics and Measurement. Annu Rev Phytopathol 1993; 31:191-216.

Stephens, P.M., Crowley, J.J., O'Connel, C. Selection of pseudomonad strains inhibiting *Pythium ultimum* on sugarbeet seed in soil. Soil Biol Biochem 1993; 25:1283-1288.

Sturz, A.V., Carter, M.R., Johnston, H.R. A review of plant disease, pathogen interactions and microbial antagonism under conservation tillage in temperate humid agriculture. Soil Till Res 1997; 41:169-189.

Sturz A.V., Christie, B.R. Beneficial microbial allelopathies in the root zone: the management of soil quality and plant disease with rhizobacteria. Soil Agro-ecosystems: Impacts of Management on Soil Health and Crop Diseases - Special Issue. Soil Till Res 2003; 72:107-123.

Sturz, A.V., Christie, B.R., Arsenault, W.J. Red clover-potato cultivar combinations for improved potato yield. Agron J 2003; 95:1089-1092.

Sturz, A.V., Christie, B.R., Nowak, J. Bacterial Endophytes: A Critical Component of Sustainable Crop Production, CRC Press, Crit Rev Plant Sci 2000; 19:1-30.

Sturz, A.V., Christie, B.R., Matheson, B.G., Arsenault, W.J., Buchanan, N.A. Endophytic bacteria, antibiosis and community ecology in potato tubers, and implications for induced resistance against plant pathogens. Plant Pathol 1999; 48:360-370.

Sturz, A.V., Matheson, B.G., Arsenault, W.J., Kimpinski, J., Christie, B.R. Weeds as a source of beneficial rhizobacteria in agricultural soils. Can J Microbiol 2001; 47:1013-1024.

Sturz, A.V., Nowak, J. Endophytic communities of rhizobacteria and the strategies required to create yield enhancing associations with crops. Appl Soil Ecol 2000; 15:183-190.

Sturz, A.V., Ryan, D.A.J., Matheson, B.G., Arsenault, W.J., Kimpinski, J., Christie, B.R. Stimulating disease suppression in soils: sulphate fertilizers can increase biodiversity and antibiosis ability of root zone bacteria against *Streptomyces scabies*. Soil Biol Biochem 2004; 36:343-352.

Surette, M.G., Bassler, B.L. Quorum sensing in *Escherichia coli* and *Salmonella typhimurium*. Proc Natl Acad Sci USA. 1998; 95:7046-7050.

Szabolcs, I. The concept of soil resilience. In, *Soil resilience and sustainable land use*. Greenland, D.J. and Szabolcs, I. eds. CAB International: Wallingford, U.K. 1994; pp. 33-39.

Tateda, K., Comte, R., Pechère, J.C., Köhler, T., Yamaguchi, K., Van Delden, C. Azithromycin inhibits quorum sensing in *Pseudomonas aeruginosa*. Antimicrob. Agents Chemother 2001; 45:1930-33.

Thomashow, L.S., Bonsall, R.F., Weller, D.M. Antibiotic production by soil rhizosphere microbes in situ. In *Manual of environmental microbiology*. Hurst, C.J., Knudsen, G.R., McInerney, M.J., Stetzenbach L.D., Walter, M.V. eds. ASM Press: Washington, 1997; pp. 493-499.

Timms Wilson, T.M., Ellis, R.J., Renwick ,A., Rhodes, D.J., Mavrodi, D.V., Weller, D.M., Thomashow, L.S., Bailey, M.J. Chromosomal insertion of phenazine-1-carboxylic acid biosynthetic pathway enhances efficacy of damping-off disease control by *Pseudomonas fluorescens*. Mol Plant-Microbe Interact 2000; 13:1293-300.

Tremblay, J., Beauchamp, C.J. Split applications of supplementary N fertilizer following incorporation of chipped ramical wood: changes in selected biological and chemical properties of a soil cropped to potatoes. Can J Soil Sci 1998; 78:275-282.

Tu, J.C., Findlay, W.I. The effects of different green manure crops and tillage practices on pea root rots. British Crop Protection Conference - Pest and Diseases. British Crop Protection Council Publ.: Surrey, U.K. 1986; pp. 229-236

Tuzun S., Kloepper J.W. Induced systemic resistance by plant growth promoting rhizobacteria. In *Improving Plant Productivity with Rhizosphere Bacteria*. Ryder, M.H., Stephens, P.M., Bowen, G.D. eds. Proceedings of 3rd International Workshop on PGPR. CSIRO, Division of Soils: Adelaide. 1994; pp. 104-109.

Van Buren, A.M., Andre, C., Ishimaru, C.A. Biological control of the bacterial ring rot pathogen by endophytic bacteria isolated from potato. Phytopathology 1993; 83:1406.

van Peer, R., Punte, H.L.M., De Weger, L.A., Schippers, B. Characterization of root surface and endorhizosphere pseudomonads in relation to their colonization of roots. Appl Environ Microbiol 1990; 56:2462-2470.

van Peer R., Niemann, G.J., Schippers B. Induced resistance and phytoalexin accumulation in biological control of Fusarium wilt of carnation by *Pseudomonas* sp. strain WCS417r. Phytopathology 1991; 81:728-34.

van Wees, S.C.M., Luijendijk, M., Smoorenburg, I., van Loon, L.C., Pieterse, C.M.J. Rhizobacteria-mediated induced systemic resistance (ISR) in *Arabidopsis* is not associated with a direct effect on expression of known defense-related genes but stimulates the expression of the jasmonate-inducible gene Atvsp upon challenge. Plant Mol Biol 1999; 41:537-549.

van Wees, S., Pieterse, C., Trijssenaar, A., Van't Westende, Y., Hartog, F., Van Loon, L.C. Differential induction of systemic resistance in *Arabidopsis* by biocontrol bacteria. Mol Plant-Microbe Interact 1997; 10:716-24.

Walker, J.C., Snyder, W.C. Pea wilt and root rots. Wisc Agr Exp Sta Bull 1933; 424:1-16.
Wardle, D.A., Bonner, K.I., Nicholson, K.S. Biodiversity and plant litter: experimental evidence which does not support the view that enhanced species richness improves ecosystem function. Oikos 1997; 79:247-258.
Wei, G., Kloepper, J.W., Tuzun, S. Induction of systemic resistance of cucumber to *Colletotrichum orbiculare* by select strains of plant-growth promoting rhizobacteria. Phytopathology 1991; 81:1508-1512.
Weller, D.M., Raaijmakers, J.M., McSpadden Gardener, B.B., Thomashow, L. Microbial populations responsible for specific soil suppressiveness to plant pathogens Ann. Rev. Phytopathol 2002; 40:309-348.
Weller, D.M. Biological control of soilborne plant pathogens in the rhizosphere with bacteria. Ann Rev Phytopathol 1988; 26:379-407.
Whipps, J.M. Microbial interactions and biocontrol in the rhizosphere. J Exp Bot 2001; 52:487-511.
Whitehead, N.A., Welch, M., Salmond, G.P.C. Silencing the majority. Nat. Biotechnology. 2001; 19:735-736.
Withers, H., Swift, S., Williams, P. Quorum sensing as an integral component of gene regulatory networks in Gram-negative bacteria. Curr Opin Microbiol 2001; 4:186-193.
Yamada, Y., Nihira, T. Microbial hormones and microbial chemical ecology. In *Comprehensive Natural Products Chemistry* Barton, D.H.R., Nakanishi, K. eds.. Elsevier: Oxford, 1998; pp. 377-413.

ROBERT J. KREMER

THE ROLE OF ALLELOPATHIC BACTERIA IN WEED MANAGEMENT

U.S.D.A., Agricultural Research Service, Cropping Systems & Water Quality Unit, University of Missouri, Columbia, Missouri, U.S.A.
Email: KremerR@missouri.edu

Abstract. Allelopathic bacteria encompass those rhizobacteria that colonize the surfaces of plant roots, produce, and release phytotoxic metabolites, similar to allelochemicals, that detrimentally affect growth of plants. Practical application of this group of bacteria to agriculture could contribute to biological weed management systems that have less impact on the environment than conventional systems by reducing inputs of herbicides. Allelopathic bacteria have been investigated for potential as inundative-type biological control agents on several weeds. Because allelopathic bacteria generally do not attack specific biochemical sites within the plant, unlike conventional herbicides, they offer a means to control weeds without causing direct selective pressure on the weed population, therefore, development of resistance is not a major consideration. Additionally, the use of allelopathic bacteria appears to be environmentally benign relative to herbicides. These characteristics make allelopathic bacteria an attractive approach for managing crop weeds in a sustainable manner, even within the boundaries of conventional agriculture systems. However, recent evidence suggests that indigenous allelopathic bacteria might be exploited under certain crop and soil management practices that are inherently part of sustainable agricultural systems. The development of "weed-suppressive" soils in diverse sustainable systems is encouraging because indigenous populations of allelopathic bacteria may develop in several soils and environments using similar practices. The recent demonstrations of apparent weed-suppressive soils may lead to development of specific management strategies for the establishment and persistence of native allelopathic bacteria directly in soils conducive to annual weed infestations.

1. INTRODUCTION

Biological weed management is a system that incorporates the use of diverse biological organisms and biologically-based approaches including allelopathy, crop competition, and other cultural practices to significantly reduce weed densities in a manner that is similar to use of chemical herbicides alone (Cardina, 1995). Interest in developing effective biological weed management systems continues to increase because of a growing awareness of problems associated with the constant and intensive use of chemical herbicides, which include surface- and groundwater contamination, detrimental impacts on nontarget organisms, development of weeds resistant to herbicides (including those that are used in transgenic herbicide-resistant crops), and consumer concerns for residues on food (Gliessman, 2002). A component of biological weed management involves biological control, the intentional use of living organisms (insects, nematodes, fungi, and bacteria) to reduce the vigor, reproductive capacity, density, or impact of weeds (Quimby and Birdsall, 1995). A number of reviews are

Inderjit and K.G. Mukerji (eds.),
Allelochemicals: Biological Control of Plant Pathogens and Diseases, 143–155.

available that discuss the various strategies of biological control and the numerous organisms that are involved as biological agents within these strategies (TeBeest, 1991; Harley and Forno, 1992; Boland and Kuykendall, 1998). The focus of the present paper will be a broad group of bacteria that are associated with seeds and seedlings, which can be developed to suppress the establishment and growth of weeds in agroecosystems.

Unlike many herbicides and biological agents that have been developed to attack growing weeds established in crops, most bacterial agents target weed seeds residing in soil and the roots of developing weed seedlings. The microenvironments consisting of the zones of soil surrounding the seed (spermosphere) and root (rhizosphere) provide organic substrates that are readily available for soil microorganisms in contrast with the nutrient-limiting condition of the bulk soil (Kennedy, 2005). The spermosphere and rhizosphere, therefore, are ideal sites for establishing selected bacteria able to suppress weed seed germination and seedling growth because of the continuous supply of carbon and energy sources released from germinating seeds and developing root systems (Zahir et al., 2004). Soil borne bacteria adapted to competitive colonization of the spermosphere, rhizosphere, and the root can be grouped under the general term rhizobacteria (Schroth and Hancock, 1982). Although the spermosphere is defined with reference to a seed prior to root emergence, the zone of soil contains substances exuded from the seed that rapidly attracts and regulates a microbial community, which often, if the seed survives, establishes the dominant microbial communities of the longer-lived rhizosphere environment (Nelson, 2004). A component of the rhizobacteria group in the spermosphere and rhizosphere that inhibit plant growth without causing obvious disease symptoms are known as deleterious rhizobacteria. Deleterious rhizobacteria are predominantly saprophytic bacteria that live on or in plant seeds and roots, surviving on organic compounds released by plant seed root cells (Schippers et al., 1987). These bacteria do not parasitize the plant or penetrate the stele like major or true pathogens, but may colonize seed tissues or root hairs and the root tip as well as in the intercellular spaces of root cortical cells (Schippers et al., 1987), in which case they may be characterized as "endorhizal bacteria." Because many of these deleterious rhizobacteria release phytotoxic metabolites that are also considered allelochemicals that influence the growth of plants, it has been suggested that term "allelopathic bacteria" may more accurately describe these bacteria (Barazani and Friedman, 1999; Sturz and Christie, 2003). Therefore, rhizobacteria that detrimentally affect seed germination and seedling development of weeds through the production of allelochemical substances will be referred to as allelopathic bacteria (AB) in this paper.

The use of rhizobacteria in weed management has typically involved an inoculative approach whereby selected AB are applied at high rates to establish critical population densities in soil or on vegetative residues to achieve rapid initiation of growth-inhibitory activity (Kremer and Kennedy, 1996). The goal is not complete kill or eradication of the weed population but the reduction of the competitive ability of the weeds growing with the crop. Recent investigations of AB suggest that sustainable agricultural practices for many crops may be linked to management of the indigenous bacterial communities for biological control of weeds, thus reducing dependence on selected AB as inoculative biocontrol agents. Development of agroecosystems with the capacity to suppress weeds using naturally-occurring soil bacteria-weed interactions has received very little attention (Gallandt et al., 1999). Therefore, weed management strategies involving AB should also consider the development of beneficial, indigenous soil bacteria similar to a *conservation biological control* approach (Newman et al., 1998), which conceivably would address sustainable weed management with reduced or no herbicide inputs (Parr et al., 1992).

The objectives of this chapter are to demonstrate how both introduced and indigenous AB contribute to weed management in cropping systems and to identify crop and soil management strategies that promote the proliferation of AB.

2. ALLELOPATHIC BACTERIA AND INTERACTIONS IN AGROECOSYSTEMS

Biological weed control using introduced agents including selected AB has often been limited by inconsistency of efficacy when placed into practical field use. This suggests that standard screening bioassays conducted under laboratory conditions poorly represent actual environmental situations where biological activity must be expressed. For individual AB agents, therefore, selection, bioassays, and inoculum development should be based on ecological principles, accounting for characteristics of the environment in which the agents are to be introduced. The principles should further be applied to potential strategies for enhancing *indigenous* AB in the soil environment. The ecological interactions associated with biological control of weed

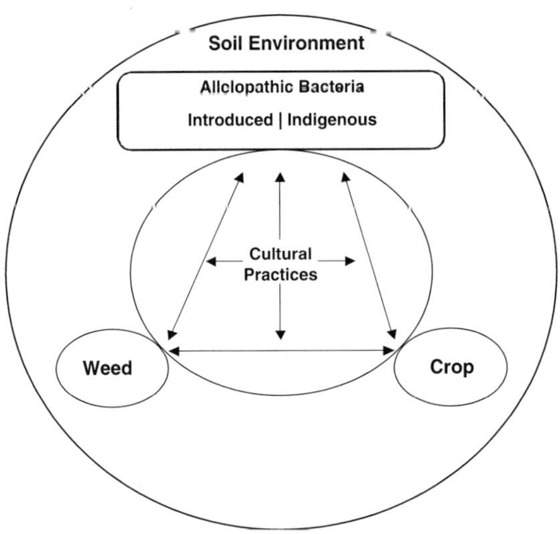

Figure 1. Environmental-interaction diagram associated with allelopathic bacteria in agroecosystems.

seeds and seedlings with AB can be expressed graphically (Figure 1) to emphasize the following key areas for consideration: a) biology and ecology of weeds; b) growth of the crop within the established cultivation system; c) characterization of AB, including those that are selected and introduced as biological control agents and those that are indigenous to the soil environment; and d) the wide range of cultural practices available for implementation of biological control within the agroecosystems.

After recognizing the need for consideration of ecological interactions, a program for development of selected AB or management of the indigenous bacterial population can be undertaken

3. CHARACTERIZATION OF ALLELOPATHIC BACTERIA

Kremer et al. (1990) identified colonizing ability, chemotactic response, and mode of action to be vital characteristics for the successful development of rhizobacteria as weed biocontrol agents. Bacteria that can rapidly colonize the root will likely be a successful biocontrol agent. Migration towards the seed or root is the first step in colonization, illustrated by movement of rhizobacterial isolates through 2-cm of soil towards velvetleaf seeds (Begonia and Kremer, 1994). As the seedling develops, movement of bacteria along roots and within the rhizosphere is influenced by root binding sites, amounts of organic material present, type of root (i.e., seminal vs. nodal roots), water movement through soil and along roots, and soil texture (Bolton et al., 1993). Compared to other bacterial groups present in non-rhizosphere soil, gram-negative bacteria readily colonize the rhizosphere, partly due to their metabolic diversity (Nehl et al., 1997). Pseudomonads are particularly adapted for rhizosphere colonization because of the ability to utilize diverse carbon sources present in root exudates. Observations with scanning electron microscopy reveal the intimate relationship of rhizoplane colonization by selected AB (Figure 2).

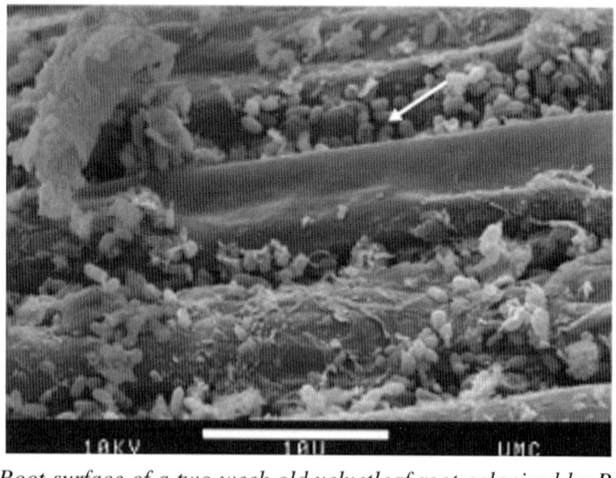

Figure 2. Root surface of a two-week old velvetleaf root colonized by Pseudomonas fluorescens strain 239 cells (arrow) aligned in intercellular spaces of root epidermal cells. Magnification is X 6,000. Scanning electron micrograph from Begonia et al. (1990).

Successful competition of bacteria living in the rhizosphere depends on several factors, including rapid growth on multiple substrates, antibiotic production, and downward growth with the root. A major factor contributing to successful competition of rhizobacteria over other microorganisms is the growth stimulation by exuded organic compounds and sloughed-off root hair and epidermal cell materials (De Weger et al., 1995). The ability to efficiently compete for these available resources and to produce siderophores for obtaining iron is important in establishment, colonization, and persistence of rhizobacteria in the rhizosphere.

A characteristic of many AB is the high specificity toward their weed host(s) with no detrimental effects on growth of nonweedy plant species (Cherrington and Elliott, 1987; Elliott and Lynch, 1985; Kennedy et al., 1991; 2001). Although effects on plants are subtle (Kremer and Kennedy, 1996), AB may be as significant as traditional bacterial pathogens in affecting plant growth (Schroth and Hancock, 1982; Suslow and Schroth, 1982). Because AB attack the seed and/or seedling rather than the growing plant, weed seed or vegetative propagule production is suppressed, a key to any weed management program, which reduces the need for repeated postemergence herbicide applications and increases the chances of success for control of a growing, competitive weed (Aldrich and Kremer, 1997).

4. MODES OF ACTION OF ALLELOPATHIC BACTERIA

Many AB strains produce secondary metabolites that are inhibitory to plants, including phytotoxins and antibiotics, which can be considered allelopathic. Phytotoxins from fluorescent *Pseudomonas* spp., a diverse group of plant pathogenic bacteria abundant in the soil and rhizosphere, have been well studied (Mitchell, 1991). There are fewer reports on phytotoxins from AB and many have not been extensively studied.

A phytotoxin from *Pseudomonas fluorescens* strain D7 was shown to be responsible for root growth inhibition of downy brome (*Bromus tectorum*) (Tranel et al., 1993). Further characterization revealed that the active fraction was a complex of chromopeptides, other peptides and fatty acid esters in a lipopolysaccharide matrix (Gurusiddaiah et al., 1994). Secondary metabolites isolated from *Pseudomonas syringae* strain 3366 inhibitory to downy brome consisted of phenazine-1-carboxylic acid, 2-aminophenoxazone and 2-aminophenol (Gealy et al., 1996). Gealy et al. (1996) showed that phenazine-type antibiotics of *Pseudomonas fluorescens* also inhibited downy brome root growth. Electron microscopy of AB colonizing the rhizoplane and endorhizal cells of leafy spurge (*Euphorbia esula*) revealed disruption of plant cell walls and membranes apparently due to production of phytotoxins and/or enzymes by the bacteria, which consequently inhibited seedling growth (Souissi et al., 1997). AB may also produce "phytotoxic antibiotics" that affect plant growth such as the broad-spectrum antibiotic, 2,4-diacetylphloroglucinol, released by *P. fluorescens* strain CHA0, which suppressed soilborne fungal plant pathogens but was also highly phytotoxic to seedlings of several plant species (Keel et al., 1992).

Plant-inhibitory effects of some AB are auxin-mediated, illustrated by direct uptake of bacterially produced indoleacetic acid (IAA). Plant response to microbially synthesized auxins is related to the concentration released into the rhizosphere. Growth inhibition of several weed and crop species was correlated with elevated IAA levels produced by AB (Sarwar and Kremer, 1995). Similar responses were observed when tryptophan, an IAA precursor, was added to soil (Sarwar and Frankenberger, 1994; Sarwar and Kremer, 1995). Presumably, the presence of extra tryptophan in the soil provided rhizobacteria with additional substrates for auxin biosynthesis.

The production of hydrogen cyanide (HCN), a volatile metabolite that negatively affects root metabolism and root growth by inhibiting cytochrome oxidase respiration, is common among rhizosphere pseudomonads (Schippers et al., 1990). The rate of HCN synthesis is affected by the availability of precursors such as glycine, methionine, proline and cyanogenic glucosides (Knowles and Bunch, 1986; Schippers et al., 1990). The amino acid composition of root exudates as well as environmental factors affecting root exudation (i.e, light intensity, soil water potential, nutrients) may be important as well (Schippers et al., 1990). Two strains of cyanogenic pseudomonads, *Pseudomonas putida* and *Acidovorax delafieldii*, significantly inhibited velvetleaf (*Abutilon theophrasti*) growth, but did not reduce corn growth in the presence of supplemental glycine (Owen and Zdor, 2001). Cyanide production by several rhizobacterial strains was a major factor in the inhibition of seedling growth of several weed species and was suggested as a trait for consideration in selecting AB as potential weed biological control agents (Kremer and Souissi, 2001).

5. ROLES OF ALLELOPATHIC BACTERIA IN WEED MANAGEMENT STRATEGIES

Strategies for the potential use of allelopathic bacteria for weed management include application of selected cultures as bioherbicides, integration of bioherbicides with other crop and soil management practices, use of AB in sustainable agricultural systems with other non-chemical means of weed control, and management of soils to enhance populations and activity of indigenous AB as a conservation biological control strategy.

5.1. Bioherbicides

The high host specificity of AB is a disadvantage for their use as a biocontrol strategy in most agroecosystems that are typically infested with multiple weed species. However, AB may have the greatest impact for management of specific weed problems in certain cropping systems. Isolates of AB specific for management of downy brome in wheat in the Pacific Northwest are under development as bioherbicides (Kennedy et al., 1991; 2001). In Kansas, selected AB suppressed early seedling growth of downy brome and Japanese brome (*Bromus japonicus*) in soils under winter wheat production (Harris and Stahlman, 1996). Several strains of rhizobacteria from over 2000 accessions isolated from prairie soils in Canada have been selected for inhibition of downy brome and green foxtail (*Setaria viridis*) seed germination and seedling growth (Boyetchko, 1999). AB that were formulated in a starch-based granular carrier and applied to soil in leafy spurge-infested sites in South Dakota suppressed growth by decreasing root weight and root carbohydrate content (Brinkman et al., 1999).

Most of the weeds targeted for biocontrol by AB infest cereal and row crops, but a few are perennial weeds of rangeland and forest ecosystems (Kremer, 2002). Selected AB are intended for soil application, however, some cultures might be effective when applied directly to growing weeds in a postemergence control strategy. Selected AB might also be applied directly to growing weeds as a postemergence control strategy. For example, cultures and cell-free supernatants of AB strains sprayed on common chickweed (*Stellaria media*), common lambsquarters (*Chenopodium album*), and field pennycress (*Thlaspi arvense*) in the greenhouse and field reduced plant biomass and survival (Weissmann and Gerhardson, 2001). Preliminary results suggest AB cultures might be used for selective weed control in growing crops through a one-time foliar application or as a follow-up to soil-incorporation of the same cultures.

5.2. Integrated Weed Management

Integrated weed management systems rely on a number of available strategies including tillage, cultural practices, herbicides, allelopathy, and biological control to reduce the weed seedbank, prevent weed emergence, and minimize competition from weeds growing with the crop (Aldrich and Kremer, 1997). These systems may be most suitable for implementing bioherbicides based on AB to counteract their limited weed host specificity. Like chemical herbicides, such bioherbicides may be most effective as a component in a multi-faceted management program rather than as a single tactic approach (Hatcher and Melander, 2003). Effective weed management offers several opportunities for integration of selected AB at the critical stages of weed development: as seeds in soil, as growing and competitive plants, and during seed production (Aldrich and Kremer, 1997). This may be the most promising situation for AB to be considered as practical management options in cropping systems.

To broaden the limited spectrum of activity of AB, several tactics have been proposed for integration of these organisms with other weed management methods. Weed growth suppression by AB combined with herbicides applied at sublethal rates has met with some success (Greaves and Sargent, 1986). AB inhibitory to downy brome and jointed goatgrass suppressed growth to a greater extent when combined with metribuzin and/or diclofop at less than label rates (Kremer and Kennedy, 1996). An understanding of the mechanisms of herbicide-AB interactions will lead to strategies where AB selected for activity toward a weed can be paired with a specific chemical that increases the susceptibility of that weed to the AB (Kremer, 1998). Use of AB in this manner may develop into a systems management approach that involves integration of bioherbicides and herbicides on a physiological basis to control economically important weeds in corn and other crops. This is currently under intensive evaluation as a potential integrated management system in Europe (Müller-Schärer et al., 2000). Successful development of these integrated strategies will increase efficacy of AB agents, reduce herbicide inputs for weed control, and decrease potential environmental contamination.

Application of selected AB with cultural practices such as tillage may be effective in integrated weed management. Greater proportions of indigenous rhizobacteria inhibitory to downy brome and jointed goatgrass were detected under either conventional or reduced tillage compared to no-till (Kremer and Kennedy, 1996). Vegetative residues at or near the soil surface may serve as substrates for production of weed-suppressive chemicals by AB applied as bioherbicides directly to the residues. Kremer (1998) suggested that application of AB to surface vegetative residues to promote phytotoxin production might suppress weed growth prior to planting the crop, similar to a preemergence herbicide tactic. Crop rotation may also be manipulated to encourage development of specific inhibitory bacteria on weed roots. A "rotation effect" in corn, due partly to certain rhizobacteria specifically associated with corn roots, illustrates the potential for using AB to achieve suppression of weeds in crop rotation systems (Turco et al., 1990). Crop rotation may be manipulated to encourage development of specific inhibitory bacteria on weed roots. Thus, weed seeds and seedlings might be attacked by selected AB directly inoculated onto crop seeds or by promoting colonization of crop roots by microbial agents applied at planting (Skipper et al., 1996). Crop roots may deliver AB to adjacent roots of weeds and also maintain or even enhance AB numbers for attack of seedlings emerging later in the season. These observations suggest that combination of biocontrol agents, including selected AB, with cultural practices enhance weed growth suppression (Kremer and Kennedy, 1996) and may be effective in controlling weeds that escape cultural control methods (Hatcher and Melander, 2003).

5.3. Sustainable Agricultural Systems

Sustainable agricultural systems involve a range of technological and management options to reduce costs, protect health and environmental quality, and enhance beneficial biological interactions and natural processes. Sustainable agricultural systems offer the greatest opportunities to study and refine nonchemical weed management (Liebman and Gallandt, 1997) and yield valuable information useful in developing improved bioherbicides, including AB, and advancing their use in broader biologically based weed management systems.

High inputs of organic amendments and green manure (cover crop residues) in sustainable agricultural systems promotes the ability of crops to compete more vigorously with weeds, which intuitively suggests that the efficacy of bioherbicides would also be enhanced when used with these amendments (Gallandt et al., 1999). Addition of organic matter to soil is one of the most effective ways to change soil environment and favor increases in populations of beneficial soil organisms. Organic amendments subjected to decomposition in soils release compounds that suppress pathogens and provide substrates for other organisms, indigenous or those added to the amendments, that may also produce compounds that inhibit pathogens and/or weed seedling growth. Inoculation of organic materials (manures, composts, mulches) has been suggested as a means for assuring establishment and efficacy of the selected biocontrol agents, however, field performance to date has been inconsistent (Sturz and Christie, 2003).

Cover crops and mulches as components of sustainable management systems may be used for integrating bioherbicides by delivering the agents on seeds and promoting their establishment in soils for attack of weeds and seedlings prior to planting. Recent research demonstrated that several cover crop species inoculated with selected AB at planting established and maintained the selected bacterial populations on their roots and in adjacent soil. When giant foxtail (*Setaria faberi*) emerged later in the season, the selected bacteria colonized seedling roots after the cover crop was terminated (Kremer, 2000). The selected AB and allelopathic cover crop residues acted synergistically to suppress the growth of the weeds.

5.4. Suppressive Soils and Conservation Biological Control

Soils under sustainable management may develop antagonistic microbial populations in rhizospheres of selected weeds that suppress their growth. This occurrence is similar to natural disease-suppressive soils in which indigenous soil microorganisms effectively protect crop plants from soilborne plant pathogens (Weller et al., 2002). Disease-suppressive soils may be defined as soils in which a pathogen does not establish or persist, establishes but causes little or no damage, or establishes and causes disease for a short time but thereafter the disease is less important even though the pathogen may persist in the soil (Baker and Cook, 1974). Suppression is due primarily to antagonistic microorganisms, however, soil physical and chemical factors also may be involved (Weller et al., 2002). Similarly, weed-suppressive soils may be defined as soils in which certain weeds do not establish or persist, or establish and grow with the crop but cause little interference due to suppressed growth and vigor caused by native AB. Native and desirable plants may also in stimulate high populations of AB in their rhizospheres that reduce growth of invasive weed species, suggesting that plant-soil interactions are also involved in development of weed-suppressive soils (Kulmatiski et al., 2004).

Evidence of apparent weed-suppressive soils has been reported for a variety of sustainable cropping systems. A study of crop management practices on claypan soils (Epiaqualfs) that involved reduced tillage, maintenance of high soil organic matter, and limited inputs of agrichemicals found increased levels of AB associated with weed seedlings that likely contributed to natural weed suppression (Li and Kremer, 2000). It was reported that agronomic practices that resulted in relatively high organic matter, such as uncultivated prairie, organic farming and integrated cropping systems, supported higher proportions of weed AB. Compost-amended soils planted to winter wheat showed 29 and 78% reductions in broadleaf and grassy weed densities, respectively, compared to soils amended with inorganic fertilizers only (Carpenter-Boggs et al., 2000). Organic amendments (composts and cover crops) increased soil microbial biomass and decreased the seedbank density and emergence of shepherd's purse (*Capsella bursa-pastoris*) and burning nettle (*Urtica urens*) in a California vegetable production field (Fennimore and Jackson, 2003). The natural soil suppressiveness of the parasitic weed *Striga hermonthica* in Nigeria appears to be related to soils under rotations of cereal and leguminous crops that promote antagonist microbial populations that destroys *Striga* seeds before germination or kills the germinated seedlings (Berner et al., 1996; Dashiel et al., 1991). Each of the above systems strongly suggests that AB growth can be exploited as a sustainable weed control strategy using relatively simple management practices.

Tactics and approaches for manipulating the field environment to enhance survival, physiological behavior, and performance of AB might easily be incorporated into diverse sustainable crop production systems. Such a strategy for natural weed suppression, also known as conservation biological control (Newman et al., 1998) or endemic soil-based control (Kulmatiski et al., 2004) relies on establishment of populations of indigenous or endemic, weed-suppressive microorganisms in soil. As demonstrated previously, many of these indigenous microorganisms are AB (Kremer and Li, 2003; Li and Kremer, 2000). Management practices including tillage, crop rotation, residue manipulation, and organic amendments enhance or induce favorable factors in the habitat for sustaining effective populations of natural AB. Crop management practices that involve reduced tillage, maintain high soil organic matter, and limit inputs of agrichemicals increased levels of deleterious rhizobacteria associated with weed seedlings and contribute to natural weed suppression (Li and Kremer, 2000). Deliberate use of management practices that benefit natural weed-antagonistic AB can adversely affect weed population dynamics in production fields through seed and seedling mortality and growth suppression.

6. SUMMARY

The future use of AB to manage weeds in both conventional and sustainable agriculture seems promising. Because AB generally do not attack specific biochemical sites within the plant, unlike conventional herbicides, they offer a means to control weeds without causing direct selective pressure on the weed population, therefore, development of resistance is not a major consideration. Additionally, the use of AB appears to be environmentally benign relative to herbicides. These characteristics make AB an attractive approach for managing crop weeds in a sustainable manner, even within the boundaries of conventional agriculture systems. The recent demonstrations of apparent weed-suppressive soils may lead to development of specific management strategies for the establishment and persistence of native AB directly in soils conducive to annual weed infestations.

7. REFERENCES

Aldrich, R.J., Kremer, R.J. *Principles in Weed Management*. Iowa State University Press: Ames, IA, 1997

Baker, K.F., Cook, R.J. *Biological Control of Plant Pathogens*. Freeman: San Francisco, 1974.

Barazani, O., Friedman, J. Allelopathic bacteria and their impact on higher plants. Crit Rev Plant Sci 1999; 18:741-755.

Begonia, M.F.T., Kremer, R.J. Chemotaxis of deleterious rhizobacteria to velvetleaf (*Abutilon theophrasti*) seeds and seedlings. FEMS Microbiol Ecol 1994; 15:227-236.

Begonia, M.F.T., Kremer, R.J., Stanley, L., Jamshedi, A. Association of bacteria with velvetleaf roots. Trans Mo Acad Sci 1990; 24:17-26.

Berner, D., Carsky, R., Dashiell, K., Kling, J., Manyong, V. A land management based approach to integrated Striga hermonthica control in sub-Saharan Africa. Outlook Agric 1996; 25:157-164.

Boland, G.J., Kuykendall, L.D. Plant-Microbe Interactions and Biological Control. Marcel Dekker: New York, 1998.

Bolton, Jr., H., Fredrickson, J.K., Elliott, L.F. Microbial ecology of the rhizosphere. In: *Soil Microbial Ecology*. Blaine Metting, Jr., F. ed., Marcel Dekker, Inc.: New York., 1993; pp. 27-63.

Boyetchko, S.M. Innovative applications of microbial agents for biological weed control. In: *Biotechnological Approaches in Biocontrol of Plant Pathogens*. Mukerji, K.G., Chamola, B.G., Upadhyay, R.K., eds., Plenum Press: London, 1999; pp. 73-97.

Brinkman, M.A., Clay, S.A., Kremer, R.J. Influence of deleterious rhizobacteria on leafy spurge (*Euphorbia esula*) roots. Weed Technol 1999; 13:835-839.

Cardina, J. Biological weed management. In, *Handbook of Weed Management Systems*, Smith, A.E. ed. Marcel Dekker: New York, 1995; pp. 279-341.

Carpenter-Boggs, L., Reganold, J.P., Kennedy, A.C. Biodynamic preparations: short-term effects on crops, soils, and weed populations. Am J Altern Agric 2000; 15:110-118.

Cherrington, C.A., Elliott, L.F. Incidence of inhibitory pseudomonads in the Pacific Northwest. Plant Soil 1987; 101:159-165.

Dashiell, K.E., Jackai, L.E.N., Hartman, G.L., Ogundipe, H.O., Asafo-Adjei, B. Soybean germplasm diversity, uses and prospects for crop improvement in Africa. In, *Crop Genetic Resources of Africa*, Ng, N.O., Perrino, P., Attere, F., Zedan, S. eds. Vol II. Proc. International Conference on Crop Genetic Resources of Africa (pp. 203-212), IITA: Ibadan, Nigeria, 1991; pp. 203-212.

De Weger, L.A., van der Bij, A.J., Dekkers, L.C., Simons, M., Wijffelman, C.A., Lugtenberg, B.J. Colonization of the rhizosphere of crop plants by plant-beneficial pseudomonads. FEMS Microbiol Ecol 1995; 17:221-228.

Elliott, L.F., Lynch, J.M. Plant growth-inhibitory pseudomonads colonizing winter wheat (*Triticum aestivum* L.) roots. Plant Soil 1985; 84:57-65.

Fennimore, S.A., Jackson, L.E. Organic amendment and tillage effects on vegetable field weed emergence and seedbanks. Weed Technol 2003; 17:42-50.

Gallandt, E.R., Liebman, M., Huggins, D.R. Improving soil quality: implications for weed management. J Crop Product 1999; 2:95-121.

Gealy, D. R., Gurusiddaiah, S., Ogg, Jr., A.G. Isolation and characterization of metabolites from *Pseudomonas syringae* strain 3366 and their phytotoxicity against certain weed and crop species. Weed Sci 1996; 44:383-392.

Gliessman, S.R. Allelopathy and agroecology. In *Chemical Ecology of Plants: Allelopathy in Aquatic and Terrestial Ecosystems*, Inderjit, A.U. Mallik eds., Birkhauser Verlag: Zurich, 2002; pp. 173-185.

Greaves, M. P., Sargent, J.A. Herbicide-induced microbial invasion of plant roots. Weed Sci 1986; 34:50-53.

Gurusiddaiah, S., Gealy, D.R., Kennedy, A.C., Ogg, Jr., A.G. Isolation and characterization of metabolites from *Pseudomonas fluorescens*-D7 for control of downy brome (*Bromus tectorum*). Weed Sci 1994; 42:492-501.

Harley, K.L.S., Forno, I.W. Biological Control of Weeds: A Handbook for Practitioners and Students. Inkata Press: Melbourne, 1992.

Harris, P.A., Stahlman, P.W. Soil bacteria as selective biological control agents of winter annual grass weeds in winter wheat. Appl Soil Ecol 1996; 3:275-281.

Hatcher, P.E., Melander, B. Combining physical, cultural and biological methods: prospects for integrated non-chemical weed management strategies. Weed Res 2003; 43:303-322.

Keel, C., Schnider, U., Maurhofer, M. Voisard, C., Laville, J., Burger, U., Wirthner, P., Haas, D., DeFago, G. Suppression of root diseases by *Pseudomonas fluorescens* CHA0: importance of the bacterial secondary metabolite 2,4-diacetylphloroglucinol. Molec Plant Microbe Interact 1992; 5:4-13.

Kennedy, A.C. Rhizosphere. In, *Principles and Applications of Soil Microbiology*, Sylvia, D.M., Fuhrmann, J.J., Hartel, P.G, Zuberer D.A., eds. Pearson Prentice Hall: Upper Saddle River, NJ, 2005; pp. 242-262.

Kennedy, A.C., Johnson, B.N., Stubbs, T.L. Host range of a deleterious rhizobacterium for biological control of downy brome. Weed Sci 2001; 49:792-797.

Kennedy, A.C., Elliott, L.F., Young, F.L., Douglas, C.L. Rhizobacteria suppressive to the weed downy brome. Soil Sci Soc Am J 1991; 55:722-727.

Knowles, C. J., Bunch, A.W. Microbial cyanide metabolism. Adv Microbiol Physiol 1986; 27:73-111.

Kremer, R.J., Li, J. Developing weed-suppressive soils through improved soil quality management. Soil Till Res 2003; 72:193-202.

Kremer, R.J. Bioherbicides: potential successful strategies for weed control. In, *Microbial Biopesticides*, Koul O., Dhaliwal B., eds., Taylor & Francis: London, 2002, pp. 307-323.

Kremer, R.J., Souissi, T. Cyanide production by rhizobacteria and potential for suppression of weed seedling growth. Curr Microbiol 2001; 43:182-186.

Kremer, R.J. Growth suppression of annual weeds by deleterious rhizobacteria integrated with cover crops. In, *Proceedings of the Xth International Symposium on Biological Control of Weeds*, Spencer, N.R. ed. USDA-ARS and Montana State University: Bozeman, MT, 2000; pp. 931-940.

Kremer, R.J. Microbial interactions with weed seeds and seedlings and its potential for weed management. In, *Integrated Weed and Soil Management*, Hatfield, J.L., Buhler, D.D., Stewart, B.L. eds., Ann Arbor Press: Chelsea, MI, 1998; pp. 161-179.

Kremer, R.J., Kennedy, A.C. Rhizobacteria as biocontrol agents of weeds. Weed Technol 1996; 10:601-609.

Kremer, R.J., Begonia, M.F.T., Stanley, L., Lanham, E.T. Characterization of rhizobacteria associated with weed seedlings. Appl Environ Microbiol 1990; 56:1649-1655.

Kulmatiski, A., Beard, K.H., Stark, J.M. Finding endemic soil-based controls for weed growth. Weed Technol 2004; 18:1353-1358.

Li, J., Kremer, R.J. Rhizobacteria associated with weed seedlings in different cropping systems. Weed Sci 2000; 48:734-741.

Liebman, M., Gallandt, E.R. Many little hammers: ecological management of crop-weed interactions. In, *Ecology in Agriculture*, Jackson, L.E. ed. Academic Press: San Diego, 1997; pp. 291-343.

Müller-Schärer, H., Scheepens, P.C., Greaves, M.P. Biological control of weeds in European crops: recent achievements and future work. Weed Res 2000; 40:83-98.

Nehl, D.B., Allen, S.J., Brown, J.F. Deleterious rhizosphere bacteria: an integrating perspective. Appl Soil Ecol 1997; 5:1-20.

Nelson, E.B. Microbial dynamics and interactions in the spermosphere. Annu Rev Phytopathol 2004; 42:271-309.

Newman, R.M., Thompson, D.C., Richman, D.B. Conservation strategies for the biological control of weeds. In, *Conservation Biological Control*, Barbosa P. ed. Academic Press: San Diego, 1998; pp. 371-396.

Owen, A., Zdor, R. Effect of cyanogenic bacteria on the growth of velvetleaf *(Abutilon theophrasti)* and corn *(Zea mays)* in autoclaved soil and the influence of supplemental glycine. Soil Biol Biochem 2001; 33:801-809.

Parr, J.R., Papendick, R.I., Hornick, S.B., Meyer, R.B. Soil quality: attributes and relationship to alternative and sustainable agriculture. Amer J Alt Agric 1992; 7:5-11.

Quimby, P.C., Birdsall, J.L. Fungal agents for biological control of weeds: classical and augmentative approaches. In, *Novel Approaches to Integrated Pest Management.* Reuveni R. ed., CRC Press: Boca Raton, FL, 1995, pp. 293-308.

Sarwar, M., Kremer, R.J. Enhanced suppression of plant growth through the production of L-tryptophan derived compounds by deleterious rhizobacteria. Plant Soil 1995; 172:261-269.

Sarwar, M., Frankenberger, Jr., W.T Influence of L-tryptophan and auxins applied to the rhizosphere on the vegetative growth of *Zea mays* L. Plant Soil 1994; 160:97-104.

Schippers, B., Bakker, A.W., Bakker, P.A.H.M., van Peer, R. Beneficial and deleterious effects of HCN-producing pseudomonads on rhizosphere interactions. Plant Soil 1990; 129:75-83.

Schippers, B., Bakker, A.W., Bakker, P.A. Interactions of deleterious and beneficial rhizosphere microorganisms and the effect of cropping practices. Annu. Rev. Phytopathol 1987; 25:339-358.

Schroth, M.N., Hancock, J.G. Disease-suppressive soil and root-colonizing bacteria. Science 1982; 216:1376-1381.

Skipper, H.D., Ogg, Jr., A.G., Kennedy, A.C. Root biology of grasses and ecology of rhizobacteria for biological control. Weed Technol 1996; 10:610-620.

Souissi, T., Kremer, R.J., White, J.A. Scanning and transmission electron microscopy of root colonization of leafy spurge (*Euphorbia esula* L.) seedlings by rhizobacteria. Phytomorphology 1997; 47:177-193.

Sturz, A.V., Christie, B.R. Beneficial microbial allelopathies in the root zone: the management of soil quality and plant disease with rhizobacteria. Soil Till Res 2003; 72:107-123.

Suslow, T.V. Role of root-colonizing bacteria in plant growth. In, *Phytopathogenic Prokaryotes*, Mount, M.S., Lacy, G.H. eds. Volume 1. Academic Press: New York, 1982, pp. 187-223.

Suslow, T.V., Schroth, M.N. Role of deleterious rhizobacteria as minor pathogens in reducing crop growth. Phytopathology 1982; 72:111-115.

TeBeest, D.O. Microbial Control of Weeds. Chapman and Hall: New York, 1991.

Tranel, P.J., Gealy, D.R., Kennedy, A.C. Inhibition of downy brome *(Bromus tectorum)* root growth by a phytotoxin from *Pseudomonas fluorescens* strain D7. Weed Technol. 1993; 7:134-139.

Turco, R.F., Bischoff, M., Breakwell, D.P., Griffith, G.R. Contribution of soil-borne bacteria to the rotation effect in corn. Plant Soil 1990; 122:115-120.

Weissmann, R., Gerhardson, B. Selective plant growth suppression by shoot application of soil bacteria. Plant Soil 2001; 234:159-170.

Weller, D.M., Raaijmakers, J.M., McSpadden-Gardner, B.B., Thomashow, L.S. Microbial populations responsible for specific soil suppressiveness to plant pathogens. Annu Rev Phytopathol 2002; 40:309-348.

Zahir, Z.A., Arshad, M., Frankenberger, Jr., W.T. Plant growth promoting rhizobacteria: applications and perspectives in agriculture. Adv Agron 2004; 81:97-168.

MARK A. BERNARDS[1], LINA F. YOUSEF[1], ROBERT W. NICOL[2]

THE ALLELOPATHIC POTENTIAL OF GINSENOSIDES

[1]*Department of Biology, University of Western Ontario, London, ON, Canada, N6A 5B7*
[2]*NovoBiotic Pharmaceuticals, Cambridge MA, 02138, USA*
Email: bernards@uwo.ca

Abstract. American ginseng (*Panax quinquefolius* L.) is a perennial herb valued for the medicinal properties of its large, fleshy tap root. These medicinal properties are purported to be due to the triterpenoid saponins, or ginsenosides, that accumulate to 3-6% of the root dry weight. We asked the question: what is the ecological role of ginsenosides in *Panax* species? In addressing this question, we have determined that ginsenosides, like other saponins, possess fungitoxic properties, although different fungi and oomycotan organisms appear to be differentially affected by them *in vitro*. In order to play an allelopathic role, however, ginsenosides must be present in the soil at biologically active (i.e., ecologically relevant) concentrations. Results to date support the hypothesis that ginsenosides are phytoanticipins and serve as host resistance factors. The success of certain pathogens (e.g., *Pythium cactorum, Pythium irregulare, Cylindrocarpon destructans*) on ginseng may arise from their ability to detoxify or otherwise utilize ginsenosides as a nutritive source or growth stimulating factor, while other soil borne organisms appear susceptible to their fungitoxic properties. Ginsenosides have been isolated from rhizosphere soil and root exudates suggesting that these compounds are involved in allelopathic interactions between the host plant and soil fungi. Ultimately this allelopathic interaction may influence the fungal diseases of ginseng.

1. INTRODUCTION

American ginseng (*Panax quinquefolius* L.) produces triterpenoid saponins called ginsenosides, which are slowly released into the rhizosphere. Ginsenosides, like other saponins, are fungitoxic, and as such may modify the balance of microorganisms in the soil; i.e., they are potentially allelopathic. From a disease management point of view, the extent to which ginsenosides alter the soil microbial community may have profound consequences, especially when disease causing organisms are favoured in the new balance. Understanding how ginsenosides affect soilborne microbes is important in understanding disease cycles in this crop.

2. SAPONINS

2.1. Saponins

Saponins, are glycosylated natural products with surfactant and soap-like properties that tend to froth in aqueous solution, even at low concentration (Dewick, 1997).

Inderjit and K.G. Mukerji (eds.),
Allelochemicals: Biological Control of Plant Pathogens and Diseases, 157–175.

They also cause haemolysis, though are generally non-toxic when taken orally. Saponins are broadly categorised as being either triterpenoid- or steroid-derived, despite the common triterpenoid origin of both. Triterpenoid saponins are widely distributed in many dicot families, and are generally based on pentacyclic triterpene parent carbon skeletons (e.g., α- and β-amyrin), or tetracyclic parent compounds (e.g., the dammaranes). By contrast, steroidal saponins are commonly found in monocots, and are characterised by a spiroketal moiety at C-22. Plants of the Solanaceae contain steroidal glycoalkaloid saponins, which are nitrogen analogues of steroidal saponins.

In general, saponins are glycosylated at the C-3 hydroxyl group, typically with a glucose molecule. However, there are other potential glycosylation sites on the parent carbon skeletons, and multi-site glycosylation can occur. Accordingly, saponins are subdivided into monodesmosidic (one glycosylation site) and bisdesmosidic (glycosylation sites at both ends of the compound) categories. The majority of saponins, however, are monodesmosidic. While glucose is the most common sugar in saponins, arabinose, galactose, rhamnose and xylose, as well as sugar acids are also present. The glycosylation patterns of saponins can lead to small families of compounds based on the same parental carbon skeleton (i.e., sapogenin or aglycone), differing only in the composition and arrangement of sugars.

2.2. Saponins as Fungitoxic and Plant Defense Compounds

There are numerous reports in the literature that describe the anti-fungal activity of saponins (Zimmer et al., 1967; Levy et al., 1986; 1989; Marston et al., 1988; Takechi and Tanaka, 1990; Ohtani et al., 1993; Ouf et al., 1994; Favel et al., 1994; Escalante et al., 2002; Nicol et al., 2002). These reports coupled with molecular data (Bowyer et al., 1995; Papadopoulou et al., 1999) suggest that these phytochemicals are constitutive plant defenses against fungi (Osbourn, 1996; 2003), that is phytoanticipins (Van Etten et al., 1994). For example, the role that the triterpenoid saponin avenacin A-1 plays in conferring resistance to "take-all disease", caused by *Gaeumannomyces graminis* in oats, is well established (Turner, 1956; Burkhardt et al., 1964; Maizel et al., 1964; Bowyer et al., 1995; Papadopoulou et al., 1999).

While the molecular mechanism of saponin fungitoxicity is not known, avenacin A-1 as well as steroidal glycoalkaloid saponins such as α-tomatine and solanine have been demonstrated to form complexes with membrane sterols, thereby causing a reduction in membrane integrity (Roddick, 1979; Steel and Drysdale, 1988; Keukens et al., 1992; 1995; Armah et al., 1999). Two models have been proposed to explain the consequences of saponin-sterol aggregation in target membranes (Morrissey and Osbourn, 1999). One suggests that the deleterious effects of saponins are related to the formation of transmembrane pores (Armah et al., 1999), whereas the other suggests that membrane integrity is compromised due to the extraction of sterols (Keukens et al., 1992; 1995). Regardless of the exact mechanism, saponins probably disrupts fungal membranes through complexation with ergosterol, the major membrane sterol in higher fungi (Evans and Gealt 1985; Weete 1989; Griffiths et al., 2003).

Saponins may act as host chemical defenses, but because fungi successfully attack plants containing these defenses, there must be means of tolerating or avoiding saponin toxicity. Fungal resistance to defensive chemicals in general can either involve enzymatic detoxification of antifungal compounds or the more ambiguous "innate resistance" (Morrissey and Osbourn, 1999). Some fungi are able to detoxify phytoalexins (Van Etten et al., 1995) as well as phytoanticipins such as saponins (Osbourn et al., 1995a; Van Etten et al., 1995; Weltring et al., 1997). Although saponin detoxification generally occurs via enzymatic cleavage of the saccharides to form the aglycone parent compound, different fungi employ different strategies and enzymes. For example, fungal detoxification of α-tomatine by *Botrytis cinerea* (Quidde et al., 1998) *Septoria lycopersici* (Arneson and Durbin, 1967) and *Verticillium albo-atrum* (Pegg and Woodward, 1986) involves cleavage of one terminal monosaccharide from the tetrasaccharide chain at carbon three of the sapogenin, whereas *Fusarium oxysporum* f. sp. *lycopersici* removes the entire tetrasaccharide (Ford et al., 1977). The saponin-detoxifying enzymes tomatinase from *S. lycopersici* and avenacinase from *G. graminis* exhibit high sequence similarity (Osbourn et al., 1995b), whereas the tomatinase from *F. oxysporum* f. sp. *lycopersici* appears to be more related to a different family of glycosyl hydrolases (Roldán-Arjona et al., 1999). The ability to efficiently detoxify host chemical defenses determines virulence and host range in fungal pathogens such as *G. graminis*, *Rhizoctonia solani* and *Phoma lingam* (Bowyer et al., 1995; Pedras et al., 2000a,b). Efficient detoxification is taken an extra step by *S. lycopersici* as the hydrolysis products of the saponin-based defense subsequently inhibit inducible defenses of the host by interfering with signal transduction pathways (Bouarab et al., 2002).

Transformation of secondary metabolites (e.g., cleaving sugars from saponins) has been suggested as a way of providing fungi with a carbon source in addition to achieving detoxification (Van Etten et al., 1995; Roldán-Arjona et al., 1999). Since some fungal detoxification enzymes are repressed by glucose (Straney and Van Etten, 1994; Roldán-Arjona et al., 1999) they may have a dual function. This would be analogous to non-pathogenic fungi that obtain carbon from phenolic monomers produced during lignin decomposition (Henderson and Farmer, 1955; Rahouti et al., 1989). Presumably it would be advantageous for pathogens to circumvent host defenses and simultaneously derive nutrition.

Pathogens in the Pythiaceae (Oomycota) appear to possess an innate resistance to the toxic effects of saponins. Members in this family are reported to be relatively unaffected by aescin (Olsen, 1971) and other saponins (Assa et al., 1972) *in vitro*, and probably the lack of ergosterol in these species (Olsen, 1973a; Weete, 1989) allows them to avoid saponin toxicity (Arneson and Durbin 1968; Olsen, 1971; Weltring, 1997). Although *Pythium* spp. appear to be unaffected by saponins, saponin toxicity can be induced through the addition of sterols to the growth medium (Olsen, 1973b; Steel and Drysdale, 1988). Members of the Pythiaceae can incorporate sterols by the binding and transport action of small protein carriers called elicitins (Mikes et al., 1997; Panabières et al., 1997; Capasso et al., 2001) and this uptake of exogenous sterols could be the cause of the observed increase in the deleterious effects of saponins on these organisms.

are 6-*O* and 20-*O* diglycosides (Figure 1). Ginsenosides can represent approximately 3-7% dry weight of the root (Court et al., 1996; Li et al., 1996; Nicol et al., 2002) and are found in the foliage at lower levels (Chen and Staba, 1978). In Asian ginseng, *Panax ginseng* C.A. Meyer, ginsenosides are found in the outer cortex and periderm of roots (Tani et al., 1981).

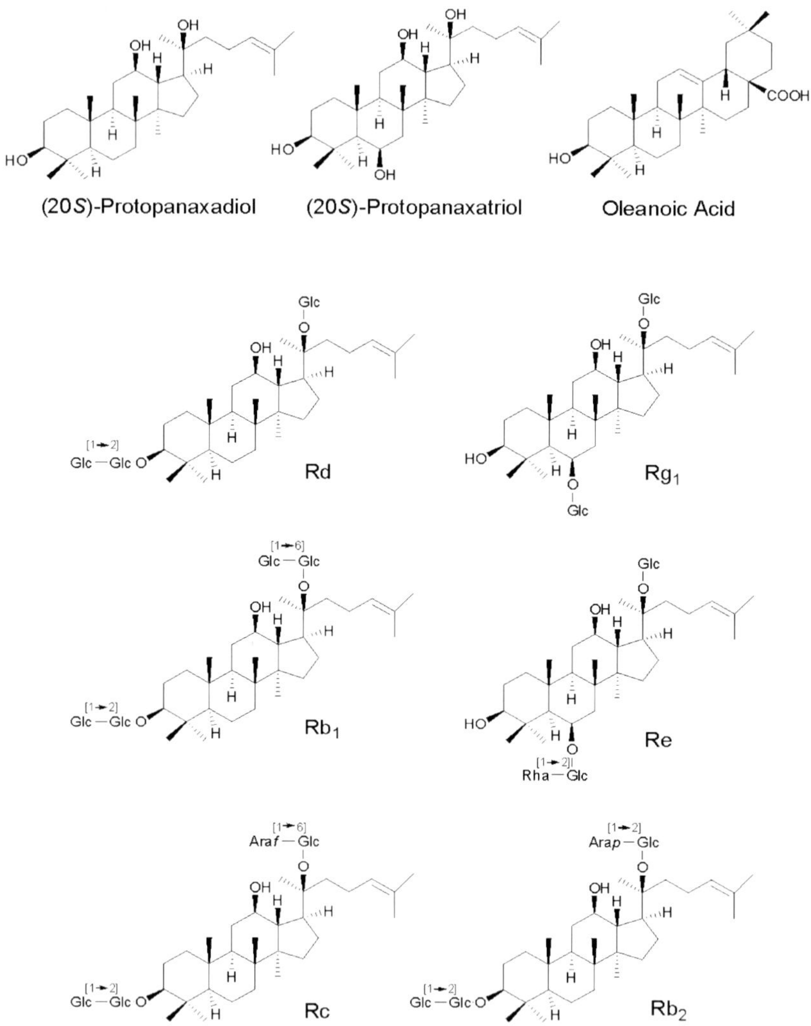

Figure 1. The Major Ginsenosides of Panax spp. The three main triterpenoid aglycones (20S)-protopanaxadiol, (20S)-protopanaxatriol, oleanoic acid), as well as the six major ginsenosides of Panax spp. are shown. Ginsenosides are named according to their migration properties on TLC, with no regard for whether they are (20S)-protopanaxadiol-, (20S)-protopanaxatriol- or oleanoic acid-derived. Ginsenoside Rf (not shown) is a major ginsenoside of P. ginseng but is not found in P. quinquefolius.

Triterpenoids are derived from the cytosolic mevalonate pathway via the common 30-carbon intermediate squalene (Dewick, 1997; Haralampidis et al., 2002). Briefly, oxidation of squalene to 2,3-oxidosqualene generates an intermediate that can be folded and cyclised (by 2,3-oxidosqualene cyclases) into any one of the different classes of triterpenes found in plants. That is, if the 2,3-oxidosqualene is folded into a *chair-boat-chair-boat* configuration, protonation of the 2,3-epoxide (which ultimately generates the 3-OH common to all triterpenoids and steroids) results in the concerted cyclisation of 2,3-oxidosqualene into a protosteryl cation which upon protonation yields either cycloartenol (plants) or lanosterol (animals, fungi), both of which are precursors to the steroidal saponins and steroids. Alternative folding into a *chair-chair-chair-boat* configuration gives rise to a dammarenyl cation (Figure 2), which

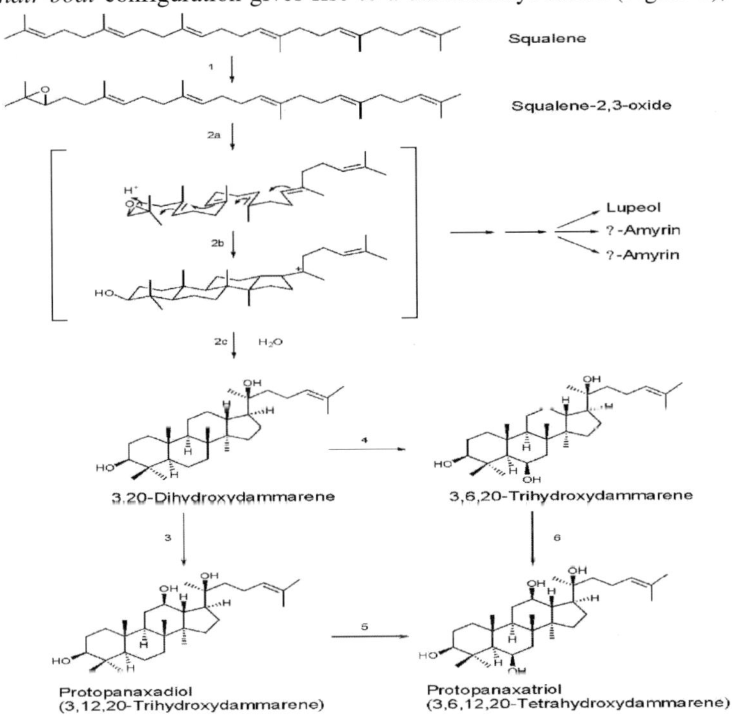

Figure 2. Hypothetical Biosynthesis of Ginsenoside Aglycones. The biosynthesis of ginsenoside aglycones begins with the formation of squalene-2,3-oxide by a flavoprotein-containing epoxide synthase (1). A putative dammarane synthase (2a-c) is thought to bind squalene-2,3-oxide and act as a template for folding it into a chair-chair-chair-boat configuration (2a). Protonation of the epoxide initiates a concerted series of ring closures (2b), terminating in a dammarenyl cation with the appropriate stereochemistry. In Panax spp., the dammarenyl cation is expected to be quenched by water (2c) yielding 3,20-dihydroxydammarene. Subsequent hydroxylation reactions would yield the protopanaxadiol and protopanaxatriol parent carbon skeletons of the common ginsenosides. Two potential hydroxylation pathways are shown (see text for details). All enzymes involved in the production of the tetracyclic ginsenosides are hypothetical for Panax spp. (based on Dewick, 1997; Haralampidis et al., 2002).

can undergo subsequent 1,2-hydride and/or 1,2 alkyl shifts and further annulation. Loss of a proton yields pentacyclic saponin precursors such as α- or β- amyrin or lupeol.

In ginseng, however, only one relatively low abundant ginsenoside (Ro) has a pentacyclic parent structure similar to most other non-ginseng triterpenoid saponins. Instead, most ginsenosides are based on the (20*S*)-protopanaxadiol and (20*S*)-protopanaxatriol parent carbon skeletons derived from the dammarenediol structure that results from the quenching of the dammarenyl cation by water. This reaction (i.e., *chair-chair-chair-boat* template folding, concerted annulation and final cation quenching) is hypothesized to be catalysed by dammarenediol synthase (Dewick, 1997; Haralampidis et al., 2002), although this has not been proven. Other enzymes involved in triterpenoid biosynthesis have been identified in Asian ginseng (*P. ginseng*) including an oxidosqualene cyclase (â-amyrin synthase, Kushiro et al., 1998) as well as other candidate oxidosqualene cyclases and enzymes involved in modification of the sapogenin (Jung et al., 2003). Further hydroxylation reactions to yield the (20*S*)-protopanaxadiol and (20*S*)-protopanaxatriol parent carbon skeletons from 3,20-dihydroxydammarene have not been described. However, two routes are theoretically possible (Figure 2). In the first, 3,20-dihydroxydammarene is hydroxylated at C-12 to yield the (20*S*)-protopanaxadiol skeleton directly. Subsequent hydroxylation at C-6 yields the (20*S*)-protopanaxatriol skeleton. Alternatively, the (20*S*)-protopanaxatriol skeleton could arise independent of the (20*S*)-protopanaxadiol skeleton via hydroxylation at C-6 first, followed by C-12 hydroxylation. In either case, the assumption is made that the parent carbon skeletons are assembled first, before glycosylation. No details are yet available with respect to the glycosylation of (20*S*)-protopanaxadiol and (20*S*)-protopanaxatriol parent carbon skeletons. Clearly glycosylation of the unique C-20 hydroxyl group, which is glycosylated in all ginsenosides except two (20S)-protopanaxadiols (Rh_2, Rg_3) and Ro, represents a novel glycosyl transferase activity, as does glycosylation of C-6 of the (20*S*)-protopanaxatriols.

4. GINSENOSIDE FUNGITOXICITY *IN VITRO*

4.1. Differential Response of Fungi and Oomycetes to the Addition of Ginsenosides to Culture Media

In commercial gardens, ginseng is attacked by the foliar pathogens *Alternaria panax*, *A. alternata*, and *Botrytis cinerea*, and the root pathogens *Cylindrocarpon destructans*, *Rhizoctonia solani*, *P. cactorum*, *Py. irregulare*, *Py. ultimum* and several species of *Fusarium* including *F. oxysporum* and *F. solani* (Brammall, 1994a, b; Reeleder and Brammall, 1995; Punja, 1997). Although the pathogens in the Pythiaceae (i.e., *Phytophthora cactorum*, *Pythium irregulare*, *Py. ultimum*) are superficially similar, and share the same nutritional mode as the other pathogens, they are not true fungi, but rather belong in the kingdom Straminipila (Burnett, 2003). Most other ginseng pathogens, are ascomycetes. The activity of ginsenosides against these pathogens, as well as saprotrophic *Trichoderma* spp., was evaluated *in vitro* (Nicol et al., 2002; unpublished data). The *Trichoderma* spp. are potential antagonists towards soilborne pathogens, evident by the greater levels of *Trichoderma* spp. found in healthy ginseng fields than in replanted ginseng fields with high disease incidence (Shin and Lee, 1986). *In vitro* anti-fungal bioassays were completed by adding ginsenosides to growth media and comparing the relative growth to that of controls (i.e., with no ginsenoside addition to the growth media). Under these conditions, the growth of six of the nine tested organisms was inhibited. The highest growth inhibition was observed in the saprotrophs *T. harzianum* and *T. hamatum* followed by the leaf pathogen *A. panax* and *T. viride* (Table 1).

Table 1. Growth Response of Selected Fungal and Oomycotan Species to Ginsenosides. The relative growth of the microorganisms was compared in vitro with and without the addition of ginsenosides isolated from Panax quinquefolius roots. Growth was standardized to that of controls (i.e., no ginsenosides added) to allow comparisons across species, even though different assay conditions were used for different organisms. Growth data is shown as a mean ± standard deviation. With the exception of F. oxysporum, the growth of all organisms in the presence of ginsenosides was significantly different from the mean growth of the respective controls (data not shown). Data compiled from Nicol et al. 2002, 2003 and unpublished data.

Fungal Species	% Growth relative to control
Trichoderma harzianum	-26.4 ± 0.9
Trichoderma hamatum	-22.2 ± 2.8
Trichoderma viride	-9.4 ± 1.1
Alternaria panax	-17.1 ± 3.8
Fusarium solani	-3.3 ± 0.6
Fusarium oxysporum	-3.0 ± 0.8
Cylindrocarpon destructans	+7.6 ± 2.9
Phytophthora cactorum	+324.9 ± 1.0
Pythium irregulare	+392.8 ± 0.5

The growth of the two *Fusarium* species, *F. oxysporum* and *F. solani*, was consistently found to be slightly lower, but not always significantly different from control (Table 1). Conversely, the growth of the causal organisms of some of the most devastating diseases in the North American ginseng industry (i.e., *C. destructans*, *P. cactorum* and *Py. irregulare*) was significantly stimulated over that of control. When analysed as a group, significantly different growth responses to the ginsenosides were observed across the fungal and oomycotan species tested (Table 1). That is, the organisms tested were generally inhibited (*Trichoderma* spp. and *A. panax*), unaffected (*Fusarium* spp.) or stimulated (*C. destructans*, *P. cactorum* and *Py. irregulare*) by ginseng saponins. By comparison, greater antifungal activity was found with aescin, a mixture of saponins from the horse chestnut tree (Nicol et al. 2002) and consequently, ginsenosides can only be considered to be mildly antifungal.

4.2. 'Detoxification" of Ginsenosides by Oomycetes

As noted in Section 1.2., two mechanisms of resistance have been proposed to explain the insensitivity of Oomycetes to saponins, (i) innate resistance, due to little or no sterols in the membranes of Oomycete species and (ii) detoxification, via an enzymatic degradation of saponins into less bioactive derivatives. With respect to the observed growth stimulation of the Oomycetes *Py. irregulare* and *P. cactorum* in the presence of ginsenosides *in vitro*, it is not clear which of these mechanisms may be involved. However, preliminary results suggest that the "detoxification" of ginsenosides (exemplified by *Py. irregulare*) is the likely mechanism. For example, we initially hypothesized that if the fungitoxicity of ginsenosides resulted from their interaction with sterols, then the addition of ergosterol to the growth medium of *Pythium* spp. and *Phytophthora* spp. would render these organism more susceptible (as has been demonstrated with aescin, e.g., Olsen, 1973b). However, we have found that the addition of ergosterol and ginsenosides to the growth medium of the two ginseng pathogens, *Py. irregulare* and *P. cactorum* resulted in a cumulative biomass increase (~ 200% and 150% compared to control colony weights) relative to cultures grown with either ergosterol or ginsenosides alone (Nicol et al., 2003). This observation suggests that the resistance of *Pythium* spp. and *Phytophthora* spp. to ginsenosides cannot simply be due to a lack of sterol content in their membranes.

Recently, a glycoside hydrolase (BGX1) was described for *P. infestans* (Brunner et al., 2002), the first glycosidase of its kind for an oomycotan organism. With the assumptions that (i) the pathogenesis mechanisms of Oomycetes are similar to those of fungi (Latijnhouwers et al., 2003), and (ii) ginsenosides act as preformed defense compounds (Nicol et al., 2002), then the production of glycosidases by members of the Pythiaceae could be an important factor in explaining their pathogenicity to American ginseng. Current attempts to rationalize the effect of ginsenosides on the growth of *Py. irregulare* involve exploring the ability of this organism to modify ginsenosides *in vitro*. Thus, when ginsenosides were re-extracted from spent broth that had been supplemented with 0.1% ginsenosides prior to inoculation with *Py. irregulare*, only the (20*S*)-protopanaxatriol-derived ginsenosides Re and Rg_1 were recovered (Figure 3, lower trace). That is, the (20*S*)-protopanaxadiol ginsenosides (Rb_1, Rb_2, Rc, Rd) were all substantially, if not completely depleted from the medium compared to the initial supplement (compare Figure 3 upper trace with Figure 3 lower trace).

Moreover, the decline in (20*S*)-protopanaxadiol ginsenosides in the spent broth of *Py. irregulare* was linear over the five day culture period (data not shown).

The disappearance of protopanaxadiol ginsenoside peaks from HPLC chromatograms was coincident with the appearance of an unidentified peak eluting at approximately 45 minutes (Figure 3, lower trace). It is tempting to speculate that the 45 minute peak is the (20*S*)-protopanaxadiol aglycone resulting from the de-glycosylation of ginsenosides by *Py. irregulare*, but this requires experimental verification. By contrast, the profile of ginsenosides recovered from broth inoculated with *T. hamatum,* (a control organism sensitive to ginsenosides; Table 1) was unaltered relative to that of broth in which no organism had been cultured (compare Figure 3 upper trace with

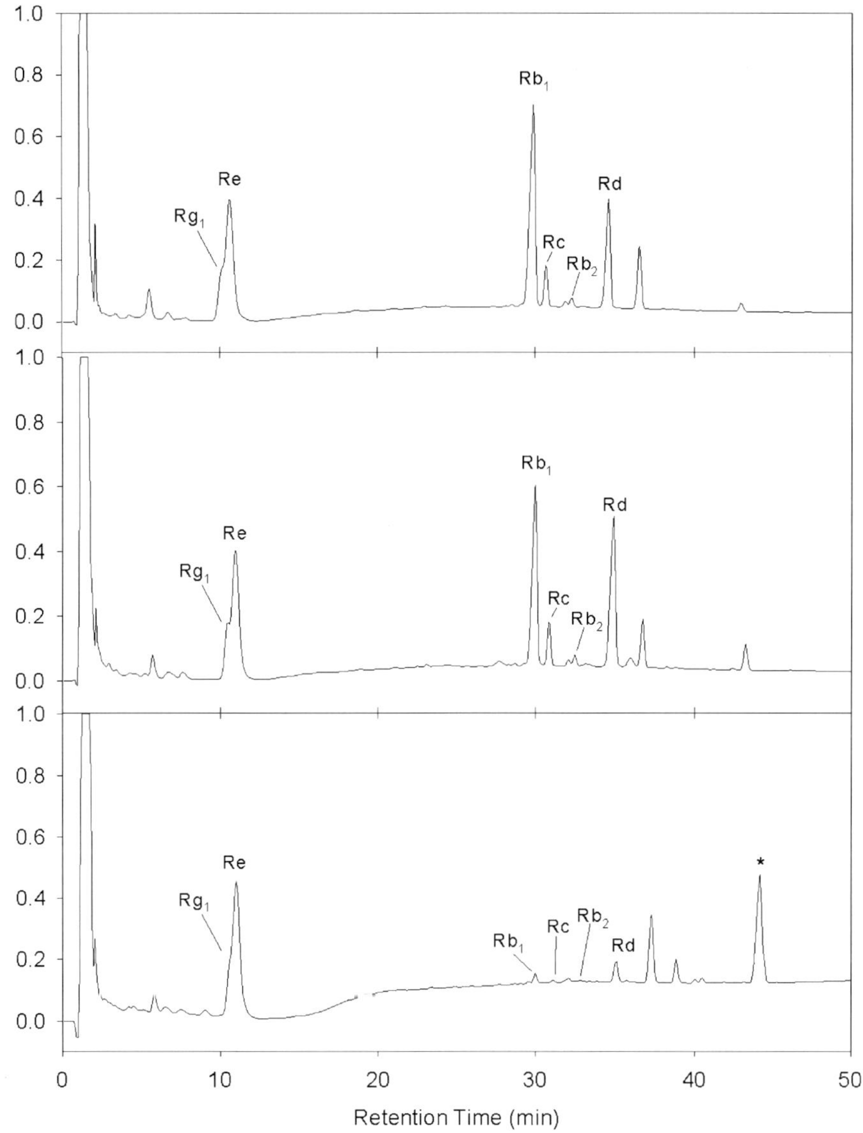

*Figure 3. HPLC Analysis of Ginsenosides. Ginsenosides were isolated from spent broth in which either no organism (upper trace), Trichoderma hamatum (middle trace) or Pythium irregulare (lower trace) had been cultured for five days at 25 °C in the dark. Ginsenosides were chromatographed on a Microsorb-MV C-18 column (150 x 4.6 mm, 5 mm) using an acetonitrile:H_2O gradient (Nicol et al., 2002), and detected at 203 nm. The * in the lower trace indicates the unknown metabolite found in the spent broth of Py. irregulare.*

Figure 3 middle trace). Therefore, the differential response of these two organisms to the presence of ginsenosides in their growth medium is coincident with differences in the profile of ginsenosides that can be recovered from their spent medium.

Interestingly, preliminary observations also show that the recovery of ginsenosides from the spent broth of *Py. irregulare* is dependent on the presence of sucrose in the original culture medium, since significantly smaller amounts of ginsenosides were recovered from a spent broth lacking sucrose (Yousef and Bernards, unpublished data). This observation implies that *Py. irregulare* is using ginsenosides as a source for carbon. However, the additional observation that mineral broth supplemented with a concentration of glucose equimolar to that expected to be released into the medium by ginsenoside de-glycosylation, does not support the same degree of growth increase in *Py. irregulare* biomass observed when the same broth is supplemented with ginsenosides (Yousef and Bernards, unpublished data), argues against a simple carbon source mechanism. Since *Pythium* spp. have been reported to incorporate sterols into their membranes (Olsen, 1973a) and because sterols are known to mediate growth (Nes, 1987), we now hypothesize that *Py. irregulare* secretes saponinases to (partially) deglycosylate ginsenosides, the latter of which are then incorporated as "sterol" triterpenoids into its membrane. It is further assumed that under the experimental conditions employed, this incorporation favours growth.

5. GINSENOSIDES IN THE RHIZOSPHERE

5.1. RETS and Soil Analysis

When the soil associated with three-year-old ginseng roots was extracted and analysed by HPLC, six major ginsenosides were tentatively identified by co-elution with standards. While much of the soil chemical HPLC profile remains unidentified, HPLC-MS analysis confirmed the presence of the six major ginsenosides (Rb_1, Rb_2, Rc, Rd, Re and Rg_1) plus pseudoginsenoside F_{11} and another protopanaxadiol ginsenoside in the soil extracts (Nicol et al., 2003). The amount of ginsenosides as percent weight of dry soil was calculated to range from 0.02% to 0.098% (average 0.06%).

In order to confirm that ginsenosides were present in the exudate of intact ginseng roots, (and not isolated from residual root tissue in our soil preparations) root exudates were collected from pot-grown ginseng plants using a root exudate trapping system, or RETS (Tang and Young, 1982). HPLC analysis of the trapped exudate revealed the presence of peaks that had the same retention times as the ginsenoside standards. These peaks were not present in the exudate collected from control pots (no ginseng plants) and were taken as evidence of ginsenosides in the exudate of pots containing ginseng plants. HPLC-MS analysis of the trapped exudate confirmed the presence of the same suite of ginsenosides as found in the soil (Nicol et al., 2003). After quantification of the ginsenoside content of the root exudates, the individual ginseng plants were determined to be losing approximately 25 μg of ginsenosides per day (i.e., amount of recovered ginsenoside / number of plants / number of days the experiment ran).

5.2. Bioactivity of Ginsenosides at Ecologically Relevant Concentrations

As emphasized in Section 1.3, before ginsenosides can be considered allelochemicals, it has to be demonstrated that they are in fact present in the rhizosphere soil at biologically active concentrations. Therefore, after demonstrating the presence of ginsenosides in the soil, *in vitro* bioassays were conducted using ginsenosides at an ecologically relevant concentration of 0.06%. At this level, ginsenosides were shown to remain bioactive. That is, the growth of *Py. irregulare* was significantly greater than control, while that of *T. hamatum* remained unchanged (Nicol et al., 2003).

5.3. Plant-Fungal Allelopathy in Ginseng Gardens and Implications for Disease

The discovery of ginsenosides in ginseng root exudates and rhizosphere soil at biologically active concentrations suggests that plant-fungal allelopathy could occur within this crop. This in turn could influence disease levels as the ginseng secondary chemicals potentially, and differentially, influence the growth of different groups of soilborne organisms (i.e., pathogens and antagonistic saprotrophs).

Naturally-occurring soilborne antagonists can play a role in preventing or reducing disease levels in crops (Azad et al., 1987; Miller et al., 1989; Larkin et al., 1993). However, the applied use of *Trichoderma* spp. as a biocontrol agent often does not exert the desired level of disease control. One reason is that isolates of these biocontrol fungi often suffer from poor rhizosphere competence (Papavizas, 1982; Chao et al., 1986). Consequently, poor (natural) *Trichoderma* spp. competence in the rhizosphere of ginseng could lead to increased disease levels in the plant. Interestingly, in an attempt to address the issue of the lack of *Trichoderma* spp. rhizosphere competence, it was found that exudates from healthy plant roots did not support the growth of *T. harzianum*, but did in fact support that of *Py. ultimum* (Green et al., 2001). More evidence for the involvement of a soil factor in preventing successful biocontrol by *Trichoderma* spp. is found in the work of Hong et al. (2000) where several *Trichoderma* isolates were shown to inhibit *C. destructans in vitro*, but this biocontrol effect failed to materialize when applied to potted ginseng plants. Again, the specific mechanism involved in both of these reports was not identified, but our results suggest that plant-fungal allelopathy is one possible explanation.

Ginsenoside-mediated stimulation of several major pathogens is likely to be involved in the eventual establishment of ginseng diseases. Three important root pathogens were repeatedly observed to grow better after the addition of ginsenosides to their growth medium (i.e., *C. destructans*, *P. cactorum* and *Py. irregulare*). Therefore, it can be hypothesized that the chemical environment of the ginseng rhizosphere favours fungal root pathogens over potential antagonists of these pathogens and that this has a direct consequence on the year-to-year disease levels in ginseng gardens. We are still working on the exact mechanism of growth stimulation in pathogens, but our preliminary results lead us to believe that in one case (i.e., *Py. irregulare*) the activity is extra-nutritional. For *Py. irregulare*, increased growth may be due to the ginsenosides or transformed ginsenosides acting as sterol analogs. Also, rhizosphere secondary chemicals in general may be serving as signal molecules for plant pathogens in a manner analogous to *Rhizobium*-legume and tree-mycorrhizae symbioses (Peters and Long, 1988; Larange et al., 2001). The ability of rhizosphere phytochemicals to fulfill a molecular need for the pathogen or to induce cellular pathways important for pathogenesis or virulence is a relatively unexplored avenue of plant-microbe allelopathy, even though this has obvious implications for diseases in economically important plants.

Anecdotal evidence suggests that ginseng crops suffer from a replant syndrome and a common farming practice is to avoid planting consecutive ginseng gardens in the same area. Although there is no data on the cause of the replant problem in ginseng, it has been suggested that pathogens could be involved (Reeleder et al., 1999). In fact the ginseng replant problem bears a similarity to the situation in apple orchards. Using our results, a replant syndrome in ginseng gardens could be postulated to have the same basic ingredients (i.e., phytochemicals, pathogens and antagonists) present in the model of the apple replant syndrome. For example, it is well established that the "dominant causal agents" (Mazzola, 1998) of the apple replant syndrome are a complex of soilborne fungal pathogens, especially species in the genera *Cylindrocarpon*, *Fusarium*, *Phytophthora* and *Pythium* (Mazzola 1999; Isutsa and Merwin, 2000) including *C. destructans*, *P. cactorum* and *Py. irregulare* (Braun, 1991; 1995; Mazzola, 1998). Although pathogens play a primary role, antagonists have also been implicated through results obtained from their application as biocontrol microbes (Utkhede et al., 2001) or surveys of orchard soil (Mazzola, 1999). Finally, it has been suggested that an important component of apple replant syndrome is the induction of virulence in microbes by root exudates (Szabo and Wittenmayer, 2000). Similarly, unidentified allelochemicals may also be involved in the asparagus replant syndrome (Peirce and Colby, 1987; Hartung and Stephens, 1983; Pedersen et al., 1991). Therefore, a ginseng replant syndrome, like that of apple and asparagus, could be due to a phytochemical-mediated alteration in the soil microbial population resulting in the stimulation of a complex of pathogens coupled with an inhibition of beneficial microbes. In other words, allelopathy may play a role in year-to-year soilborne crop disease cycles as well as replant syndromes.

6. CONCLUDING REMARKS

Plant disease researchers often focus on causal agents in relative isolation. Our research further demonstrates that other host-derived factors may also play an important role in the etiology and severity of plant diseases and replant syndromes. That is, allelopathy may contribute to plant diseases through the ability of certain microbes to grow and/or survive preferentially in the phytochemical profile of the rhizosphere surrounding specific crop roots. Because pathogenicity is a common nutritional mode in microbes, it is not surprising that pathogens are adapted to the chemical environment of their host plants. Manipulation of rhizosphere chemistry, therefore, either through genetic modification of plants, identification and use of microbe-influencing genotypes or the planting of mixed crops, represents a potential, but as yet underdeveloped, opportunity to reduce disease levels in economically important plants.

7. REFERENCES

Anderson, R.C., Fralish, J.S., Armstrong, J.E., Benjamin, P.K. The ecology and biology of *Panax quinquefolium* L. (Araliaceae) in Illinois. Am Midl Nat 1993; 129:357-372.

Armah, C.N., Mackie, A.R., Roy, C., Price, K., Osbourn, A.E., Bowyer, P., Ladha, S. The membrane-permeabilizing effect of avenacin A-1 involves the reorganization of bilayer cholesterol. Biophys J 1999; 76:281-290.

Arneson, P.A., Durbin, R.D. The sensitivity of fungi to α-tomatine. Phytopathology 1968; 58:536-537

Assa, Y., Gestetner, B., Chet, I., Henis, Y. Fungistatic activity of lucerne saponins and digitonin as related to sterols. Life Sci 1972; 11:637-647.

Azad, H.R., Davis, J.R., Schnathorst, W.C., Kado, C.I. Influence of *Verticillium* wilt resistant and susceptible potato genotypes on populations of antagonistic rhizosphere and rhizoplane bacteria and free nitrogen fixers. Appl Microbiol Biotechnol 1987; 26:99-104.

Bécard, G., Douds, D.D., Pfeffer, P.E. Extensive in vitro hyphal growth of VAM fungi in the presence of CO_2 and flavonols. Appl Environ Microbiol 1992; 58:821-825.

Black, R.L.B., Dix, N.J. Utilization of ferulic acid by microfungi from litter and soil. Trans Br Mycol Soc 1976; 66:313-317.

Bouarab, K., Melton, R., Peart, J., Baulcombe, D., Osbourn, A. A saponin-detoxifying enzyme mediates suppression of plant defences. Nature 2002; 418:889-892.

Bowen, G.D., Rovira, A.D. The rhizosphere. In, *Plant Roots, the Hidden Half.* Waisel, Y., Eshel, A., Kafkafi, U., eds. New York: Marcel Dekker Inc. 1991; pp. 641-670.

Bowyer, P., Clarke, B.R., Lunness, P., Daniels, M.J., Osbourn, A.E. Host range of a plant pathogenic fungus determined by a saponin detoxifying enzyme. Science 1995; 267:371-374.

Brammall, R.A. Alternaria blight. In *Diseases and Pests of Vegetable Crops in Canada.* Howard, R.J, Garland, J.A., Seaman, W.L. ed., The Canadian Phytopathological Society and the Entomological Society of Canada, Ottawa, ON. 1994a; pp. 294-295.

Brammall, R.A. Disappearing root rot. In, *Diseases and Pests of Vegetable Crops in Canada.* Howard, R.J., Garland, J.A., Seaman, W.L., ed., The Canadian Phytopathological Society and the Entomological Society of Canada, Ottawa, ON. 1994b; pp. 296-297.

Braun, P.G. The combination of *Cylindrocarpon lucidum* and *Pythium irregulare* as a possible cause of apple replant disease in Nova Scotia. Can J Plant Pathol 1991; 13:291-297.

Braun, P.G. Effects of *Cylindrocarpon* and *Pythium* species on apple seedlings and potential role in apple replant disease. Can J Plant Pathol 1995; 17:336-341.

Brunner, F., Wirtz, W., Rose, J.K.C., Darvill, A.G., Govers, F., Scheel, D., Nuernberger, T. A beta-glucosidase/xylosidase from the phytopathogenic oomycete, *Phytophthora infestans.* Phytochemistry 2002; 59:689-696.

Burkhardt, H.J., Maizel, J.V., Mitchell, H.K. Avenacin, an antimicrobial substance isolated from *Avena sativa.* II. Structure. Biochemistry 1964; 3:426-431.

Campbell, C.D., Grayston, S.J., Hirst, D.J. Use of rhizosphere carbon sources in sole carbon utilisation tests to discriminate soil microbial communities. J Microbiol Methods 1997; 30:33-41

Capasso, R., Cristinzio, G., Di Maro, A., Ferranti, P., Parente, A. Syringicin, a new alpha-elicitin from an isolate of *Phytophthora syringae*, pathogenic to citrus fruit. Phytochemistry 2001; 58:257-262.

Catling, P.M., Spicer, K.W. Pollen vectors in an American ginseng crop. Econ Bot, 1995; 49:99-102.

Chao, W.L., Nelson, E.B., Harman, G.E., Hoch, H.C. Colonization of the rhizosphere by biological control agents applied to seeds. Phytopathology 1986; 76:60-65.

Chen, S.E., Staba, E.J. American ginseng I. Large scale isolation of ginsenosides from leaves and stems. Lloydia 1978; 41:361-366.

Court, W.A., Reynolds, L.B., Hendel, J.G. Influence of root age on the concentration of ginsenosides of American ginseng (*Panax quinquefolium*). Can J Plant Sci 1996; 76:853-855.

D'Arcy Lameta, A., Jay, M. Study of soybean and lentil root exudates: 3. Influence of soybean isoflavonoids on the growth of rhizobia and some rhizosphere microorganisms. Plant Soil 1987; 101:269-272.

Dewick, P.M. Medicinal natural products: A biosynthetic approach. Chichester: John Wiley and Sons, 1997.

Escalante, A.M., Santecchia, C.B., López, S.N., Gattuso, M.A., Ravelo, A.G., Monache, F.D., Sierra, M.G., Zacchino, S.A. Isolation of antifungal saponins from *Phytolacca tetramera*, an Argentinean species in critic risk. J Ethnopharm 2002; 82:29-34.

Evans, J.L., Gealt, M.A. The sterols of growth and stationary phases of *Aspergillus nidulans* cultures. J Gen Microbiol 1985; 131:279-284.

Favel, A., Steinmetz, M.D., Regli, P., Vidal-Olivier, E., Elias, R., Balansard, G. In vitro antifungal activity of triterpenoid saponins. Planta Med 1994; 60:50-53.

Ford, J.E., McChance, D.J., Drysdale, R.B. The detoxification of α-tomatine by *Fusarium oxysporum* f. sp. *lycopersici*. Phytochemistry 1977; 16:544-546.

Fountain, M.S. Vegetation associated with natural populations of ginseng *Panax quinquefolium* in Arkansas USA. Castanea 1986; 51:42-48.

Fuzzati, N., Gabetta, B., Jayakar, K., Pace, R. Peterlongo, F. Liquid chromatography-electrospray mass spectrometric identification of ginsenosides in *Panax ginseng* roots. J Chromatography A, 1999; 854:69-79.

Grayer, R.J., Harborne, J.B. A survey of antifungal compounds from higher plants, 1982-1993. Phytochemistry 1994; 37:19-42.

Grayston, S.J., Griffith, G.S., Mawdsley, J.L., Campbell, C.D., Bardgett, R.D. Accounting for variability in soil microbial communities of temperate upland grassland ecosystems. Soil Biol Biochem 2001; 33:533-551.

Green, H. Heiberg, N., Lejbolle, K., Jensen, D.F. The use of a GUS transformant of *Trichoderma harzianum*, strain T3a, to study metabolic activity in the spermosphere and rhizosphere related to biocontrol of Pythium damping-off and root rot. Eur J Plant Pathol 2001; 107:349-359.

Griffiths, K.M., Bacic, A., Howlett, B.J. Sterol composition of mycelia of the plant pathogenic ascomycete *Leptosphaeria maculans*. Phytochemistry 2003; 62:147-153.

Haralampidis, K., Trojanowska, M., Osbourn, A.E. Biosynthesis of triterpenoid saponins in plants. In *Advances in Biochemical Engineering/Biotechnology*. Scheper Th. ed., Berlin: Springer-Verlag, 2002; pp.31-49.

Hartung, A.C., Stephens, C.T. Effects of allelopathic substances produced by asparagus on incidence and severity of asparagus decline due to *Fusarium* crown rot. J Chem Ecol 1983; 9:1163-1174.

Helal, H.M., Sauerbeck, D.R. Influence of plant roots on C and P metabolism in soil. Plant Soil 1984; 76: 175-182.

Henderson, M.E.K., Farmer, V.C. Utilization by soil fungi of p-hydroxybenzaldehyde, ferulic acid, syringaldehyde and vanillin. J Gen Microbiol 1955; 12:37-46.

Hong, S.C., Gray, A.B., Asiedu, S.K., Ju, H.Y. The evaluation of *Trichoderma* isolates, benomyl, and propiconazole against *Cylindrocarpon destructans*. Can J Plant Sci 2000; 80:231.

Isutsa, D.K., Merwin, I.A. *Malus* germplasm varies in resistance or tolerance to apple replant disease in a mixture of New York orchard soils. Hortscience 2000; 35:262-268.

Johnson, N.C., Tillman, D., Wedin, D. Plant and soil controls on mycorrhizal fungal communities. Ecology 1992; 73:2034-2042.

Jung, J.D., Park, H.W., Hahn, Y., Hur, C.G, In, D.S., Chung, H.J., Liu, J.R., Choi, D.W. Discovery of genes for ginsenoside biosynthesis by analysis of ginseng expressed sequence tags. Plant Cell Reports 2003; 22:224-230.

Keukens, E.A.J., De Vrije, T., Fabrie, C.H.J.P., Demel, R.A., Jongen, W.M.F., De Kruijff, B. Dual specificity of sterol-mediated glycoalkaloid induced membrane disruption. Biochim Biophys Acta 1992; 1110:127-136.

Keukens, E.A.J., De Vrije, T., Van Den Boom, C., De Waard, P., Plasman, H.H., Thiel, F., Chupin, V., Jongen, W.M.F., De Kruijff, B. Molecular basis of glycoalkaloid induced membrane disruption. Biochim Biophys Acta 1995; 1240:216-228.

Kushiro, T., Shibuya, M. Ebizuka, Y. beta-Amyrin synthase: Cloning of oxidosqualene cyclase that catalyzes the formation of the most popular triterpene among higher plants. Eur J Biochem 1998; 256:238-244.

Lagrange, H., Jay-Allemand, C., Lapeyrie, F. Rutin, the phenolglycoside from *Eucalyptus* root exudates, stimulates *Pisolithus* hyphal growth at picomolar concentrations. New Phytol 2001; 150:349-355.

Larkin, R.P., Hopkins, D.L., Martin, F.N. Effect of successive watermelon plantings on *Fusarium oxysporum* and other microorganisms in soils suppressive and conducive to Fusarium wilt of watermelon. Phytopathology 1993; 83:1097-1105.

Latijnhouwers, M., de Wit, P.J.G.M., Govers, F. Oomycetes and fungi: similar weaponry to attack plants. Trends Microbiol, 2003; 11:462-469.

Lawrence, G.H.M. Taxonomy of Vascular Plants. Macmillan Press, New York, 1951.

Levy, M., Zehavi, U., Naim, M., Itzhack, P. An improved procedure for the isolation of medicagenic acid 3-O-ß-D-glucopyranoside from alfalfa roots and its antifungal activity on plant pathogens. J Agric Food Chem 1986; 34:960-963.

Levy, M., Zehavi, U., Naim, M., Polacheck, I. Isolation, structure determination, synthesis, and anti-fungal activity of a new native alfalfa-root saponin. Carbohydr Res 1989; 193:115-123.

Lewis, W.H., Zenger, V.E. Population dynamics of the American ginseng *Panax quinquefolium* (Araliaceae). Am J Bot 1982; 69:1483-1490.

Li, T.S.C., Mazza, G., Cottrell, A.C., Gao, L. Ginsenosides in roots and leaves of American ginseng. J Agric Food Chem 1996; 44:717-720.

Maizel, J.V., Burkhardt, H.J., Mitchell, H.K. Avenacin, an antimicrobial substance isolated from *Avena sativa*. I. Isolation and antimicrobial activity. Biochemistry 1964; 3:424-426.

Marston, A., Gafner, F., Dossaji, S.F., Hostettmann, K. Fungicidal and molluscicidal saponins from *Dolichos kilimandscharicus*. Phytochemistry 1988; 27:1325-1326.

Mazzola, M. Elucidation of the microbial complex having a causal role in the development of apple replant disease in Washington. Phytopathology 1998; 88:930-938.

Mazzola, M. Transformation of soil microbial community structure and *Rhizoctonia*-suppressive potential in response to apple roots. Phytopathology 1999 89:920-927.

Miethling, R., Wieland, G., Backhaus, H., Tebbe, C.C. Variation of microbial rhizosphere communities in response to crop species, soil origin, and inoculation with *Sinorhizobium meliloti* L33. Microbial Ecology 2000; 40:43-56.

Mikes, V., Milat, M.L., Ponchet, M., Ricci, P., Blein, J.P. The fungal elicitor cryptogein is a sterol carrier protein. FEBS Lett 1997; 416:190-192.

Miller, H.J., Henken, G., van Veen, J.A. Variation and composition of bacterial populations in the rhizosphere of maize, wheat and grass cultivars. Can J Microbiol 1989; 35:656-660.

Morrissey, J.P., Osbourn, A.E. Fungal resistance to plant antibiotics as a mechanism of pathogenesis. Microbiol. Mol Biol Rev 1999; 63:708-724.

Narasimhan, K., Basheer, C., Bajic, V.B., Swarup, S. Enhancement of plant-microbe interactions using a rhizosphere metabolomics-driven approach and its application in the removal of polychlorinated biphenyls. Plant Physiol 2003; 132:146-153.

Nes, D.W. Biosynthesis and requirement for sterols in the growth and reproduction of Oomycetes. In *Ecology and Metabolism of Plant Lipids*. Fuller, G., Nes, W.D. ed., American Chemical Society, Washington, DC. 1987; pp. 304-328.

Nicol, R.W., Yousef, L., Traquair, J.A., Bernards, M.A. Ginsenosides stimulate the growth of soilborne pathogens of American ginseng. Phytochemistry 2003; 64:257-264.

Nicol, R.W., Traquair, J.A., Bernards, M.A. Ginsenosides as host resistance factors in American Ginseng *(Panax quinquefolius)*. Can J Bot 2002; 80:557-562.

Ohtani, K., Mavi, S., Hostettmann, K. Molluscicidal and antifungal triterpenoid saponins from *Rapanea melanophloeos* leaves. Phytochemistry 1993; 33:83-86.

Olsen, R.A. Triterpene glycosides as inhibitors of fungal growth and metabolism. 1. effect on growth, endogenous respiration and leakage of UV-absorbing material from various fungi. Physiol Plant 1971; 24:534-543.

Olsen, R.A. Triterpene glycosides as inhibitors of fungal growth and metabolism. 5. Role of sterol contents of some fungi. Physiol Plant 1973a ; 28:507-515.

Olsen, R.A. Triterpene glycosides as inhibitors of fungal growth and metabolism. 6. The effect of aescin on fungi with reduced sterol contents. Physiol Plant 1973b ; 29:145-149.

OMAF. Ontario ginseng exports, 1996-2002. Queen's Printer for Ontario, 2003.

Osbourn, A.E., Wubben, J.P., Daniels, M.J. Saponin detoxification by phytopathogenic fungi. In *Plant-Microbe Interactions* Volume 2. Stacey, G., Keen, N.T. ed., New York: Chapman and Hall, 1995a, pp. 99-124.

Osbourn, A. Bowyer, P., Lunness, P., Clarke, B., Daniels, M. Fungal pathogens of oat roots and tomato leaves employ closely related enzymes to detoxify different host plant saponins. Mol Plant Microb Interact 1995b ; 8:971-978.

Osbourn, A.E. Preformed antimicrobial compounds and plant defense against fungal attack. Plant Cell 1996; 8:1821-1831.

Osbourn, A.E. Saponins in cereals. Phytochemistry 2003; 62:1-4.

Ouf, S.A., Hady, F.K., Abdel, E.M.H., Shaker, K.H. Isolation of antifungal compounds from some *Zygophyllum* species and their bioassay against two soil-borne plant pathogens. Folia Microbiol 1994; 39:215-221.

Panabières, F., Ponchet, M., Allasia, V., Cardin, L., Ricci, P. Characterization of border species among Pythiaceae: several *Pythium* isolates produce elicitins, typical proteins from *Phytophthora* spp. Mycol Res 1997; 101:1459-1468.

Papadopoulou, K., Melton, R.E., Leggett, M. Daniels, M.J., Osbourn, A.E. Compromised disease resistance in saponin-deficient plants. Proc Natl Acad Sci USA 1999; 96:12923-12928.

Papavizas, G.C. Survival of Trichoderma harzianum in soil and in pea and bean rhizosphere. Phytopathology 1982; 71:121-125.

Pedersen, C.T., Safir, G.R., Siqueira, J.O., Parent, S. Effect of phenolic compounds on asparagus mycorrhizae. Soil Biol Biochem 1991; 23:491-494.

Pedras, M.S.C., Khan, A.Q. Biotransformation of the phytoalexin camalexin by the phytopathogen *Rhizoctonia solani*. Phytochemistry 2000a ; 53:59-69.

Pedras, M.S.C., Okanga, F.I., Zaharia, I.L., Khan, A.Q. Phytoalexins from crucifers: synthesis, biosynthesis, and biotransformation. Phytochemistry 2000b; 53:161-176.

Pegg, G.F., Woodward, S. Synthesis and metabolism of alpha tomatine in tomato *Lycopersicon esculentum* isolines in relation to resistance to *Verticillium albo-atrum*. Physiol Mol Plant Pathol 1986; 28:187-202.

Peirce, L.C., Colby, L.W. Interaction of asparagus root filtrate with *Fusarium oxysporum* f.sp. *asparagi*. J Am Soc Hort Sci 1987; 112:35-40.

Peters, N.K., Long, S.R. Alfalfa root exudates and compounds which promote or inhibit induction of *Rhizobium meliloti* nodulation genes. Plant Physiol 1988; 88:396-400.

Proctor J.T.A., Bailey, W.G. Ginseng, industry botany and culture. Hort Rev 1987; 9:187-236.

Punja, Z.K. Fungal pathogens of American ginseng (*Panax quinquefolium*) in British Columbia. Can J Plant Pathol 1997; 19:301-306.

Quidde, T., Osbourn, A.E., Tudzynski, P. Detoxification of alpha-tomatine by Botrytis cinerea. Physiol Mol Plant Pathol 1998; 52:151-165.

Rahouti, M. Seigle-Murandi, F., Steiman, R., Eriksson, K.E. Metabolism of ferulic acid by *Paecilomyces variotii* and *Pestalotia palmarum*. Appl Env Microbiol 1989; 55:2391-2398.

Reeleder, R.D., Brammall, R.A. Pathogenicity of *Pythium* species, *Cylindrocarpon destructans*, and *Rhizoctonia solani* to ginseng seedlings in Ontario. Can J Plant Pathol 1995; 16:311-316.

Reeleder, R.D., Roy, R., Capell, B. Seed and root rots of ginseng (*Panax quinquefolius*) caused by *Cylindrocarpon destructans* and *Fusarium* spp. Phytopathology 1999; 89:S65.

Rice, E.L. Allelopathy. Academic Press Inc., New York, 1984.

Roddick, J.G. Complex formation between solanaceous steroidal glycoalkaloids and free sterols in vitro. Phytochemistry 1979; 18:1467-1470.

Roldán-Arjona, T., Pérez-Espinosa, A., Ruiz-Rubio, M. Tomatinase from *Fusarium oxysporum* f. sp. *lycopersici* defines a new class of saponinases. Mol Plant Microb Interact 1999; 12:852-861.

Sandrock, R.W., Van Etten, H.D. Fungal sensitivity to and enzymatic degradation of the phytoanticipin a-tomatine. Phytopathology 1998; 88:137-143.

Schlessman, M.A. Floral biology of American ginseng (*Panax quinquefolium*). Bull Torrey Bot Club 1985; 112:129-133.

Schmidt, S.K., Ley, R.E. Microbial competition and soil structure limit the expression of allelochemicals in nature. In *Principles and Practices in Plant Ecology: Allelochemical Interactions,* Inderjit, K.M.M., Dakshini, Foy, C.L. ed. Boca Raton: CRC Press, 1999, pp. 339-351.

Scow, K.M. Soil microbial communities and carbon flow in agroecosystems. In *Ecology in Agriculture*. Jackson, L.E. ed. Academic Press, N.Y. 1997, pp. 367-413.

Shepherd, T. The microchemistry of plant/microorganism interactions. In *Ecology of Plant Pathogens*. Blakeman, J.P., Williamson, B. ed.,Wallingford: CAB International, 1994, pp. 39-62.

Shin, H.S., Lee, M.W. Environmental factors and the distribution of soil microorganisms in ginseng field. Kor J Microbiol 1986; 24:184-193.

Singh, H.P., Batish, D.R., Kohli, R.K. Autotoxicity: concepts, organisms, and ecological significance. Crit Rev Plant Sci 1999; 18:757-772.

Small, E., Catling, P.M. Canadian medicinal crops. NRC Research Press, Ottawa, ON, 1999.

Sparling, G.P., Ord, B.G., Vaughan, D. Microbial biomass and activity in soils amended with glucose. Soil Biol Biochem 1981; 13:99-104.

Steel, C.C., Drysdale, R.B. Electrolyte leakage from plant and fungal tissues and disruption of liposome membranes by a-tomatine. Phytochemistry 1988; 27:1025-1030.

Straney, D.C., Van Etten, H.D. Characterization of the PDA1 promoter of *Nectria haematococca* and identification of a region that binds a pisatin-responsive DNA binding factor. Mol Plant Microb Interact 1994; 7:256-266.

Suleman, P., Tohamy, A.M., Saleh, A.A., Madkour, M.A., Straney, D.C. Variation in sensitivity to tomatine and rishitin among isolates of *Fusarium oxysporum* f. sp. *lycopersici*, and strains not pathogenic on tomato. Physiol Mol Plant Pathol 1996; 48:131-144.

Szabo, K., Wittenmayer, L. Plant specific root exudations as possible cause for specific replant diseases in Rosaceen. J Appl Bot 2000; 74:191-197.

Takechi, M., Tanaka, Y. Structure-activity relationships of the saponin α-hederin. Phytochemistry 1990; 29:451-452.

Tang, C.S., Young, C.S. Collection and identification of allelopathic compounds from the undisturbed root system of bigalta limpograss (*Hemarthria altissima*). Plant Physiol 1982; 69:155-160.

Tani, T., Kubo, M., Katsuki, T., Higashino, M., Hayashi, T., Arichi, S. Histochemistry II. Ginsenosides in ginseng (*Panax ginseng* root). J Nat Prod 1981; 44:401-407.

Turner, E.M.C. The nature of the resistance of oats to the take-all fungus. II. Inhibition of growth and respiration of *Ophiobolus graminis* Sacc. and other fungi by a constituent of oat sap. J Exp Bot 1956; 7:80-92.

Utkhede, R.S., Sholberg, P.L., Smirle, M.J. Effects of chemical and biological treatments on growth and yield of apple trees planted in *Phytophthora cactorum* infested soil. Can J Plant Pathol 2001; 23:163-167.

Van Etten, H.D., Mansfield, J.W., Bailey, J.A., Farmer, E.E. Two classes of plant antibiotics: phytoalexins versus "phytoanticipins". Plant Cell 1994; 6:1191-1192.

Van Etten, H.D., Sandrock, R.W., Wasmann, C.C., Soby, S.D., McCluskey, K., Wang, P. Detoxification of phytoanticipins and phytoalexins by phytopathogenic fungi. Can J Bot 1995; 73:518-525.

Wardle, D.A. A comparative assessment of factors which influence microbial biomass carbon and nitrogen levels in soil. Biol Rev 1992; 67:321-358.

Weete, J.D. Structure and function of sterols in fungi. Adv Lipid Res 1989; 23:115-167.

Weltring, K.M., Wessels, J., Geyer, R. Metabolism of the potato saponins α-chaconine and α-solanine by Gibberella pulicaris. Phytochemistry 1997; 46:1005-1009.

Zimmer, D.E., Pedersen, M.W., McGuire, D.F. A bioassay for alfalfa saponins using the fungus *Trichoderma viride* Pers. ex Fr. Crop Sci 1967; 7:223-224.

HIROYUKI NISHIMURA and ATSUSHI SATOH

ANTIMICROBIAL AND NEMATICIDAL SUBSTANCES FROM THE ROOT OF CHICORY (*Cichorium intybus*)

Department of Biosciences and Technology, School of Engineering, Hokkaido Tokai University, Sapporo 005-8601, Japan
Email: nishimura@db.htokai.ac.jp

Abstract. Antimicrobial sesquiterpenoids, 8a-angeloyloxycichoralexin and guaianolides such as cichoralexin and 10a-hydroxycichopumilide were isolated and identified from the root of chicory (*Cichorium intybus*). These sesquiterpenoids exhibited antifungal activities against *Pyricularia oryzae, Pellicularia sasaki* and *Alternaria kikuchiana*. Ether soluble phenolics from the chicory root were found to exhibit nematicidal activity. The addition of dry chicory root powder to noodles, a boiled fish paste, and cocoa- and coffeecakes, during food processing, provided some protection from food spoilage organisms, and this product may have value as a natural food preservative.

1. INTRODUCTION

Green plants produce numerous secondary compounds, that are not involved in primary metabolism (Kaur et al., 2000). The essential role that secondary metabolites, such as terpenoid and phenolic compounds, play in complex interactions among living organisms in the natural environment is gradually being determined. Although some of these natural products have been found to be pollinator or feeding attractants, many of them seem to function as chemical weapons against insects, pathogenic organisms, and competing plants (Whittaker and Feeny, 1971; Seister, 1977; Berenhaum, 1995; Inderjit and Duke, 2003). Such secondary metabolites may take part in plant-animal, plant-microorganism and plant-plant interactions, and are generally termed allelochemicals.

Chicory (*Cichorium intybus*) is cultivated in cool regions such as Northern Europe. Recently, this vegetable has arisen out of claims that it is able to promote "good health" since no pesticides are used to cultivate chicory in the field, while the plant remains noticeably free from herbivore and microbial attack. The bitter substances, lactupicrin, 8-deoxylactucin and some phenolics had previously been shown to possess insect antifeedant properties in chicory (Rees and Harborne, 1985). Specifically, scsquiterpenoid lactones from chicory leaves, such as 8-deoxylactucin and lactupicrin (Figure 1), were identified as insect antifeedants against desert locust, *Schistocerca gregaria*. Similarly, we found some biologically active secondary metabolites in the

Inderjit and K.G. Mukerji (eds.),
Allelochemicals: Biological Control of Plant Pathogens and Diseases, 177–180.

chicory roots and reported their activity against microorganisms and small animals living in soil (Nishimura et al., 2000). In addition, chicory dry-powder has a potential application in preserving processed foods.

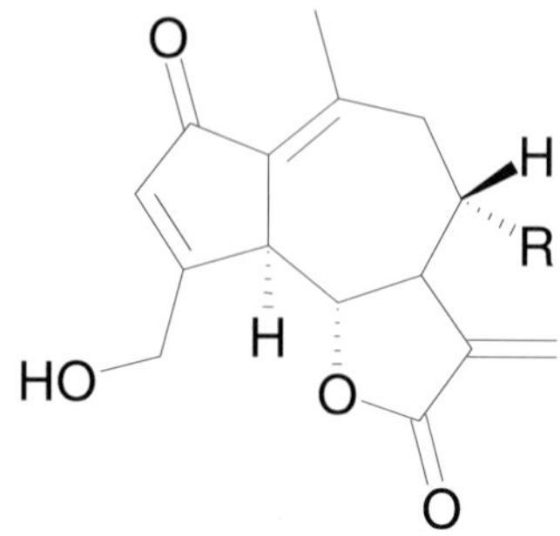

a : R = H, 8-deoxylactucin
b : R = $OCOCH_2C_6H_4OH$(p), lactupicrin

Figure 1. Insect antifeedants in chicory leaves. Source: Nishimura et al. (2000). Reproduced after permission from Kluwer Academic Publisher (Springer).

2. ANTIPATHOGENIC SUBSTANCES

2.1. Antimicrobial Substances in Chicory Root

White shoots germinated from cultivated chicory roots have been used for French and Italian dishes for a long time. We noticed that fungi were not observed on the chicory root in spite of being placed in moist conditions favorable for the development of rots. We found that chicory root extracts, recovered with hexane or ether, exhibited antifungal activities against *Cladosporium herbarum, Pyricularia oryzae, Pellicularia sasaki* and *Alternaria kikuchiana.* The ether extract was fractionated by SiO_2 column chromatography. Rechromatography in a SiO_2 column and HPLC with a $CHCl_3$-MeOH solvent system revealed three active compounds. Two of these were sesquiterpenoids (C-1 and C-2) identified as cichoralexin (Monde et al., 1990) and 10α-hydroxycichopumilide, respectively, by the interpretation of spectral data. From the interpretation of its spectral data, the third compound (C-3) was identified as 8α-angeloyloxy-cichoralexin (Figure 2) (Nishimura et al., 2000). These three sesquiterpenoids, cichoralexin, 10α-hydroxycichopumilide and 8α-angeloyloxycichoralexin from the chicory root were found to have significant antimicrobial properties.

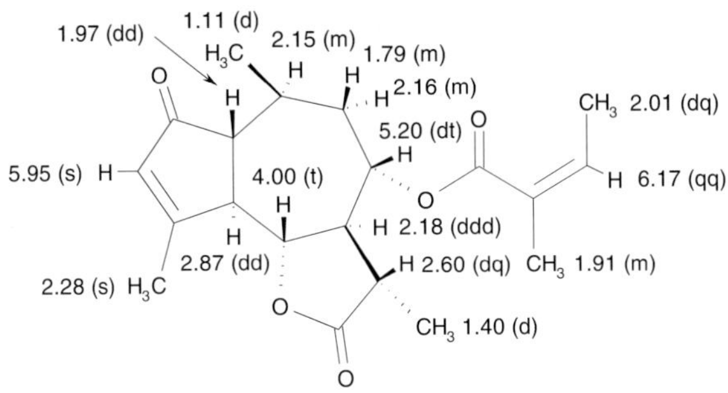

Figure 2. Chemical structure of 8α-angeloyloxycichoralexin. Source: Nishimura et al. (2000). Reproduced after permission from Kluwer Academic Publisher (Springer).

2.2. Nematicidal Substances in Chicory Root

We found that the ether and ethyl acetate extracts of chicory roots exhibited the best nematicidal activities. The extracts were separated according to their acidity to give organo-acidic, phenolic, basic and neutral fractions. The phenolic fraction was found to have the highest activity.

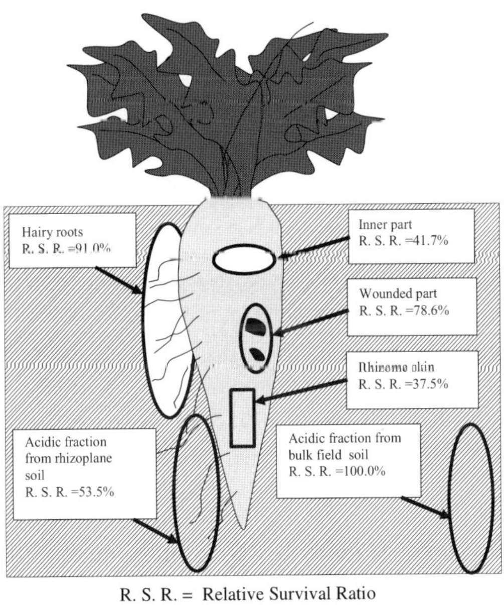

Figure 3. The relative survival ratios (%) of the golden nematode (Globodera rostochiensis) following exposure to chicory plant extracts recovered from the skin, inner part, wounded part and hairy roots of chicory and from the acidic fraction of neighboring soil.
Source: Nishimura et al. (2000). Reproduced with permission from Kluwer Academic Publisher (Springer).

The nematicidal activities of extracts from different parts of a chicory root were examined. The relative survival ratios (R. S. R.) at the root skin, inner root tissue, wounded root tissue and hairy roots were measured (Figure 3). The extract from the root skin exhibited the highest nematicidal activity. Interestingly, the nematicidal activity of an acidic fraction from rhizoplane soil was much higher than that from bulk field soil. We found that some phenolics from the root also exhibited nematicidal activity. Thus, it seems that secondary metabolites such as terpenoids and phenolics can play an important role in chicory defense.

3. UTILIZATION OF ALLELOCHEMICALS IN PROCESSED FOODS

Traditional Japanese foods - noodles, a boiled fish paste, and cocoa- and coffee cakes - were cooked without chicory dry powder. After five days, food spoilage fungi such as *Mucor* and *Rhizopus* species were observed growing on the test foods. Interestingly, when 0.1 – 5% (w/w) of the dried chicory powder was added to the processed foods, no fungi were observed even after ten days storage. It seems that antimicrobial sesquiterpenoids in chicory root can inhibit the growth of spoilage fungi.

4. REFERENCES

Berenhaum M.R. Turnabout is fair play: secondary roles for primary compounds. J Chem Ecol 1995; 21: 925-940.

Inderjit, Duke S.O. Ecophysiological aspects of allelopathy. Planta 2003; 217:529-539.

Kaur H., Inderjit, Keating K.I. Do allelochemicals operate independent of substratum factors? In: *Chemical ecology of plants: allelopathy in aquatic and terrestrial ecosystems*. Inderjit, Mallik AU eds. Birkhauser-Verlag AG, Basal, Switzerland, 2002; pp 99-107.

Monde, K., Oya, T., Shirata, A., Takasugi, M. A guaianolide phytoalexin, cichoralexin, from *Cichorium intybus*., Phytochemistry 1990; 29:3449-3451.

Nishimura, H., Kondo, Y., Nagasaka, T., Satoh, A. Allelochemicals in chicory and utilization in processed foods, J Chem Ecol 2000; 26:2233-2241.

Rees, S.B., Harborne, J.B. The role of sesquiterpene lactones and phenolics in the chemical defense of the chicory plant. Phytochemistry 1985; 24:2225-2231.

Seigler D.S. Primary roles for secondary compounds. Biochem Syst Ecol 1977; 5:195-199

Whittaker R.H., Feeny P.P. Allelochemics: chemical interactions between species. Science 1971; 171:757-770.

REN-SEN ZENG

DISEASE RESISTANCE IN PLANTS THROUGH MYCORRHIZAL FUNGI INDUCED ALLELOCHEMICALS

Chemical Ecology Lab., Institute of Tropical & Subtropical Ecology, Agricultural College, South China Agricultural University Wushan, Tianhe District, Guangzhou, 510642 P.R. China
Email: rszeng@scau.edu.cn; zengrs8@hotmail.com

Abstract. Allelochemals induced in mycorrhizal plants play an important role in disease resistance. Mycorrhizal associations are the most important symbiosis systems in terrestrial ecosystems and offer many benefits to the host plant. Arbuscular mycorrhizal associations can reduce damage caused by soil and root - borne pathogens.

1. INTRODUCTION

As plants are non-motile orgnaisms and lack the sophisticated immune system that animals have, they had to develop their own defence system against pathogens and predators, and systems to lure motile creatures for fertilization and dissemination. Plants are capable of synthesizing an overwhelming variety of low-molecular-weight organic compounds, called secondary methbolities, many with very unique and complex structures. More than 100,000 different secondary metabolites have been found from plants (Buckingham, 1998; Dixon, 2001), with many more yet to be discovered; estimated of the total number in plants exceed 500,000 (Mendelsohn and Balick, 1995). Many of these compounds are differentially distributed among limited taxonomic groups within the plant kingdom and, conversely, each plant species has a distinct profile of secondary metabolites.

Many of these secondary products are bactericidal, fungicidal, repellent, or even poisonous to herbivores. Plant secondary metabolites have for thousands of years played an important role in medicine, crop disease and weed control. Nowadays, they are used either directly or after chemical modification. Examples of these compounds include flavonoids, phenols and phenolic glycosides, unsaturated lactones, sulphur compounds, saponins, cyanogenic glycosides and glucosinolates (Go'mez Garibay et al., 1990; Bennett and Wallsgrove, 1994; Grayer and Harborne, 1994).

Plant discases cause serious losses in crop production. Pesticide application is currently the primary way to control crop disease, but it has raised an array of environmental problems. Achieving sustainable agriculture will require avoiding a

Inderjit and K.G. Mukerji (eds.),
Allelochemicals: Biological Control of Plant Pathogens and Diseases, 181–192.

heavy reliance on pesticides. Enhancing crop resistance against diseases and herbivores is an ideal approach to reduce pesticide application.

Plants possess both constitutive and inducible chemical defense mechanisms. Before pathogen infection, plants may contain significant amounts of constitutive secondary metabolites including phenolics, terpenoids, and steroids, which are toxic to invading organisms (Levin, 1976; Mauch-Mani and Metraux, 1998). Plants may also activate their production of certain defensive chemicals after pathogen infection. These inducible defense compounds are usually only produced and accumulated after specific recognition of the invading organism, and are known as phytoalexins (Dixon, 1986; Hammerschmidt, 1999). Plants can acquire induced resistance to pathogens after infection with necrotrophic attackers, nonpathogenic root-colonizing pseudomonads, salicylic acid, beta-aminobutyric acid and many other natural or synthetic compounds (Conrath et al., 2002; Benhamou, 1996).

Mycorrhizal fungi provide an effective alternative method of disease control especially for those pathogens which effect the below ground plant parts. In mycorrhizal fungi lies enormous potential for use as biological control agent for soil-borne diseases as the root diseases are of the most difficult to manage and lead to losses in disturbing proportions.

The mycorrhizal symbiosis substantially influence plant growth under a variety of stressful conditions and their role in biological control of soil/root - borne pathogens is of immense importance both in the agricultural system as well as in the forestry (Linderman, 1994).

Mycorrhizal associations increase growth and yield of many crop plants by enhancing nutrient uptake, resistance to drought and salinity and increases tolerance to pathogens (Gianinazzi-Pearson, 1996; Mukerji, 1999; Singh et al., 2000; Ludwig-Müller, 2004). Of the seven types of mycorrhiza known (Srivastava et al., 1996; Mukerji et al., 1997; Raina et al., 2000; Redecker et al., 2000), ectomycorrhiza and vesicular-arbuscular mycorrhiza (VAM) are more important in agriculture and forestry. Ecotmyocrrhizal associations are more prevalent in temperate and sub-temperate regions, while VAM/AM associations are common features of tropical and substropical regions of the world. During colonization, distinct structures are formed by the arbuscular mycorrhizal (AM) fungi, with in the host roots - internal hyphae, arbuscules and vesicles (Walker, 1992). The complex cellular relationship between host roots and AM fungi requires a continuous exchange of signals, for proper development of mycorrhiza in the roots of a host plant (Gianinazzi-Pearson, 1996). Plant hormones may be a suitable candidate for the regulation of such a symbiosis. There is little information about the function of plant harmones during the colonization process although there is evidence that they are involved in signaling events between AM fungi and host plants (Barker and Tagu, 2000; Ludwig - Müller, 2000a,b). In addition, it has been suggested that phytohormones, such as IAA and cytokinins, released by mycorrhizal fungi may also contribute to the enhancement of plant growth. (Frankenberger Jr. and Arsad, 1995).

This review describes the role of mycorrhizal associations in disease control.

2. MYCORRIZA IN DISEASE RESISTANCE

Safir (1968) found that inoculation of onion with *Glomus mosseae* could significantly reduce pink root disease due to *Pyrenochaeta terrestris*. Later studies indicate that AM fungi can induce resistance or increase tolerance to some root-borne pathogens (Azcon - Aguilar and Barea, 1996, 1997; Caron et al., 1986; Caron, 1989; Cordier et al., 1997; Dehne, 1982; Hooker et al., 1994; Trotta et al., 1996). *Glomus mosseae* protected peanut plants from infection by pod rot fungal pathogens *Fusarium solani* and *Rhizoctonia solani* (Abdalla and Abdel-Fateh, 2000).

The *Glomus intraradices* increased P uptake and reduced disease development of *Aphanomyces euteiches* in pea roots (Bodker et al., 1998) Mycorrhization with *Glomus mosseae* and *G. intraradices* induced local or systemic resistance to *Phytophthora parasitica* in tomato roots (Cordier et al., 1996, 1998; Pozo et al., 2002). Decreased pathogen development in mycorrhizal and non-mycorrhizal parts of inoculated roots is associated with accumulation of phenolics and plant cell defense response. The protective effects induced by AM fungi against a phytoplasma is reported in tomato (Lingua et al., 2002).

AM protects an annual grass from root pathogenic fungi in the field (Newsham et al., 1995). Inoculation of onion with *Glomus* sp. Zac-19 delayed onion white rot epidemics caused by *Sclerotium cepivorum* Berk by two weeks and increased the yield by 22% under field conditions (Torres-Barragan et al., 1996). *Diospyros lotus* inoculated with *Glomus mosseae, Glomus intraradices, Glomus versiforme* significantly increased the plant growth and decreased the disease caused by *Cercospora kaki* under field conditions. The AM fungal inoculum even suppressed postharvest disease of potato dry rot *(Fusarium sambucinum)* in prenuclear minitubers (Niemira et al., 1996).

Root rot caused by *Fusarium solani* significantly contributes to crop yield decline, up to 50%. The inoculation of common bean *(Phaseolus vulgaris)* with *Glomus mosseae*, besides decreasing propagule number of *F. solani* in the rhizosphere, decreased root rot by 34 to 77% (Dar et al., 1997). In the presence of the root nodulating symbiont *Rhizobium leguminosarum,* mycorrhizal inoculated plants were more tolerant to the fungal root pathogen. This indicates that interactions between mycorrhizal fungi and other rhizosphere microbes might have greater effects on soil-borne pathogens than mycorrhizal fungi alone. Davis and Menge (1980) found that *Glomus fasciculatum* reduced *Phytophthora* root rot of citrus at low level of soil phosphorus but had no effect in high phosphorus soil. The VAM fungi has also been employed as biocontrol agents for *Macrophomina* root rot of cowpea and *Fusarium* wilt of tomato (Ramaraj et al., 1988). The understanding of the mechanisms of plant disease resistance in mycorrhizal plants would provide better directions towards the development of efficient crop production and sustainable agriculture.

3. MECHANISM OF DISEASE CONTROL BY MYCORRHIZAL FUNGI

The mycorrhizal symbiosis involves several mechanisms in control of plant diseases. (i) Creating a mechanical barrier for the pathogen penetration and subsequent spread as in the case of sheathing mycorrhiza. In ectomycorrhizal associations forming highly interwoven mycelial network cover on the root (fungal mantle) and internally with cortical cells whose cell walls are surrounded by fungal hyphae i.e., hartigs' net (Ingham, 1991; Maronek, 1981).

(ii) Thickening of cell wall through lignifications and production of other polysaccharides which in turn hinder the entry of root pathogen (Dehne and Schonbeck, 1979a,b).

(iii) Stimulating the host roots to produce and accumulate sufficient concentrations of metabolities (terpenes, phenols etc.) which impart resistance to the host tissue against pathogen invasion (Krupa et al., 1973; Sampangi, 1989).

(iv) Stimulating flavonolic wall infusions as in the case of *Laccaria bicolor* which prevented lesion formation by the pathogen *Fusarium oxgsporum* in roots of Douglas fir (Strobel and Sinclair, 1991).

(v) Increasing the concentraton of orthodihydroxy phenols in roots which deter the activity of pathogens (Krishna and Bagyarj, 1984).

(vi) Producing antifungal and antibacterial antibiotics and toxins that act against pathogenic organisms (Marx, 1972).

(vii) Competing with the pathogens for the uptake of essential nutrients in the rhizosphere and the root surface (Reid, 1990).

(viii) Stimulating the microbial activity and competitions in the root zone (rhizosphere, rhizoplane) and thus preventing the pathogen to get access to the roots (Rambelli, 1973; Singh and Mukerji, 2005). Roots colonized by VAM/AM fungi may also harbour more actinomycetes antagonistic to root pathogens (Secilia and Bagyaraj, 1987; Dixon, 2000; Bansal et al., 2000; Singh et al., 2000; Mukerji, 2002).

(ix) Compensating the nutrient absorption system from damage to roots by pathogens (Simth and Reid, 1997).

(x) Changing the amount and type of plant root exudates. Pathogens dependent on certain exudates will be at a disadvantage as the exudates change (Matsubara et al., 1995; Newsham et al., 1995).

4. PLANT DEFERENCE REACTIONS

There is considerable evidence for the role of VAM/AM fungi in control of root pathogens (St-Arnaud et al., 1995, 1997; Mukerji, 1999). In AM associations some plant resistance marker molecules are formed in the tissue which interfere with the pathogens to attack (Gianinazzi-Pearson et al., 1996). These are phytoalexins, callose, peroxidase, chitinase, β-1,3 glucanase, PR-1 protein. Plants which constitutively over-express defence-related genes provide interesting material for determing how AM fungi contend with plant defence, or their modifications may occur in the expression of other genes in such plants is still unclear.

4.1. Mycorrhiza induced secondary Metabolites

In AM plants, accumulation of secondary metabolites has been reviewed (Azcon-Aguilar and Barea, 1996; Morandi, 1996; Vierheilig et al., 1998; Mukerji, 1999).

Phytoalexins are produced *de novo* in plants in response to infection or attempted invasion by microbial pathogens or treatment with abiotic agents (Harborne and Ingham, 1978). They are low molecular weight, anti-microbial compounds that are both synthesized by and accumlated in plants after contact (exposure) with microbial pathogens (Paxton, 1981). Enhancement of disease resistance in mycorrhizal plants is due to accumulation of soluble antimicrobial phytoalexins. Rishitin and solavetivone are well-known phytoalexins produced by potato plants in response to pathogen infection (Engström et al., 1999). Mycorrhizal colonization of potato by *Glomus etunicatum* stimulated the accumulation of the phytoalexins rishitin and solavetivone in the roots of plantlets challenged with *Rhizoctonia solani* but did not change phytoalexin levels in non-challenged plantlet roots (Yao et al., 2002, 2003). The phytoalexin concentrations in mycorrhizal roots were significantly higher than those roots only challenged with *Rhizoctonia solani.* The result shows that mycorrhization can amplify phytoalexin production. Accumulation of phenolpropanoid compounds such as phytoalexins, and hydroxyproline-rich glycoproteins, which play important roles in plant defense, have been shown to increase either temporarily during mycorrhiza formation (Lambais, 2000; Garcia- Garrido and Ocampo, 2002). The major phytoalexin in alfalfa *(Medicago truncatula)* is the isoflavonoid (-)-medicarpin. Mycorrhizal colonization resulted in medicarpin accumulation between 7 and 13 days (Harrison and Dixon, 1993). The isoflavonoids glyceollin and coumestrol which are fungitoxic and nematostatic, respectively, accumulated in the mycorrhizal roots (Morandi and Le Querre, 1991). Isoflavonoids also has been showed to accumulate in mycorrhizal soybean roots (Morandi, 1984).

Consistent increases in formononetin levels and transient increases in medicarpin-3-0-glycoside and formononetin conjugates were found in inoculated alfafa roots when mycorrrizal colonization began (Volpin et al., 1995). It is interesting that concentrations of formononetin increases when an AM fungus was in the rhizosphere, even when the plants growing there were not yet colonized by the fungus (Volpin and Kapulnik, 1994). Formononetin has not been shown to have any antimicrobial activities, but is a precursor of isoflavonoid phytoalexin produced in alfalfa in response to microbial infections

Mycorrhizal roots of many plants develop so-called 'yellow pigment' , which has been used as an indicator to estimate the degree of mycorrhizal colonization (Fyson and Oaks, 1992). The chromophore of the 'yellow pigment' is an acyclic C_{14} carotenoid-derived polyene called mycorradicin (Klingner et al., 1995). Other components of the highly complex mixture of apocarotenoids accumulating in mycorrhizal roots are glycosylated C_{13} cyclohexenone derivatives (Peipp et al., 1997). Mycorrhizal roots of barley, wheat, rye and oat were found to accumulate the blumenin-a terpenoid glycoside (Maier et al., 1995). The level of the compound was directly related to the degree of root colonization. Upon colonization by AM fungi, roots of many plant families accumulate certain apocarotenoids (Fester et al., 1999, 2002; Maier et al., 1995; 1998, 1999, 2000; Walter et al., 2000). Colonization of barley, wheat and maize roots by different arbuscular mycorrhizal fungi, i.e.

Glomus intraradices, Glomus mosseae, and *Gigaspora rosea* leads to the accumulation of similar cyclohexenone derivatives (Vierheilig et al., 2000). However, no fungus-specific induction of these compounds are known. Pathogens and endophyte did not induce the formation of cyclohexenone derivatives in barley roots (Maier et al., 1997). The role of cyclohexenone derivatives in disease resistance is unknown.

In response to pathogen attack, plants activate an array of inducible defense reactions, many of which involve the transcriptional activation of the corresponding defense genes, including genes that encode enzymes involved in the synthesis of lignin and phytoalexins (Dixon et al., 1984; Dixon and Harrison 1990). Transcript levels of some pivotal enzymes of defense response of plants significantly increase after mycorrhizal fungal infection (Harrison and Dixon, 1993; 1994). Several inducible defence-related genes, including those encoding isoflavonoid phytoalexins such as phenylalanine ammonia lyase (PAL), chalcone synthase (CHS), chalcone isomerase (CHI) and for the cell wall structural protein HRGP, have been reported to be induced during mycorrhiz establishment (Tagu and Martin, 1996). Mycorrhization resulted in a local and systemic induction of plant defence-related enzymes chitinase, chitosanase and beta-1,3-glucanase, as well as superoxide dismutase in tomato plants (Pozo et al., 2002). Chalcone isomerase (CHI) and chitinase activities increased in inoculated roots prior to mycorrhizal colonization (Volpin et al., 1994, Kapulink et al., 1996), whereas the increase in PAL activity coincided with colonization. Production of some new compounds, and increase in the activity of the enzymes peroxidase and polyphenol oxidase, was observed following inoculation with AM fungi (Charitha Devi and Reddy, 2002). Dumas-Gaudot et al. (1992a,b) found new chitinase isoforms that were specifically induced in several AM associations and were different from those elicited by root fungal pathogens, indicating a different pattern of plant response to pathogenic and mutualistic fungi. *Glomus mosseae* induced new chitinase isoforms in tomato roots (Pozo et al., 1996). Expression of genes encoding enzymes that synthesize phenolpropanoid compounds has been detected in mycorrhizal roots (Garcia-Garrido and Ocampo, 2002). Other defense related genes shown to be upregulated in mycorrhizal symbioses include: genes involved in metabolism of reactive oxygen species, chitinase and beta I ,3-glucanase, and genes involved in senescence, including glutathione-S-transferase. Mycorrhiza also induced changes in PR protein expression in tobacco leaves (Shaul et al., 1999).

Colonization of barley, wheat and maize and rice roots by *Glomus intraradices* resulted in strong induction of transcript levels of the pivotal enzymes of methylerythritol phosphate pathway of isoprenoid biosynthes i.e., 1 -deoxy-D-xylulose 5-phosphate synthase (DXS) and 1 -deoxy-D-xylulose 5-phosphate reductoisomerase (DXR) (Walter et al., 2000). At the same time six cyclohexenone derivatives were characterized from mycorrhizal wheat and maize roots. DXS2 transcript levels are low in most tissues but are strongly stimulated in roots upon colonization by mycorrhizal fungi, correlated with accumulation of carotenoids and apocarotenoids (Walter et al., 2002). Some reports show that the AM symbiosis may cause an increase, decrease, or no change in the plant defense reactions (Guenoune et al., 2001; Mohr et al., 1998).

DIMBOA (2,4-dihydroxy-7-methoxy-1 ,4-benzoxazin-3-one) play an important role in the chemical defense of cereals against insects, pathogenic fungi and bacteria (Niemeyer, 1988). Researches with maize indicate that mycorrizal colonization induced accumulation of DIMBOA and increase in transcript levels of *Bx9*. Concentrations of DIMBOA in maize roots inoculated both *G. mosseae* and *Rhizoctonia solani* were significantly higher than those roots inoculated separately with *G. mosseae* or *Rhizoctonia solani.*

Phenolic compounds in plants play a role in disease resistance (Morandi, 1996). Mycorrhizal plants of maize accumulated more phenolic compounds p-hydrocinnamic acid and ferulic acid in roots than non-mycorrhizal plants (Huang, 2003). Mycorrhizal Ri T-DNA transformed carrot roots accumulated more phenolic compounds when challenged by *Fusarium oxysporum* (Benhamou et al., 1994). All cultivars of pea colonized with *G. mosseae* accumulated more phenolic acids, and total phenolic accumulation were closely correlated to disease intensity (Singh et al., 2004).

Many studies show that AM fungi initiate a host defence response which is subsequently suppressed (Lambais and Mehdy, 1993; Volpin et al. 1994, 1995). The decreases were accompanied by differential reductions in the levels of mRNAs encoding for different endochitinase and endoglucanase isoforms. But the activation of specific plant defence reactions by AM fungi could predispose the plant to an early response to attack by a root pathogen (Gianinazzi-Pearson et al., 1994).

Although studies on growth inhibition of fungi by isolated plant compounds suggest a role in plant defense, such *in vitro* tests may not always give an accurate indication of the significance of these compounds in restricting fungal growth in the plant. Despite increasing efforts in research on metabolic changes in mycorrhizal plants, the precise understanding of the mechanisms is poorly understood, and the role of secondary metabolites induced by AM fungi in disease resistance is still obscure.

5. CONCLUSIONS

Mycorrhizal fungi protect plant roots from diseases in several ways : (i) by providing a physical barrier to the invading pathogens. Physical protection is more likely to exclude soil insects and nematode than bacteria or fungi in ectomycorrhizal plants. However some nematodes can penetrate the fungal mantle, (ii) by competing with the pathogen; (iii) by producing allelopathic chemicals like secondary metabolites, antagonistic chemicals like antibiotics, toxins etc., and amount and type of the root exudates.

For effective and persistent disease management and biocontrol the need is to evaluate the mycorrhizal symbionts in the natural system under field conditions. The use of mixed inoculum of mycorrhizal symbionts can be more effective and give better results than use of a single species. Selection of superior indigenous mycorrhizal symbionts may have an adaptive advantage to the soils and environment in which pathogen and host co-occur as compared to non-indigenous mycorrhizal symbionts.

6. REFERENCES

Abdalla M.E., Abdel-Fattah G.M. Influence of the endomycorrhizal fungus *glomus* mosseae on the development of peanut pod rot disease in Egypt. Mycorrhiza 2000; 10: 29-35.

Mzcon-Aguilar C., Barea J.M. Arbuscular mycorrhizas and biological control of soil-borne plant pathogens - an overview of the mechanisms involved. Mycorrhiza 1996; 6: 457-464.

Azcon-Aguilar C., Barea J.M. Applying mycorrhizal biotechnology to horticulture: significance and potentials. Sci Hortic 1997; 68: 1-24.

Bansal, M., Chamola, B.P., Sarwar, N., Mukerji, K.G. Mycorrhizosphere : Interaction between Rhizosphere microflora and VAM fungi. In, *Myocrrhizal Biology*, K.G. Mukerji, B.P. Chamola, Jagjit Singh eds., Kluwer Academic/Plenum Publishers, New York, Dordrecht, London; 2000; pp. 143-152.

Barker, S.J., Tagu, D. The role of auxins and cytokinins in mycorrhizal symbiosis. J. Plant Growth Regul 2000; 19: 144-154.

Benhamou N. Elicitro-induced pant defense pathways. Trends in Plant Science 1996; 1:233-240.

Benhamou N., Fortin J.A., Hamel C., St.-Arnaud M, Shatilla A. Resistance responses of mycorrhizal Ri T-DNA transformed carrot roots to infection by *Fusarium oxysporum* f. sp. *chrysanthemi*. Phytopathology 1994; 84: 958-968.

Bennett R.N., Wallsgrove R.M. Secondary metabolites in plant defense mechanisms. New Phytol 1994; 127: 617-633.

Bodker L., Kjoller R. Rosendah S. Effect of phosophate and the arbuscular mycorrhizal fungus *Glomus intraradices* on disease severity of root rot of peas (*Pisum sativum*) caused by *Aphanomyces euteiches.* Mycorrhiza 1998; 8: 169-174.

Buckingham J. Dictionary of natural products on CD-ROM. Chapman & Hall, London, 1998.

Caron M. Potential use of mycorrhizae in control of soil-borne disease. Can J Pl Pathol 1989; 11: 177-179.

Caron M., Richard C., Fortin, J.A. Effect of prteinfestation of the soil by a vesicular-arbuscular mycorrhizal fungus, *Glomus intraradices*, on *Fusarium crown* and root rot of tomatoes. Phytoprotection. 1986; 67: 15-19.

Charitha Devi M., Readdy M.N. Phenolic acid metabolish of groundnut (*Arachis hypogaea* L.) plants inoculated with VAM fungus and *Rhizobium*. Pl Gr Regul 2002; 37: 151-156.

Conrath U., Pieterse C.M.J., Mauch Mani B. Priming in plant pathogen interactions. Trends Pl Sci 2002; 7: 210-216.

Cordier C., Gianianzzi S., Gianinazzi-Perason V. Colonization patterns of root tissues by *Phytophthora nicotianae* var. *parasitica* related to reduced disease in mycorrhizal tomato. Plant Soil 1996; 185: 223-232.

Cordier C., Pozo M.J. Barea J.M., Gianinazzi S. Gianinazzi-Pearson V. Cell defense responses associated with localized and systemic resistance to *Phytophthora parasitica* induced in tomato by an arbuscular mycorrhizal fungus. Mol Plant-Microbe Interactions 1998; 11:1017-1028.

Cordier C., Trouvelot A., Gianinazzi S., Gianinazzi-Pearson V. Arbuscular mycorrhizal technology applied to micropropagated *Prunus avium* and to protection against *Phytophthora cinnamomi.* Agronomie 1997; 17: 256-265.

Dar G.H. Zargar M.Y. Beigh G.M. Biocontrol of *Fusarium* root rot in the common bean *(Phaseolus vulgaris* L.) by using symbiotic *Glomus mosseae* and *Rhizobium leguminosarum.* Microb Ecol 1997; 34: 74-80.

Davis R.M., Menge J.A. Influence of *Glomus fasciculatum* and soil phosphorus on phytopththora root rot of citrus. Phytopathology 1980; 70:447-452.

Dehne H.W. Interaction between vesicular arbuscular mycorrhizal fungi and plant pathogens. Phytopathology 1982; 72: 1115-1119.

Dehne, H.W., Schonbeck, F. Unterschungen Zum Einflu-β-der endotrophen Mykorrhiza auf Pflanzenkrankheiten, II. Phenol stoffwechsel and Lignitiezierung. Phytopathol Z 1979b, 19: 210-216.

Dehne, H.W., Schonbeck, F. The infleunce of endotrophic mycorrhizae on plant disease. I colonization to tomato plants by *Fusarium oxysproum* f. sp. *lyeopersici.* Phytopatol Z 1979a; 95: 105-109.

Dixon R.A. The phytoalexin response: elicitation, signaling and control of host gene expression. Biol Rev 1986; 61: 239-291.

Dixon R.A. Natural productions and plant disease resistance. Nature 2001; 411: 843-847.

Dixon R.A. and Gerrish C. Lamb C.J., Robbins M.P. Elicitor-mediated induction of chalone isomerase in *Phaseolus vulgaris* cell suspension cultures. Planta 1984; 159: 561-569.

Dixon R.A. and Harrison M.J. Activation, structure and organization of genes involved in microbial defense in plants. Adv Gentics 1990; 28: 165-233.

Dixon, R.K. The Global Carbon Cycle and Global change : Resposns and fedcdback from the Mycorrhizaosphre. In, *Mycorrhizal Biology*, K.G Mukerji, B.P. Chamola, Jaggit Singh eds. Kluwer Academic/Plenum Publishers, New York, Dordrecht, London. 2000; pp. 85-100.

Dumas-Gaudot E., Furlan V., Grenier J., Asselin A. New acidic chitinase isoforms induced in tobacco roots by vesicular-arbuscular mycorrhizal fungi. Mycorrhiza 1992a; 1: 133-136.

Dumas-Gaudot E., Renier J. Furlan V., Asselin A. Chitinase, chitosanase and b-1, 3-glucanase activites in *Allium* and *Pisum* roots colonized by *Glomus* species. Plant Sci 1992b; 84: 17-24.

Engström K., Widmark A.K., Brishammar S., Helmersoon S., Antifungal activity to *Phytophthora infestans* of sesquiterpendoids from infectd potato tyubers. Potato Res 1999; 42: 43-50.

Fester T., Maier W., Strack D. Accumulation of secondary compounds in barley and wheat roots in response to inoculation with an arbuscular mycorrhizal fungus and co-inoculation with rhizsophsere bacteria. Mycorrhiza 1999; 8: 241-246.

Fester T., Schmidt D., Lohse S., Walter M.H., Giuliano G., Bramley P.M., Fraser P.d., Hause B., Strack D. Stimualtion of carotenoid metabolism in arbuscular mycorrhizal roots. Planta 2002; 216: 148-154.

Frankenberger Jr. W.T., Arshad M. Microbial synthesis of auxins. In *Phytohormones in soils*, W.T. Frankberger Jr., M. Arshad eds., Marcel Deeker Inc., N.Y. 1995, pp. 35-71.

Fyson A., Oaks A. Rapid methods for quantifying VAM fungal infections in maize roots. Plant Soil 1992; 147: 317-319.

Garcia-Garrido J.M., Ocampo J.A. Regulation of the plant defence response in arbuscular mycorrhizal symbiosis. J Exp Bot 2002; 53: 1377-1386.

Gianinazzi-Pearson V. Plant Cell responses to arbuscular mycorrhizal fungi getting to the roots of the symbiosis. Plant cell 1996; 8: 1871-1883.

Giaminazzi-Pearson V., Dumas-Gaudot E., Gollote A., Tahiri-Alaoui A., Gianinazzi S. Cellular and molecular defence related root responses to invasion by arbuscular mycorrhizal fungi. New Phytol 1996; 133: 45-57.

Gianinazzi-Pearson V., Gollotte A. Dumas-Gaudo E. Franken P., Gianinazzi S. Gene expression and molecular modifications associated with plant responses to infection by arbuscular mycorrhizal fungi. In: *Advnaces in molecular genetics of plantmicrobe interactions*. M.J. Daniels, J.A. Downic, A.E. Obsourn. eds. Kluwer, Boston, 1994; pp 179-186.

Go'mez Garibay F., Reyes Chilpa R., Quijano L., Caldero'n Pardo J.S. Ray'os'Castillo T., Methoxifurans auranols with fungostatic activity from *Lonchocarpus castilloi*. Phytochemistry 1990; 29: 159-163.

Grayer R.J. Harborne J.B., A survey of antifungal compounds from higher plants, 1982-1993. Phytochemistry 1994; 37. 19-42.

Guenoune D., Galili S., Phillips D.A., Volpin H. Chet I., Okon Y., Kapulnik Y. The defense response elicted by the pathogen *Rhizoctonia solani* is suppressed by colonization of the AM-fungus *Glomus intraradices*. Pl Sci 2001; 160: 925-932.

Hammerschmidt R. Induced disease resistance: How do induced plants stop pathogens? Physiol Molecular Pl Pathol 1999; 55: 77-84.

Harborne J.B., Ingham J.L. Biochemical aspects of the co-evaluation of higher plants with their fungal parasites. In *Biochemical Aspects of Plant and Animal Co-evaluation*. J.B. Harborne. ed. Academic Press, London, 1978; 343-405.

Harrison M.J., Dixon A. Isoflavonid accumulation and expression of defense gene transcripts during the establishment of vesicular arbuscular mycorrhizal association in roots of *Medicago truncatula*. Mol Plant-Microbe Interactions 1993; 6: 643-659.

Harrison M.H., Dixon R.A. Spatial patterns of expression of flavonoid/isoflavonoid pathway genes during interactions between roots of *Medicago truncatula* and the mycorrhizal fungus *Glomus versiforme*. Plant J 1994; 6: 9-20.

Hooker J.E., Jaizme-Vega M., Atkinson D. Biocontrol of plant pathogens using arbuscular mycorrhizal fungi. In : *Impact of Arbuscular Mycorrhizas on Sustainable Agriculture* and *Natural Ecosystems*. ALS, Birkhauser, Basel 1994; pp. 191-200.

Huang J.H., Chemical mechanisms of plant disease resistance induced by arbuscular mycorrhizal fungi toward fungal pathogens. Doctoral dissertation submitted to South China Agricultural University, Guangzhou, 2003.

Ingham E.R. Interactions among mycorrhizal fungi, rhizosphere orgnaims, and plants. In, *Microbial mediation of Plant-herbivore Interactions.* John Wiley and Sons, Inc. 1991; p. 530.

Kapulink Y., Volpin H., Itzhaki H., Ganon D., Galili S., David R., Shaul O., Elad Y., Chet I., Okon Y. Suppression of defence responses in mycorrhizal alfalfa and tobacco roots. New Phytol 1996; 133: 59-64.

Klinger A., Bothe H., Wray V., Marner F.J. Identification of a yellow pigment formed in maize roots upon mycorrhizal colonization. Photochemistry 1995; 38: 53-55.

Krishna K.R., Bagyaraj D.J. Interaction between *Glomus fasciculatum* and *Sclerotium rolfsii* in peanut. Can J Bot 1984; 61: 2349.

Krupa S., Anderson J., Marx D.H. Studies on ectomycorrhizae of pine. IV. Volatile organic compounds in mycorrhizal and non mycorrhizal root systems of *Pinus echinata* Eur J For Pathol 1973; 3: 194-200.

Lambais M.R., Mehdy M.C. Suppression of endochitinase, β-1, 3-endoglucanase and chalcone isomerase expression in bean vesicular-arbuscular mycorrhizal roots under different soil phosphate conditions. Mol Plant-Microbe Interactions 1993; 6: 75-83.

Lambais M.R. Regulation of plant defense-related genes in arbuscular mycorrhizae. In: *Current advances in mycorrhizae research.* G.K. Podila, D.D. Douds eds. APS Press, Minnesota U.S.A. 2000.

Levin D. Alkaloid-bearing plants: An ecogeographi perspective. Am Natur 1976; 110: 261-284.

Linderman R.G. Role in VAM fungi in biocontrol. In: *Mycorrhiza and Plant Health*, F.L. Pfleger, R.G. Linderman eds. APS Press, St. Paul, 1994; pp. 1-26.

Lingua G. D'Agostino G. Massa N., Antosiano M., Berta G. Mycorrhiza induced differential response to a yellows disease in tomato. Mycorrhiza 2002; 12: 191-198.

Ludwig-Müller, J. Indole-3-butyric acid in plant growth and development. Plant Growth Regul 2000a; 32: 219-230.

Ludwig-Müller J. Hormonal balance in plants during colonization by mycorrhial fungi. In, *Arbuscular Mycorrhizas : Physiology and function,* D.D. Douds, Y. Kapulnik eds. Kluwer Academic Publishers, Dordrecht, the Netherlands. 2000b; pp. 263-285.

Ludwig-Müller J. From auxin homeostasis to understanding plant pathogen and plant symbiont interaction : editor's research interests JPGR 2004; 23: 1-8.

Maier W., Hammer K. Dammann U., Schulz B., Strack D., Accumulation of sesquiterpenoid cyclohexenone dervatives induced by an arbuscular fungus in members of the Poaceae. Planta 1997; 202: 36-42.

Maier W., Peipp H., Schmidt J., Wray V., Strack D. Levels of a terpenoid glycoside (blumenin) and cell wall-bound phenolics in some cereal mycorrhizas. Pl Physiol 1995; 109: 465-470.

Maier W., Schmidt J., Nimtz M., Wray V., Strcak D. Secondary products in mycorrizal roots of tobacco and tomato. Phytochemistry 2000; 54: 473-479.

Maier W., Schmidt J. Wray V., Walter M.H. Strack D. The arbuscular mycorrhizal fungus *Glomus intraradices* induces the accumulation of cyclohexenone derivatives in tobacco roots. Planta 1999; 207: 620-623.

Maier W., Schneier B., Strack D. Biosynthesis of sesquiterpenoid cyclohezenone derivatives in mycorrhizal barley roots proceeds via the glyceraldehydes 3-phosphate/pyruvate pathway. Tetrahedron Letter 1998; 39: 521-524.

Maronek, D.M. Mycorrhizal fungi and their importance in horticulture crop production. Hort Rev 1981; 3: 172-213.

Marx D.H. Ectomyocrrhizae as biological deterents to pathogenic root infections. Ann Rev Phytopath 1972; 10: 429-441.

Matsubara Y., Tamura H., Harada T. Growth enhancement and *Verticillium* wilt control vesicular-arbuscular mycorrhiza fungus inoculation in eggplant. J Japan Soc Hort Sci. 1995; 555-561.

Mauch-Mani B., Metraux J.P. Salicylic acid and systemic acquired resistance to pathogen attack. Ann Bot 1998; 82: 535-561.

Mendelsohn R., Balick M.J. The value of undiscovered pharmaceuticals in tropical forests. Eco Bot 1995; 49: 223-228.

Mohr U., Lange J., Boller T., Wiemken A., Vogeli-Lange R. Plant defence genes are induced in teh pathogenic interaction between bean roots and *Fusarium solani*, but not in the symbiotic interaction with the arbuscular mycorrhizal fungus *Glomus mosseae*. New Phytol 1998; 138: 589-598.

Morandi D. Isoflavonoid accumulation in soybean roots infected with vesicular-arbuscular mycorrhizal fungus. Physiol Pl Pathol 1984; 24: 357-364.

Morandi D. Occurrence of phytoalexins and phenolic compounds in endomyocrrhizal interactions, and their potential role in biological control. Plant Soil. 1996; 185: 241-251.

Morandi D., LeQuerre J.L. Influence of nitrogen on accumulation of isosojagol (a new deterred coumestan in soybean) and associated isoflavonoids in roots and nodules of mycorrhizal and non-mycorrhizal soybean. New Phytol 1991; 117: 75-79.

Mukerji K.G. Mycorrhiza in control of plant pathogens : Molecular approaches. In, *Biotechnological Approaches in Biocontrol of Plant Pathogens*, K.G. Mukerji, B.P. Chamola, R.K. Upadhyay eds. Kluwer Academic/ Plenum Publishers, New York, Dordrecht, London, 1999; pp. 135-155.

Mukerji K.G. Rhizosphere Biology. In, *Techniques in Mycorrhizal Studies,* K.G. Mukerji, C. Manoharachary, B.P. Chamola eds. Kluwer Academic Publishers, Dordrecht, The Netherlands, 2002; pp. 87-101.

Mukerji K.G., Upadhyay R.K., Kaushik A. Mycorrhiza and Integrated Disease Management. In, *IPM Systems in Agriculture. Vol 2. Biocontrol in Emerging Biotechnology.* R.K. Upadhyay, K.G. Mukerji, R.L. Rajak eds. Aditya Books Pvt. Ltd., New Delhi, India, 1997; pp. 423-452.

Neswham K.K., Fitter A.H. Watkinson A.R. Arbuscular mycorhiza protect an annual gras from root pathogenic fungi in the field. J Ecol 1995; 83: 991-1000.

Niemeyer H.M. 1988. Hydroxamic acids (4-hydroxy-1, 4-benzoxazin-3-ones), defense chemicals in the GRamineae. Phytochemistry 1988; 27: 3349-3358.

Niemira B.A., Hammerschmidt R., Sar G.R. Postharvest suppression of potato dry rot (Fusarium sambucinum) in prenuclear minitubers by arbuscular mycorrhizal fungal inoculum. An Potato J 1996; 73: 509-515.

Paxton J.D. Phytoalexins-a working redefinition. Phytopathol 1981; 101: 106-109.

Peipp H., Maier W., Schmidt J. Wray V. Strack D. Arbuscular mycorrhizal fungus-induced changes in the accumulation of secondary compounds in barely roots. Phytochemistry 1997; 44: 581-587.

Pozo M.J., Cordier C., Dumas-Gaudot E., Gianinazzi S., Barea J.M., Azcon-Aguilar C. Localized versus systemic effect of arbuscular mycorrhizal fungi on defence responses to *Phytophthora* infection in tomato plants. J Exp Bot 2002; 525-534.

Pozo M.T., Dumas-Gaudot e., Slezack S., Cordier C., Asselin A., Gianinazzi S., Gianinazzi-Pearson V., Azcon-Aguilar C., Barea J.M. Infuction of new chitinase isoforms in tomato roots during interactions with *Glomus mosseae* and/or *Phytophthora nicotianae* var *parasitica*. Agronomie 1996; 16: 689-697.

Raina S., Chamola B.P., Mukerji K.G. Evolution of Mycorrhiza. In, *Mycorhizal Biology,* K.G. Mukerji, B.P. Chamola, J. Singh eds. Kluwer Academic / Plenum Publishers, New York, pp. 1-26.

Ramaraj B., Shanmugam N., Reddy D.A. Biocontrol of *Macrophomina* root rot of cowpea and *Fusarium* wilt of tomato by using VAM fungi. In: *Mycorrhizae for green Asia*. A Mahadevan ed. Proc. 1st Asian Conf Mycorrhizae, University of Madras, Madras, India 1988; pp. 250-251.

Rambelli A. The rhizosphere of mycorrhizae. In, Ectomycorrhizae, G.L. Marks, T.t. Koslowski eds. Academic Press, New York, 1973; pp. 299-343.

Redecker D., Kodner R., Graham L.E. Glomalean fungi from the Ordovician. Science 2000; 289: 1920-1921.

Reid C.P.P. Mycorrhizas. In, *The Rhizosphere*, J.M. Lynch ed. John Wiley and Sons, Chickester, England; 1990; pp. 281-315.

Safir G. The influence of vesicular mycorrhiza on the resistance of onion of *Pyrenochaeta terrestris.* M.S. Thesis, Univernsty of Illinois, Urbana. 1968.

Sampangi R.K. Some recent advances in the study of fungal root diseases. Ind Phytopth 1989; 22: 1-17.

Secilia J., Bagyaraj D.J. Bacteria and actinomycetes associated with pot cultures of vesicular arbuscular mycorrhiza. Can J Microbol 1987; 33: 1069-1073.

Shaul O., Galili S., Volpin H., Ginzberg I.I., Elad Y., Chet I. I., Kapulnik Y. Mycorrhiza - induced changes in disease severity and PR protein expression in tobacco laves. Mol Plant-Microbe Interact 1999; 12: 1000-1007.

Singh D.P., Srivastava J.S., Bahadur A., Singh U.P., Singh S.K. Arbuscular mycorrhizal fungi induced biochemical changes in pea (*Pisum sativaum*) and their effect on powdery mildew (*Erysiphe pisi*). Phytopathol Z fur Pelanzekrankheiten und Pelanzenschutz 2004; 111: 266-272.

Singh G., Mukerji K.G. Root exudates as determinant of rhizospheric microbial diversity. In, *Microbial activity in the rhizosphere*. K.G. Mukerji, C. Manoharachary, J. Singh eds. Springer Verlag, Berlin, Heidelberg. 2006; 39-55.

Singh R., Adholeya A., Mukerji K.G. Mycorrhiza in control of soiolborne pathogens. In: *Mycorrhizal biology.* Mukerji K.G., Chamola B.P., Singh J. eds. Kluwer Academic / Plenum Publishers, New York, Dordrecht, London, 2000; pp. 173-196.

Smith S.E., Reid D.J. Mycorrhizal Symbiosis. Academic Press, London, New York. 1997; p. 605.

Srivastava D., Kapoor R., Srivastava S.K., Mukerji K.G. Vesicular-arbuscular mycorrhiza - an overview. In, *Concept in mycorrhizal research*, K.G. Mukerji ed. Kluwer Academic Publishers, Dordrecht, The Netherlands, 1996; pp. 1-39.

St-Arnaud M., Hamel C., Carom M, Fortin J.A. Inhibition of *Pythium ultimum* in roots and growth substrate of mycorrhizal *Tagetes patula* colonized with *Glomus intraradices*. Can J Path 1995; 16: 187-194.

St-Arnaud M., Hamel C., Vimard B., Caron M., Fortin J.A. Inhibition of *Fusarium oxysporum* f. sp. *dianthi* in the non-VAM species Dianthus caryophyllus by co-culture with *Tagetes patula* companion plants colonized by *Glomus intraradices*. Can J Bot 1997; 75: 998-1005.

Strobel N.E., Sinclair W.A. Role of flavonilic wall inclusions in the resistance induced by *Laccaria bicolor* to *Fusarium oxysporum* in primary roots of Donglas fir. Phytopathology 1991; 81: 420-425.

Tagu D., Martin F. Molecular anlaysis of cell wall proteins expressed during early steps of ectomycorrhiza development. New Phytol 1996; 133: 73-85.

Torres-Barragan A., Zavaleta-Mejia E., Gonzalez-Chavez C., Ferrera-Cerrato R. The use of arbuscular mycorrhizae to control onion white rot (*Sclerotium cepivorum* Berk.) under field conditions. Mycorrhiza 1996; 6: 253-257.

Trotta A., Varese G.C., UNAVI E., Fusconi A., Sampo S., Berta G. Interactions between the soil-borne root pathogen *Phytophthora nicotianae* var *parasitica* and arbuscular mycorrhizal fungus *Glomus mosseae* in tomato plants. Plant Soil 1996; 185: 199-209.

Vierhilig H., Bago B., Albrecht C., Poulin M.J., Piche Y. Flavonoids and arbuscular mycorrhizal fungi. In : *Falvonoids in the living system*. J. Manthey, B. Buslig eds. Plenum, New York, 1998; pp. 9-33.

Vierheilig H., Gagnon H., Strack D., Maier W. Accumulation of cyclohexenone derivatives in barley, wheat and maize roots in response to inoculation with different arbuscular mycorrhizal fungi. Mycorrhiza 2000; 9: 291-293.

Volpin H., Elkind Y. Okon Y., Kapulnik Y. A vesicular arbuscular mycorrhizal fungus (*Glomus intraradix*) induce a defense response in alfalfa roots. Pl Physiol 1994; 104: 683-689.

Volpin H., Kapulnik Y. Interaction of *Azospirillum* with beneficial soil microorganisms. In: *Azospirillum / Plant Associations*. Okon Y. ed. CRC Press Boca Baton FL 1994; pp. 111-118.

Volpin H., Phillips D.A., Okon Y., Kapulnik Y. Suppression of an isoflavonoid phytoalexin defense response in myocrrhizal alfalfa roots. Plant Physiol 1995; 108: 1449-1545.

Walker C. Systamatics and Taxonomy of the arbuscular endomycorrhizal fungi (Glomales) - a possible way forward. Agronomie 1992; 12: 887-892.

Walter M.H., Festr T. Strack D. Arbuscular mycorrhizal fungi induce the non-mevalonate methylerythritol phosphate pathway of isoprenoid biosynthesis correlated with accumulation of the 'yellow pigment' and other apocarotenoids. Plant J 2000; 21: 571-578.

Walter M.H., Hans J., Strack D. Two distantly related genes encoding 1-deoxy-D-xylulose 5-phosphate synthases: differential regulation in shoots and apocarotenoid-accumulating mycorrhizal roots. Plant J 2002; 31: 243-254.

Yao M.K. Desilets H., Charles M.T., Boulanger R. Tweddell R.J. Effect of mycorrhization on the accumulation of rishitin and solavetivone in potato plantlets challenged with *Rhizoctonia solani*. Mycorrhiza 2003; 13:333-336.

Yao M., Tweddell R., Desilets H. Effect of two vesicular-arbuscular mycorrhizal fungi on the growth of micropropagated potato plantlets and on the extent of disease caused by *Rhizotonia solani*. Mycorrhiza 2002; 12: 235-242.

CHUIHUA KONG

ALLELOCHEMICALS FROM *Ageratum conyzoides* L. AND *Oryza sativa* L. AND THEIR EFFECTS ON RELATED PATHOGENS

Institute of Applied Ecology, Chinese Academy of Sciences, Shenyang 110016, China, and South China Agricultural University, Guangzhou 510642, China.
E-mail: kongch@mail.edu.cn

Abstract. Allelochemicals play an important role in biological control of plant pathogens and diseases. Weed *Ageratum conyzoides* L. and food crop *Oryza sativa* L. can produce and release many kinds of allelochemicals participating in their defense against pathogens. The essential oil from *A. conyzoides* has been found to have significant negative effects on several plant pathogens. In the *A. conyzoides* intercropped citrus orchard, *A. conyzoides* released allelopathic flavones and agreatochromene into the soil to reduce the populations of soil pathogenic fungi *Phytophthora citrophthora*, *Pythium aphanidermatum* and *Fusarium solani*. Further research revealed that ageratochromene underwent a reversible transformation in the soils, that is, ageratochromene released from *A. conyzoides* plants was transformed into its dimers, and the dimers can be remonomerized in the soils. The reversible transformation between ageratochromene and its dimers in the *A. conyzoides* intercropped citrus orchard soil can be an important mechanism maintaining bioactive allelochemicals at an effective concentration, thus, sustaining the inhibition of pathogenic fungi in soil. Many kinds of allelochemicals in rice were identified. Among them, alkylresorcinols, flavone and cyclohexenone had high antifungal activities on *Pyricularia oryzae* and *Rhizoctonia solani*. Furthermore, these antifungal allelochemicals formed by rice can be triggered by a large number of abiotic and biotic factors. Antifungal allelochemicals from rice mainly involved two types of diterpenes and flavones, including momilactones A and B, oryzalexins A-F and S, phytocassanes A-E and sakuranetin. These compounds help rice establishing its own pathogen defense mechanism. However, it remains obscure which allelochemicals in rice are predominantly involved in defense mechanisms against the pathogens. Therefore, further clarification of the resistance mechanism and multiple functions of these compounds on rice are warranted.

1. INTRODUCTION

Plants produce many kinds of low-molecular-mass secondary metabolites that are generally non-essential for the basic metabolic processes of the plant. Among these secondary plant metabolites, some are known as allelochemicals that improved defense against other plant competition, microbial attack or insect/animal predation. Plants cannot move to escape pathogens. However, plants have evolved to successfully withstand infection by a vast majority of pathogens that attack them (Stuiver and Custers, 2001). It was found that plants biosynthesize phytoalexins as soon after pathogenic attack (Dangl and Jones, 2001). The concept of phytoalexins as induced anti-microbial allelochemicals in plant was first developed in 1940 (Muller and Borger,

Inderjit and K.G. Mukerji (eds.),
Allelochemicals: Biological Control of Plant Pathogens and Diseases, 193–206.

1940; Bailey and Mansfield, 1982). Anti-microbial allelochemicals or phytoalexins have been extensively investigated and play important roles in plant defense (Dixon, 2001). Most of anti-microbial allelochemicals have relatively broad-spectrum activity on pathogens and specificity is often occurred in cropping systems. Accordingly, an understanding of the interactions between allelochemicals and pathogenic organisms is essential in disease control in agro-ecosystems.

The structures and sources of many allelochemicals with anti-microbial activity have been documented (Dixon, 2001; Grayer and Harborne, 1994; Harborne, 1999). In this chapter, evidence for anti-microbial functions of allelochemicals from weed *Ageratum conyzoides* L. and food crop *Oryza sativa* L. has been reviewed. Effects of these allelochemicals on related pathogen management in the *A. conyzoides* intercropped citrus orchard and the paddy ecosystem were discussed.

2. ALLELOCHEMICALS OF *A. conyzoides* AND THEIR EFFECT ON RELATED PATHOGENS

2.1. Ageratum conyzoides L.

Ageratum conyzoides of the family Compositae (Asteraceae) is native to Central America (Kossmann and Groth, 1993) Caribbean and Florida (USA). It has spread to West Africa, Southeast Asia, South China, India, Australia and South America (Okunade, 2002; Stadler et al., 1998). *A. conyzoides* is an annual erect, branched herb growing 15 to 100 cm tall. Its stem is covered with fine white hairs, leaves are opposite, pubescent with long petioles and include glandular trichomes. It has a shallow tap root system. The inflorescence contains 30 to 50 pink or purple flowers arranged in a corymb and are self-incompatible (Jhansi and Ramanujam, 1987). The fruit is an achene with an aristate pappus and is easily dispersed by wind and animals fir. Seeds are positively photoblastic and remain viable upto 12 months (Okunade, 2002). The seeds germinate between 20-25°C. It prefers a moist, well drained soil but may tolerate dry conditions (Ladeira et al., 1987). This species has great morphological variations and appears highly adaptable to varying ecological conditions (Hu and Kong, 2002a). It is a pioneer plant growing in ruined sites and cultivated fields and often becomes dominant and forms a stand in natural community and is resistant to common insects or diseases (Liang and Hunag, 1994). Although it is harmful to crops and invades cultivated fields and interferes with the natural community compositions, it has been used as folk medicine in several countries and it has anti-microbial, insecticidal and nematicidal properties (Ming, 1999; Okunade, 2002). In Central America *A. conyzoides* has been bred for many colours of flowers (Stadler et al., 1998). In South China *A. conyzoides* is traditionally used as green manure in fields to increase the crop yields, and usually is intercropped as understory in citrus orchards to suppress weeds and control other pests (Liang and Hunag, 1994; Kong et al., 2004b). This species appears to be a valuable agricultural resource (Ming, 1999).

2.2 The essential oil of A. conyzoides and their biological activities on related pathogens

A. conyzoides has a wide range of secondary metabolites including flavonoids, chromenes, benzofurans and terpenoids. Among these secondary metabolites, some are allelochemicals inhibiting the growth of other organisms (Okunade, 2002; Pari et al., 1998). Usually, *A. conyzoides* can produce and release volatile chemicals into the environment. The concentration of its released volatiles is so high that the unpleasant odor can be smelled in the fields. Therefore, most investigations have focused on chemical components of its essential oil (Albersberg and Singh, 1991; Ekundayo et al., 1988; Menut et al., 1993; Wandji et al., 1996). It was found that ageratochromenes and their derivatives, monoterpenes and sesquiterpenes were the major components of the essential oil from *A. conyzoides* (Kong et al., 1999; 2002a; Pari et al., 1998).

The allelopathic potential of volatile allelochemicals from *A. conyzoides* has been reported in our previous papers (Kong et al., 1999; 2002a; 2004a). Anti-microbial effects of the essential oil from *A. conyzoides* have been confirmed for a long time (Biond et al., 1993; Dixit et al., 1995; Rao et al., 1996). Table 1 showed that several fungal pathogens, such as *Rhizoctonia solani, Botrytis cinerea, Sclerotinia sclerotiorum*, were significantly inhibited by the essential oils of *A. conyzoides* (Kong et al., 2001; 2002a). The quantity and variety of allelochemical produced by *A. conyzoides* varies depending on its growth stages and habitats and so do their growth inhibitory effects on the pathogens (Kong et al., 2002a, 2004a). *A. conyzoides* produces different volatiles in larger quantities when infected with *Erysiphe cichoracearum* (Kong et al., 2002a).

Table 1. Inhibitory effects of essential oil from the A. congzoides collected from different growth stages and habitats on fungal pathogens.

The essential oil collected from	Pathogens		
	R. solani	*B. cinerea*	*S. sclerotiorum*
Growth stages			
4-leaf	23.8±3.8[a]	19.6±5.3[a]	39.8±7.6[a]
Pre-flowering	56.3±10.1[b]	49.7±9.1[b]	60.2±7.9[b]
Peak-flowering	100[c]	82.3±11.5[c]	100[c]
Mature	60.6±8.8[b]	58.3±7.4[b]	40.8±9.5[a]
Habitats			
Cultivated field	79.6±10.2[d]	83.5±6.3[c]	93.8±7.7[c]
Under citrus canopy	58.4±4.3[b]	45.6±6.1[b]	66.2±5.5[b]
Roadside	100[c]	100[d]	100[c]
Control			
50% Carbendazin	100[c]	59.8±5.2[c]	92.3±7.9[c]

Test concentrations of the essential oil were 100 μg/ml. All data are inhibitory percentage of spore germination of fungal pathogens tested and mean of 3 replicates with standard error. Data in a column not followed by the same letter are significantly different, p=0.05, ANVOA with Ducan's multiple range test.

2.3. *Allelochemicals and pathogen management in the A. conyzoides intercropped citrus orchard*

Besides the volatiles, *A. conyzoides* can biosynthesize and release non-volatile allelochemicals into the soil, thus, inhibiting the growth of other plants and microorganisms in soils. Polymethoxyflavones, ageratochromene and its analogues are rare in natural products but they have been found in *A. conyzoides* (Adesogan and Okunade, 1979; Gonzalez et al., 1991; Horie, et al 1993; Okunade, 2002). These compounds have obvious anti-microbial activity and have been used in managed ecosystem.

In South China *A. conyzoides* is often intercropped in citrus orchards as an understory plant that quickly becomes dominant in citrus orchards. In addition, intercropping *A. conyzoides* makes the citrus orchard ecosystem more favorable for predatory mites (*Amblyseius* spp.). These mites are effective natural enemies of the citrus red mite (*Panonychus citri*). Further investigations showed that the pathogenic fungi *Phytophthora citrophthora, Pythium aphanidermatum* and *Fusarium solani* were isolated from both the *A. conyzoides* intercropped and non-intercropped citrus orchards soils. However, populations of these fungi were lower in the *A. conyzoides* intercropped citrus orchard than in the non-intercropped citrus orchard (Figure 1), indicating that intercropping with *A. conyzoides* in citrus orchards markedly decreased the population of soil pathogenic fungi. It may have resulted from the phytotoxins in the *A. conyzoides* intercropped citrus orchard soil (Kong et al., 2004c).

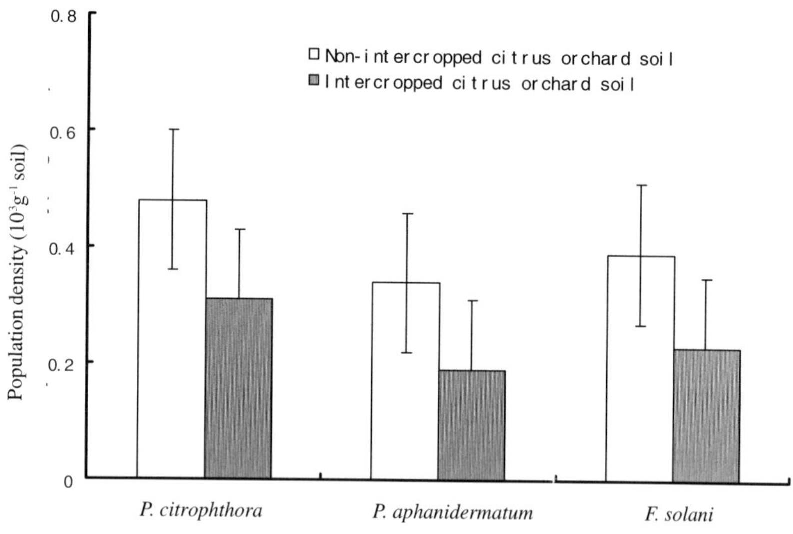

Figure 1. Population of major pathogenic fungi in the A. conyzoides intercropped and non-intercropped citrus orchards soils. Population density was mean individual in per gram soil.

Several flavones, ageratochromene and its two dimers were isolated and identified from the *A. conyzoides* intercropped citrus orchard soil (Figures 2 and 3), their amounts ranged from 11 to 93μg g^{-1} in the air dried soil. However, these chemicals could not be found in the non-intercropped citrus orchard soil. The results showed that these chemicals were primarily released from ground cover plants of *A. conyzoides* and accumulated in the soil year by year.

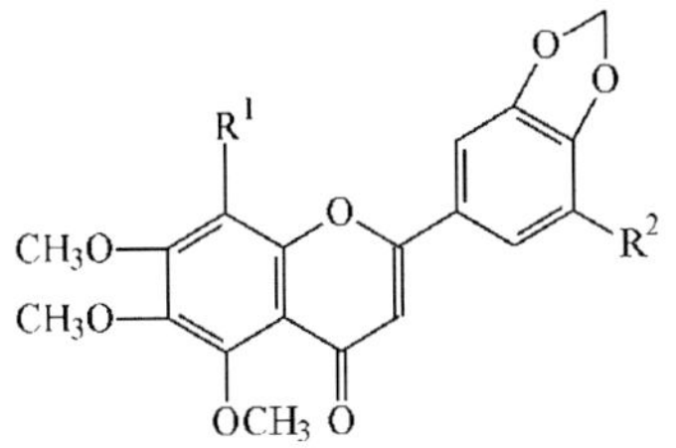

A: $R^1=R^2=OCH_3$; **B**: $R^1=H$, $R^2=OCH_3$; **C**: $R^1=OCH_3$, $R^2=H$

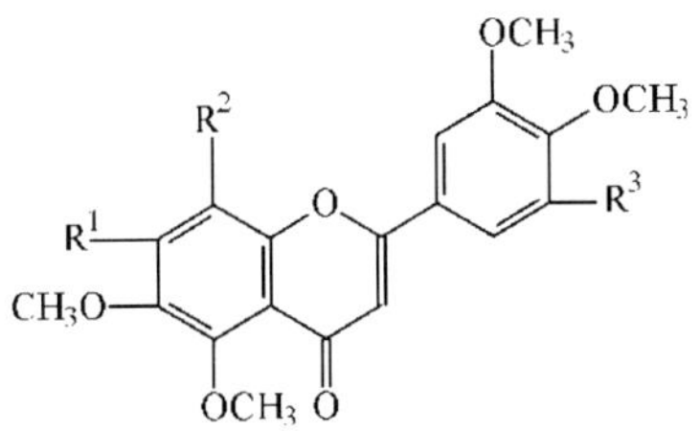

D: $R^1=R^3=OCH_3$, $R^2=H$; **E**: $R^1=R^2=R^3=OCH_3$;
F: $R^1=R^3=H$, $R^2=OCH_3$; **G**: $R^1=H$, $R^2=R^3=OCH_3$;
H: $R^1=R^3=OCH_3$, $R^2=OH$; **I**: $R^1=R^3=OCH_3$, $R^2=\alpha$- rhamnosyl.

Figure 2. Flavones produced and released from the A. conyzoides intercropped citrus orchard soil.

Bioassays showed that ageratochoromene and flavones could significantly inhibit spore germination of the pathogenic fungi *P. citrophthora, P. aphanidermatum* and *F. solani*, but two dimers of ageratochromene had no inhibitory effects on them (Table 2). Thus, the flavones and ageratochromene could be one of the key factors that *A. conyzoides* plants are able to reduce the populations of soil pathogenic fungi in the citrus orchard. Two dimers, though not biologically active, may be the products of ageratochromene transformation in soil.

Further studies revealed that ageratochromene underwent a reversible transformation in the soils, that is, ageratochromene released from ground *A. conyzoides* plants was transformed into its dimers, and the dimers can be remonomerized in the soils (Kong et al., 2004c). However, this dynamic transformation did not occur in the soil with low organic matter and fertility (Figure 3). The reversible transformation between ageratochromene and its dimers in the *A. conyzoides* intercropped citrus orchard soil can be an important mechanism maintaining bioactive

Table 2. Inhibitiory effects of allelochemicals isolated from the A. conyzoides intercropped citrus orchard soil on major pathogenic fungi.

Chemicals	P. citrophthora	P. aphanidermatum	F. solani
Flavone A	100[a]	90.6±8.6[a]	72.1±6.5[a]
Flavone B	90.5±7.6[b]	75.9±8.1[b]	39.9±5.8[b]
Flavone C	91.2±6.3[b]	76.3±6.5[b]	35.6±7.1[b]
Flavone D	100[a]	90.5±8.7[a]	59.6±7.1[c]
Flavone E	100[a]	93.2±9.2[a]	60.2±7.3[c]
Flavone f	87.5±7.9[b]	80.5±7.2[b]	38.9±4.5[b]
Flavone G	77.3±6.1[c]	74.2±6.2[b]	40.2±5.3[b]
Flavone H	98.6±8.0[b]	80.3±6.6[b]	57.7±4.1[c]
Flavone I	39.8±2.1[d]	39.7±7.5[c]	35.3±5.0[b]
Ageratochromene	43.5±6.4[d]	37.6±5.4[c]	60.8±9.9[c]
Dimer A	6.1±1.1[e]	0.9±0.3[d]	1.8±0.6[d]
Dimer B	5.0±0.9[e]	1.6±0.3[d]	7.2±1.6[d]
50% Carbendazin	100[a]	74.8±6.1[b]	58.3±3.2[c]

Test concentrations of the essential oil were 100 μg/ml. All data are inhibitory percentage and mean of 3 replicates with standard error. Data in a column not followed by the same letter are significantly different, p=0.05, ANVOA with Ducan's multiple range test.

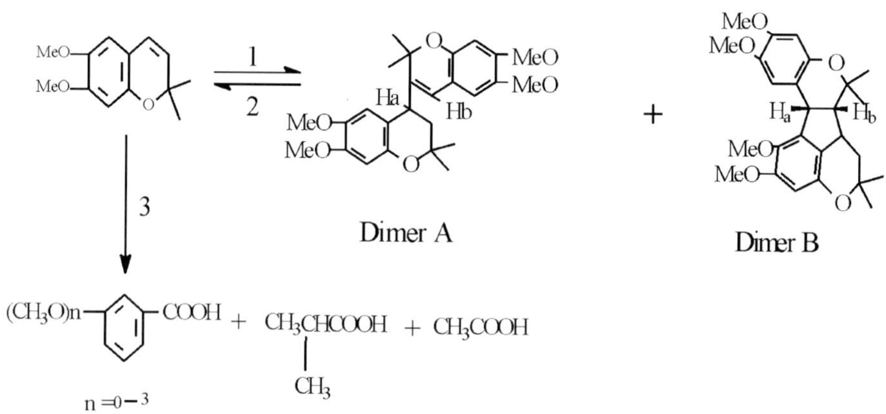

Figure 3. Transformation and degrading of ageratochormene in the soil.
Ageratochormene transformation was in soil with high organic matter content and high fertility. Transformation occurred from step 1 to step 3, only in soil with low organic content and fertility.

allelochemicals at an effective concentration, thus, sustaining the inhibition of pathogenic fungi in soil.

Therefore, *A. conyzoides* has been advocated for intercropping in citrus orchards and utilized on more than 150,000 ha of citrus orchards in South China and gave substantial ecological and economic benefits (Liang and Huang, 1994). It is an excellent example of applied aspects of allelopathy in agro-ecosystem.

3. RICE ALLELOCHEMICALS AND THEIR EFFECTS ON RELATED PATHOGENS

3.1. Rice antifungal allelochemicals

Rice (*Oryza sativa* L.) is one of the principal food crops in the world. Its production is characterized by heavy use of herbicides and fungicides that may cause environmental problems in the paddy ecosystem (Kim and Shin, 2000). Accordingly, rice allelochemicals can potentially be used to improve weed and pathogen management in rice production. Therefore, search for allelochemicals from rice has been extensively studied (Chung et al., 2001; Kong et al., 2002b; Mattice et al., 1998; Rimando et al., 2001). A range of phenolic acids was identified as potent allelochemicals from rice tissues and root exudates (Chung et al., 2001; Mattice et al., 1998; Rimando et al., 2001). However, these phenolic acids are unlikely to explain the allelopathy of rice since their soil concentrations never reach phytotoxic levels (Olofsdotter et al., 2002). More recently, an increasing number of studies have shown that a few flavones, diterpenoids and other types of compounds are also the potent allelochemicals from rice (Kato-Noguchi et al., 2002; 2003; Kong et al., 2002b; 2004d,e; Lee et al., 1999). These allelochemicals can be biosynthesized in rice seedlings and then released into their surroundings at ecologically relevant concentrations to inhibit the germination and growth of associated weeds. Similarly, rice allelochemicals may participate in the defense mechanisms of rice against pathogens.

In our laboratory, a flavone (5,7,4.-trihydroxy-3,5.-dimethoxyflavone), a cyclohexenone (3-isopropyl-5-acetoxycyclohexene-2-one-1) and a liquid mixture of low polarity, containing long-chain and cyclic hydrocarbons (Table 3), were isolated and identified from leaves of allelopathic rice accession PI 312777 (Kong et al., 2004e). Both the flavone and cyclohexenone significantly inhibited spore germination of *Pyricularia oryzae* and *Rhizoctonia solani,* but the mixture containing low polarity constituents did not show any inhibitory effect on them, even at high concentrations (Figure 4a,b). The IC_{50} values of the flavone on spore germination of *P. oryzae* and *R. solani* were ca 50 and 70 . g. g^{-1}, while the cyclohexenone were ca 75 (*P. oryzae*) and 95 (*R. solani*), respectively. At all concentrations tested, the inhibitory activity of the flavone on pathogens was slightly higher than that of the cyclohexenone. The complete inhibition of both compounds on spore germination of the pathogens was observed at 250 . g. g^{-1}. (Figure 5a, b).

a. b.

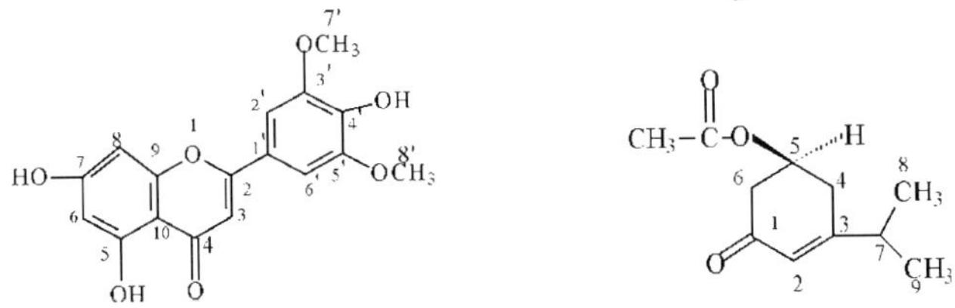

5,7,4¡Ç-trihydroxy-3¡Ç,5¡Ç-dimethoxyflavone 3-isopropyl-5-acetoxycyclohexene-2-one-1

Figure 4. a. Flavone; b. Cyclohexenone.

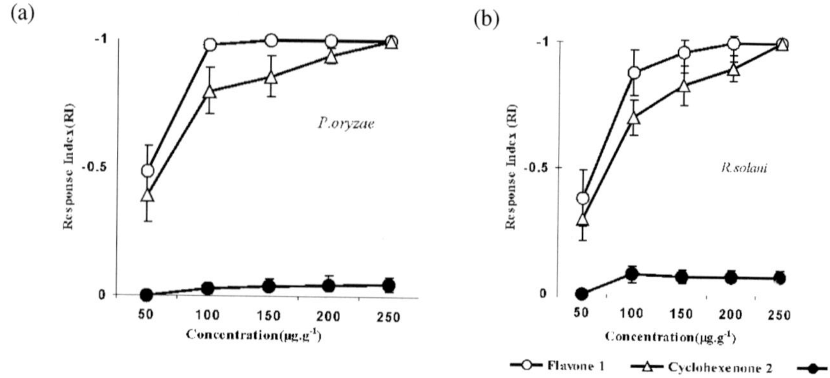

Figure 5. Inhibitory activities of flavone, cyclohexnone and mixture with low polarity against spore germination of (a) P. oryzae and (b) R. solani at different concentrations.

Table 3. Chemical constituents of the mixture with low polarity isolated from rice leaves

Retention time (min)	*Chemical constituents*	*Relative amount (%)*
1.89	2-methylhexane	8.77
5.40	4,6-dimethylundecane	2.48
6.28	2,2,6-trimethyldecane	2.35
7.26	2-hexyl-1-decanol	2.11
7.58	1-butyl-2-propylcyclopentane	8.45
8.19	2,6-dimethyl-decahydronaphthalene	22.45
8.40	*trans, trans*-1,10-dimethylspiro[4,5]decane	8.03
8.48	2,3-dimethyl-decahydronaphthalene	3.37
8.63	*trans, cis*-1,8-dimethylspiro[4,5]decane	8.19
8.78	1,2-dimethyldecahydronaphthalene	11.49
9.10	*cis, cis*-1,1-dimethylspiro[4,5]decane	11.60
10.18	1,4,6-trimethyl-1,2,3,4-tetrahydronapthalene	1.77
14.35	hexadecanoic acid, methyl ester	1.85
-	Other unknown components*	7.09

*A total of 11 components and their relative amounts was less than 1% of the mixture with low polarity

Alkylresorcinols (Figure 6) are another kind of antifungal agent that were isolated and identified from etiolated rice seedlings (Suzuki et al., 1996; 1998). Antifungal activities of alkylresorcinol against spore germination of *P. oryzae* were correlated with their concentration and structure, but complete inhibition of their spore germination occurred only at a concentration of over 100.g.ml^{-1}.

$R=-(CH_2)_{12}CH_3$, $-(CH_2)_{14}CH_3$, $-(CH_2)_{16}CH_3$, $-(CH_2)_7CH=CH(CH_2)_5CH_3$, $-(CH_2)_7CH=CH(CH_2)_7CH_3$, $-(CH_2)_7CH=CHCH_2CH=CH(CH_2)_4CH_3$

Figure 6. Alkylresorcinols.

3.2. Diterpene and flavone phytoalexins from rice

Allelochemicals play an important role in rice disease resistance (Bailey and Mansfield, 1982). Antifungal allelochemicala, phytoalexins produced by rice in response to injury, physiological stimuli or in the presence of infectious agents (Hammerschmidt, 1999). In particular, phytoalexins can be induced and accumulated by rice after fungal infection. Phytoalexins from rice mainly involve two types of diterpenes and flavones, including momilactones A and B, oryzalexins A-F and S, phytocassanes A-E and sakuranetin (Figures 7, 8).

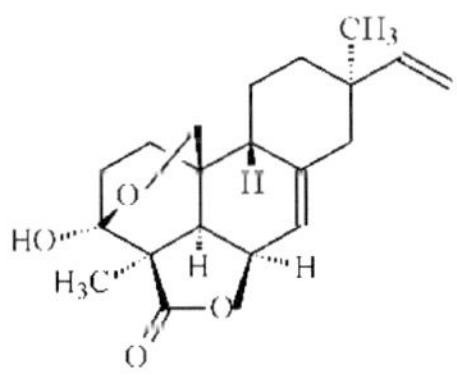

(5)

Momilactone A

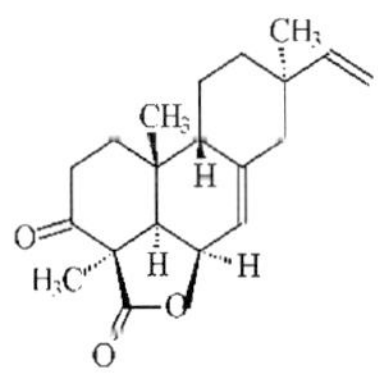

(6)

Momilactone B

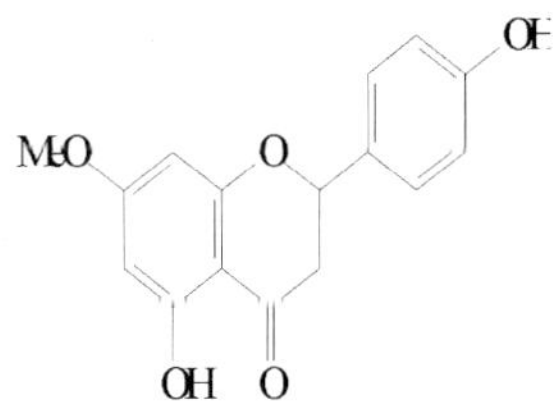

Sakuranetin

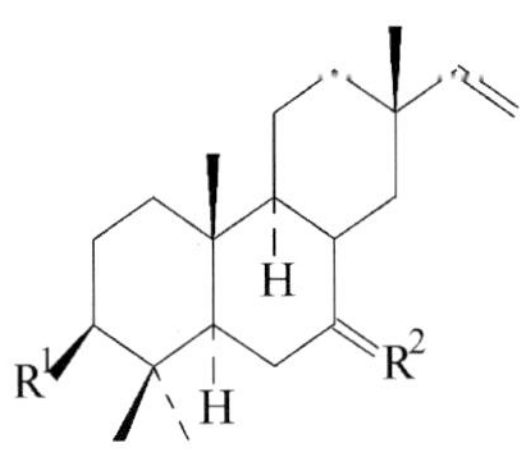

Oryzalexin A: $R^1=OH$, $R^2=CH_3$
Oryzalexin B: $R^1=O$, $R^2=OH$
Oryzalexin C: $R^1=O$, $R^2=O$
Oryzalexin D: $R^1=OH$, $R^2=O$

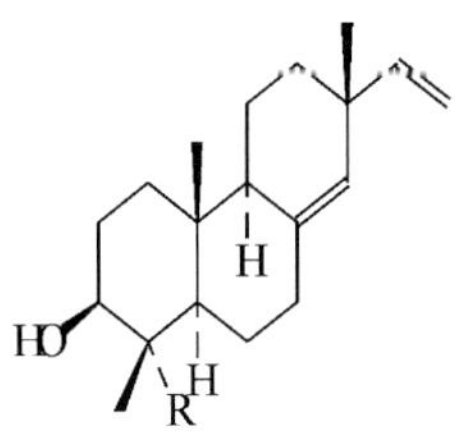

Oryzalexin E: $R=CH_3$
Oryzalexin F: $R=CH_2OH$

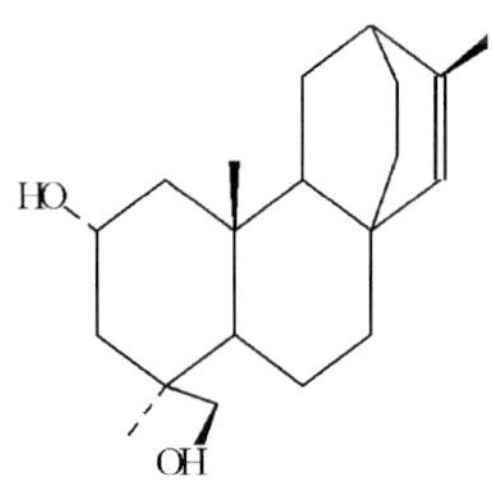

Oryzalexin S

Figure 7. Typical diterpene and flavone phytoalexins from rice.

Momilactones A and B are the first phytoalexins characterized from any member of the Gramineae. They were isolated initially as plant inhibitors from rice husk (Kato et al., 1977), and were subsequently found to be produced by rice in response to either infection by the blast disease that caused by the pathogen *Pyricularia oryzae* or irradiation with UV light (Cartwright et al., 1977; 1981). They had significant phytoalexin-like activity in *P. oryzae* and *Helminthosporium oryzae.* More recently, momilactones A and B have been found in rice root exudates, which have been found to participate in defense against weeds (Kato-Noguchi et al., 2002; 2003; Kong et al., 2004d; Lee et al., 1999).

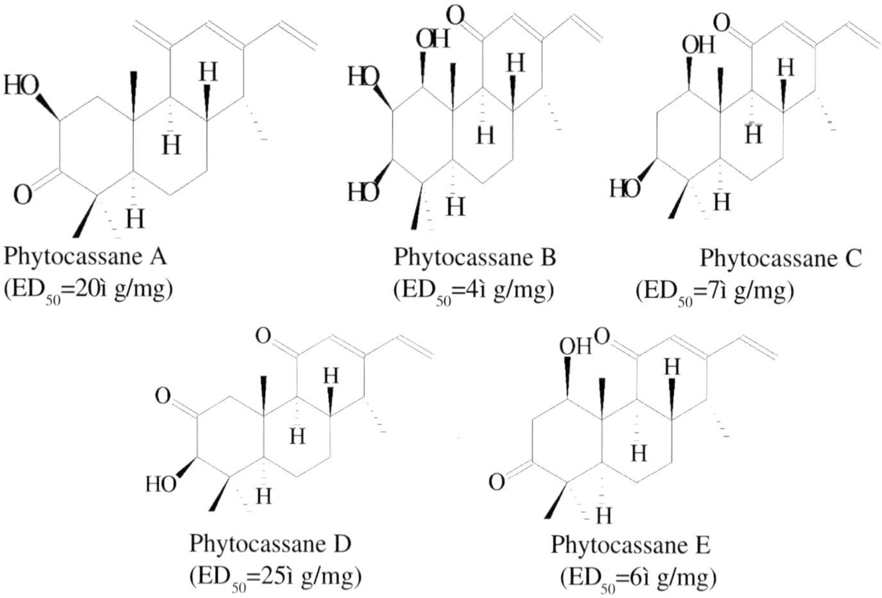

Figure 8. Structures of phytocassanes A-E and their antifungal activities (the ED_{50} values) on spore germination of Magnaporthe grisea.

Oryzalexins A-F and S, phytocassanes A-E and sakuranetin were isolated from rice leaves infected with blast fungus, *Magnaporthe grisea* (Akatsuka et al., 1985; Kato et al., 1993; 1994; Kodama et al., 1992; Koga et al., 1995; 1997; Nakazato Y et al., 2000). Oryzalexins A-C strongly inhibited the spore germination of *P. oryzae.* Their ED_{50} values were 130, 68 and 136 ppm, respectively. Complete inhibition of *P. oryzae* spore germination was observed at 200 ppm. Oryzalexins A-C also strongly suppressed the germ tube elongation of *P. oryzae.* Their ED_{50} values on *P. oryzae* germ tube elongation were 35, 18 and 35 ppm, respectively (Akatsuka et al., 1985). Similarly, oryzalexins D-F and S significantly inhibited spore germination of the rice blast fungus (Kato et al., 1993; 1994; Kodama et al., 1992). Noteworthy, oryzalexins E had slightly lower antifungal activity than oryzalexins D, but higher than Oryzalexins A-C (Kato et al., 1993). Phytocassanes A-E are the diterpenes with a cassane skeleton

(Figure 8). They were produced in rice leaves and stems with *M. grisea* and *R. solani*. Phytocassanes A-E had high antifungal activity against the pathogenic fungi, *M. grisea* and *R. solani* (Koga et al., 1995; 1997). Their ED_{50} values in prevention of spore germination and germ tube growth of *M. grisea* were very low (Figure 8). Sakuranetin is a flavonoid-type phytoalexin that was identified from rice plants. It had high antifungal activity and a large amount of accumulation in rice leaves (Nakazato et al., 2000).

It has been shown in host-pathogen interactions that resistance reactions can be triggered by a large number of abiotic and biotic factors. Among the chemical factors, macromolecules of microbial origin are very important. Plant defense responses are stimulated by very low concentrations of these molecules (Darvill and Albersheim, 1984). It was confirmed that a chemical for plant disease control might function by activating the natural resistance mechanisms of the host, for example, 2,2-dichloro-3,3-dimethyl cyclopropane carboxylic acid may exert its systemic fungicidal activity against the rice blast disease caused by *P. oryzae* (Cartwright et al., 1977). The application of methionine on wounded rice leaves induced the production of rice phytoalexins, sakuranetin and momilactone A. In the paddy field, methionine treatment has been demonstrated to reduce rice blast (Nakazato et al., 2000).

An increasing number of studies have shown that rice phytoalexins are induced by elicitor that are produced by pathogenic microorganisms and make field disease control by inducing the pathogen defense mechanism in rice (Schaffrath et al., 1995; Tamogami et al., 1997a,b). Elicitors have been investigated extensively. It has been shown that jasmonic acid and its related compounds play important roles as the signaling molecules that elicit the production of phytoalexins in rice. Sakuranetin production may be elicited by exogenously applied jasmonic acid in rice leaves. Furthermore, sakuranetin production by exogenously applied jasmonic acid was significantly counteracted by amino acid, cytokinin, kinetin and zeatin (Tamogami et al., 1997a,b).

4. CONCLUSION

Many interactions between plant pathogens and their hosts are allelopathic. Allelochemicals can be applied in biological control of weeds and plant diseases (Rice, 1995). Our research suggested that allelochemicals produced and released from *A. conyzoides* intercropping in citrus orchard did play important roles in integrated pest management. Many kinds of allelochemicals in rice not only inhibit the germination and growth of weeds, but also participate in the defense against pathogens. However, it remains unclear which allelochemicals in rice are predominantly involved in defense mechanisms against the pathogens. Therefore, further clarification of the resistance mechanism and multiple functions of rice allelochemicals are warranted.

Acknowledgments : The author thanks Dr. Inderjit and anonymous reviewers for thoughtful criticisms and correcting English on earlier versions of the manuscript. The work was supported by National Natural Science Foundation of China (NSFC No.30170182; 30430460) and Hundreds-Talent Program of Chinese Academy of Sciences.

5. REFERENCES

Adesogan, E.K., Okunade, A.L. A new flavone from *Ageratum conyzoides*. Phytochemistry 1979; 18:1863-1864.

Akatsuka, T., Kodama, Q, Sekido, H., Kono, Y, Takeuchi, S. Novel phytoalexins (Oryzalexins A, B and C) isolated from rice blast leaves infected with *Pyricularia oryxae*. Part I: Isolation, characterization and biological activities of oryzalexins. Agric Biol Chem 1985; 49:1689-1694.

Albersberg, W. G. L. Singh, Y. Essential oil of *Ageratum conyzoides*. Flavor Fragrance J 1991; 6: 117-120.

Bailey, J. A., Mansfield, J. W. eds. *Phytoalexins*. John Wiley and Sons Inc., New York, 1982.

Biond, D., Cianci, P., Geraci, C. Antimicrobial activity and chemical composition of essential oil from Sicilian aromatic plants. Flavor Fragrance 1993; 8:331-337.

Cartwright, D., Langcake, P., Pryce, R. J., Leworthy, D. P. Chemical activation of host defence mechanisms as a basis for crop protection. Nature 1977; 267:511-523.

Cartwright, D., Langcake, P., Pryce, R.J., Leworthy, D.P. Ride, J.P. Isolation and characterization of two phytoalexins from rice as momilactones A and B. Phytochemistry 1981; 20:535-537

Chung, I.M., Ahn, J.K., Yun, S.J. Identification of allelopathic compounds from rice (*Oryza sative* L.) straw and their biological activity. Can J Plant Sci 2001; 81:815-819.

Dangl, J.L., Jones, J.D.G. Plant pathogens and integrated defence responses to infection. Nature 2001; 411:826-8335.

Darvill, A.G., Albersheim, P. Phytoalexins and their elicitors-A defense against microbial infection in plants. Ann Rev Plant Pathol 1984; 35:243-275.

Dixit, S.N., Chandra, H., Tiwari, R., Dixit, V. Development of a botanical fungicide against blue mould of mandarins. J Stored Prod Res 1995; 31:165-172.

Dixon, R.A. Natural products and plant disease resistance. Nature 2001; 411:843-847.

Ekundayo, O., Laakso, I., Hiltunen, R. Essential oil of *A. conyzoid*es. Planta Med 1988; 54:55-57.

Gonzalez, A.G., Aguiar, Z.E., Grillo, T.A., Luis, J.G., Rivera, A., Calle, J. Methoxyflavones from *Ageratum conyzoides*. Phytochemistry 1991; 30:1269-1271.

Grayer, R.J. Harborne, J.B. A survey of antifungal compounds from higher plants. Phytochemistry 1994; 37:19-42.

Hammerschmidt, R. Phytoalexins: what have we learned after 60 years? Annu Rev Phytopathol 1999; 37:285-306.

Harborne, J.B. The comparative biochemistry of phytoalexin induction in plants. Biochem System Ecol 1999; 27:335-367.

Horie, T., Tominaga, H. Kawamura, Y. Revised structure of a natural flavone from *Ageratum conyzoides*. Phytochemistry 1993; 32:1078-1077

Hu, F., Kong, C.H. Allelopathy of *Ageratum conyzoides*. VI. Effects of meteorological conditions on allelopathy of *Ageratum conyzoides*. Chin J Appl Ecol 2002a ; 13:76-80. (In Chinese)

Hu, F., Kong, C.H. Inhibitory effect of flavones from *Ageratum conyzoides* on the major pathogens in citrus orchard. Chin J Appl Ecol 2002b ; 13:1166-1168 (In Chinese)

Jhansi, P., Ramanujam, C.G.K. Pollen analysis of extracted and squeezed honey of Hyderabad. India Geophytol 1987; 17:237–240.

Kato, H., Kodama, O., Akatsuka, T. Oryzalexin E, a diterpene phytoalexin from UV-irradiat rice leaves. Phytochemistry 1993; 33:79-81.

Kato, H., Kodama, O., Akatsuka, T. Oryzalexin F, a diterpene phytoalexin from UV-irradiat rice leaves. Phytochemistry 1994; 36:299-301.

Kato-Noguchi, H., Ino, T., Sata, N., Yamamura, S. Isolation and identification of a potent allelopathic substance in rice root exudates. Physiol Planta 2002; 115:401-405.

Kato-Noguchi, H., Ino, K. Rice seedlings release momilactone B into the environment. Phytochemistry 2003; 63:551-554.

Kim, K.U., Shin, D.H. eds. *Rice Allelopathy* Kyungpook National University. Taegu(Korea) 2000.

Kato, T., Tsunakawa, M., Sasaki, N. Growth and germination inhibitors in rice husks. Phytochemistry 1977; 16:45-48.

Kodama, O., Li, W. X., Tamogami, S., Akatsuka, T. Oryzalexin S, a novel stemarane-type diterpene rice phytoalexin. Biosci Biotech Biochem 1992; 56:1002-1003.

Koga, J., Ogawa, N., Yamauchi, T., Kikuchi, M., Ogasawara, N., Shimura, M. Functional moiety for the antifungal activity of phytocassane E, a diteroene phytoalexin from rice. Phytochemistry 1997; 44:249-253.

Koga, J., Shimura, M., Oshima, K., Ogawa, N., Yamauchi, T., Ogasawara, N. Phytocassanes A, B, C and D, novel diterpene phytoalexins from rice, *Oryza sativa* L. Tetrahedron 1995; 51:7907-7918.

Kong, C.H., Hu, F., Xu, T., Lu, Y.H. Allelopathic potential and chemical constituents of volatile oil from *Ageratum conyzoides*. J Chem Ecol 1999; 25:2347-2356.

Kong, C.H., Huang, S.S., Hu, F. Allelopathy of *Ageratum conyzoides*. V. Biological activities of the volatile oil from *Ageratum* on fungi, insects and plants and its chemical constituents. Acta Ecol Sin 2001; 21:584-587. (In Chinese).

Kong, C.H., Hu, F., Xu, X. H. Allelopathic potential and chemical constituents of volatile from *Ageratum conyzoides* under stress. J Chem Ecol 2002a; 28:1173-1182.

Kong, C.H., Xu, X.H., Hu, F., Cheng, X.H., Lin, B., Tan, Z.W. Using specific secondary metabolite as marker to evaluate allelopathic potential of rice variety and individual plant. Chin Sci Bull 2002b; 47:839-843.

Kong, C.H., Hu, F., Liang, W.J., Wang, P., Jiang, Y. Allelopathic potential of *Ageratum conyzoides* with different growth stages and habitats. Allelopathy J 2004a ; 13:233-240.

Kong, C.H., Hu, F., Xu, X.H., Liang, W.J., Zhang, C.X. Allelopathic plant: *Ageratum conyzoides* L. Allelopathy J 2004b; 14:5-13.

Kong, C.H., Liang, W. J., Hu, F., Xu, X. H. Wang, P., Jiang, Y. Allelochemicals and their transformations in the *Ageratum conyzoides* intercropped the citrus orchard soil. Plant Soil 2004c; 264:149-157.

Kong, C.H., Liang, W.J., Xu, X.H., Hu, F., Wang, P., Jiang, Y. Release and activity of allelochemicals from allelopathic rice seedlings. J Agric Food Chem 2004d ; 52:2861-2865.

Kong, C.H., Xu, X. H., Zhou, B., Hu, F., Zhang, C.X., Zhang, M.X. Two compounds from allelopathic rice accession and their inhibitory effects on weeds and fungal pathogens. Phytochemistry 2004e ; 65:1123-1128.

Kossmann, G., Groth, D. *Plantea Infestantes e Nocivas*. BASF Brasileira, São Paulo, 1993.

Ladeira, A.M., Zaidan, L.B.P., Figueiredo, Ribeiro, R.D.C.L. *Ageratum conyzoides* L.(Compositae): germination, flowering and occurrence of phenolic derivatives at different stages of development. Hoehnea 1987; 14:53-62.

Lee, C.W., Yoneyama, K., Takeuchi, Y., Konnai, M., Tamogami, S., Kodama, O. Momilactones A and B in rice straw harvested at different growth stages. Biosci Biotech Biochem 1999; 63:1318-1320.

Liang, W.G., Hunag, M.D. Influence of citrus orchard ground cover plants on anthropod communities in China: a review. Agric Ecosyst Environ 1994; 50:29-37.

Mattice, J., Lavy, T., Skulman, B., Dilday, R. Searching for allelochemicals in rice that control ducksalad. In *Allelopathy In Rice*; Olofsdotter, M. Ed. International Rice Research Institute: Manila, Philippines, 1998; pp. 81-97.

Menut, C., Lamaty, G., Zollo, P.H.A. Aromatic plants of tropical central Africa: Part X. Chemical composition of essential oil of *Ageratum houstonianum* and *Ageratum conyzoides* from Cameroon. Flavor Fragrance J 1993; 8:1-4.

Ming, L.C. *Ageratum conyzoides*: A tropical source of medicinal and agricultural products. In: *Perspectives on New Crops and New Uses,* Janick, J. eds. Alexandria, VA USA: ASHS Press 1999; pp 469-473.

Muller, K.O., Borger, H. Experimentelle Untersuchungen uber die Phytophthora-Resistenz der Kartoffel-zugleich ein Bewitrag zum Problem der "erworbenen Resistenz" im Pflanzenreich. Arb Biol Anst Reichsanst (Berl) 1940; 23:189-231.

Nakazato, Y., Tamogami, S., Kawai, H., Hasegawa, M., Kodama, O. Methionine-induced phytoalexin production in rice leaves. Biosci Biotech Biochem 2000; 64:577-583.

Okunade, A. L. *Ageratum conyzoides* L.(Asteraceae). Fitoterapia 2002; 73:1-16.

Olofsdotter, M., Rebulanan, M., Madrid, A., Wang, D.L., Navarez, D., Olk, D.C. Why phenolic acids are unlikely primary allelochemicals in rice. J Chem Ecol 2002; 28:229-241.

Pari, K., Rao, P. J., Subrahmanyam, B., Rasthogi, J. N., Devakumar, G. Benzofuran and other constituents of the essential oil of *Ageratum conyzoides*. Phytochemistry 1998; 49:1385-1388.

Rao, G. P., Pandey, A. K., Shukla, K. Essential oils of some higher plants vis-à-vis some legume viruses. Indian Perfumer 1986; 30:483-486.

Rice, E.L. *Biological Control of Weeds and Plant Diseases: Advances in Applied Allelopathy*. University of

Oklahoma Press, Norman, OK, 1995.

Rimando, A.M., Olofsdotter, M., Dayan, F.E., Duke, S.O. Searching for rice allelochemicals: An example of bioassay-guided isolation. Agron J 2001; 93:16-20.

Schaffrath, U., Scheinpflug, H., Reisener, H.J. An elicitor from *Pyricularia oryzae* induced resistance responses in rice: isolation, Characterization and physiological properties. Physiol Mol Plant Pathol 1995; 46:293-307.

Stadler, J., Mungai, G., Brandl, R. Weed invasion in East Africa: insights from herbarium records. African J Ecol 1998; 36:15-22.

Stuiver, M.H., Custers, J.H.H.V. Engineering disease resistance in plants. Nature 2001; 411:865-868.

Suzuki, Y., Esumi, Y., Hyakutake, H., Kono, Y., Sakurai, A. Isolation of 5-(8.Z-heptadecenyl)-resorcinol from etiolated rice seedlings as an antifungal agent. Phytochemistry 1996; 41:1485-1489.

Suzuki, Y., Esumi, Y., Saito T, Kishimoto Y, Morita T. Identification of *5-n-*(2.-oxo)alkylresorcinols from etiolated rice seedlings. Phytochemistry 1998; 47:1247-1252.

Tamogami, S., Rakwal, R., Kodama, O. Phytoalexin production by amino acid conjugates of jasmonic acis through induction of naringenin-7-o-methyltransferase, a key enzyme on phytoalexin biosynthesis in rice (*Oryza sativa* L.). FEBS Lett 1997a; 401:239-242.

Tamogami, S., Rakwal, R., Kodama, O. Phytoalexin production elicited by exogenously applied jasmonic acid in rice leaves (*Oryza sativa* L.) is under the control of cytokinins and ascorbic acid. FEBS Lett 1997b; 412:61-64.

Wandji, J., Bissangou, M. F., Ouambra, J. M., Silou, T., Abena, A. A., Keita, A. The essential oil from *Ageratum conyzoides.* Fitoterapia 1996; 67, 427-431.

Index

A

Abutilon theophrasti, 38, 83, 148
Acacia sedillense, 47
Acalypha indica, 114
Acidovorax delafieldii, 148
Adenostoma fasciculatum, 55
Aeschynomene sp., 17
Ageratum conyzoides, 42, 92, 194, 195, 196, 197, 198, 203
Aggressiveness, 93 , 94
Agrobacterium radiobacter, 63
Agrobacterium sp. 130
Alcaligenes piechaudii, 63
Allium cepa, 57
Alnus firmifolia, 51
Alternanthera philoxeroides, 111
Alternaria alternata, 111, 164
Alternaria alternata f sp. *lycopersici,* 112
Alternaria eichhorniae, 111
Alternaria kikuchiana, 178
Alternaria panax, 164, 165
Alternaria solani, 37, 53
Amaranthus hypochondriacus, 37, 46, 53
Amaranthus retroflexus, 43, 92
Amaranthus spp., 57
Amblyseius spp, 196
Anabaena sp, 110
antifungal, 165, 178, 200, 203
Aphanomyces euteiches, 183
Arabidopsis thaliana, 87, 160
Arachis hypogaea, 56, 92
Arthrobacter spp., 66
Ascochyta caulina, 83
Aspergillus flavus, 9
Aspergillus fumigatus, 9
Aspergillus niger, 9
Aspergillus parasiticus, 9
Aspergillus sp., 68
Avena fatua, 92
Avena sativa, 64, 86
Azadirachta indica, 21
Azolla nilotica, 108

B

Bacillus cereus, 66, 127
Bacillus mycoides, 61
Bacillus subtilis, 3, 66
Bacillus sp., 130
Belonolaimus longicaudatus, 65
Berula erecta, 51
Beta vulgaris, 83
Bievundimonas vesicularis, 63
biocontrol , 66, 110, 111, 118, 124, 128, 129, 130, 131, 132, 133, 134, 145, 146, 148, 149, 150, 169 , 170, 183, 187
bioherbicides, 148, 149, 150, 151
biological control, 25, 56, 70, 71, 111, 125, 131, 134, 143, 145, 146, 148, 151, 152, 182
biopesticide, 16, 15, 17
biotrophic, 80, 93, 94
biotrophs, 79
Botrytis cinerea, 3, 5, 7, 8, 9, 92, 130, 159, 164, 195
Botrytis sp., 3, 12
Brassica juncea, 42, 108
Brassica kaber, 50
Brassica napus, 87
Brassica rapa var. *pervidis,* 38
Brassica spp., 21
Bromus japonicus, 148
Bromus mollis, 92
Bromus tectorum, 147
Burkholderia cepacia, 66
Burkholderia pickettii, 63
Burkholderia sp., 20

C

Caenorhabditis elegans, 23
Callicarpa acuminata, 37
Calotropis gigantea, 106
Canavalia ensiformis, 17, 71
Capsella bursa-pastoris, 84, 152
Capsicum annuum, 42, 57
Capsicum frutescens, 9, 45
Cassia fasciculata, 17
Cassia obtusifolia, 57
Cassytha sp., 108
Centaurea solstitialis, 107
Ceratophyllum demersum, 108
Cercospora alternantherae, 111
Cercospora amboinicus, 114
Cercospora kaki, 183
Cercospora sp., 111
Chaetomella raphigera, 111
chemotaxis, 18
Chenopodium album, 43, 83, 149
Chondrilla juncea, 83
Cicer arietinum, 86
Cichorium intybus, 177
Cirsium maculosa, 45

Cirsium sp., 43
Cladosporium herbarum, 178
Cleistopholis acutatum, 12
Cleistopholis fragariae, 11
Cleistopholis patens, 9
Cleome viscosa, 114
Coleus amboinicus, 109, 114 , 115, 116, 117
Colletotrichum acutatum, 3, 5, 7, 8, 9, 12
Colletotrichum coccodes, 83
Colletotrichum fragariae , 3, 5, 8 , 9, 11
Colletotrichum gloeosporioides, 3, 5, 8, 9, 10
Colletotrichum sp., 3, 4, 8, 12
commensal, 124, 125, 127, 128
compatible, 57
contamination, 143, 149
Criconemella sp., 65
Crotalaria juncea, 17, 20, 57, 69, 70, 71
Crotalaria retusa, 71
Crotalaria spectabilis, 17, 70
Crotalaria sp.,16, 65, 67, 69, 71
Croton sparsiflorus, 114
Cucumis sativus cv. *Wisconsin*, 47
Cucumis sativus, 92
Cucurbita pepo, 47
Cuminum cyminum, 86
Curcuma longa, 114
Cyamopsis tetragonoloba, 52
Cylindrocarpon destructans, 164, 165, 169, 170
Cylindrocarpon sp., 170
Cynoches sp., 3
Cynodon dactylon, 107
Cyperus difformis,104, 106, 117
Cyperus littoralis, 106
Cyperus rotundus, 106, 111

D

Dactylaria higginsi, 111
Dactylis glomerata, 88, 92
Daucus carota, 86
Delonix regia, 35
Digitaria decumbens, 35
Diospyros lotus, 183
Drechmeria coniospora, 69
Drecshlera sp., 111

E

Echinochloa crus-galli, 37, 46, 104, 111
Echinochloa spp, 106
Eclipta alba, 114
Egeria densa, 111
Eichhornia crassipes, 108, 109, 114
elicitors, 203
Empetrum hermaphroditum, 41
Enterobacter asburiae, 63
epidemics, 88
Erwinia carotovora, 87, 126, 127
Erwinia sp, 130
Erysiphe cichoracearum, 92, 195
Eucalyptus globulus, 106
Eucalyptus sp., 106
Euphorbia esula, 147
Euphorbia hirta, 114
Exostema caribaeum, 47
Exserohilum monocerus, 111
exudate, 19, 35, 40, 69, 87, 107, 124, 148, 160, 168, 170, 184, 199

F

Fagus sylvatica, 47
Festuca arundinacea, 91
fungistasis, 84
fungitoxic, 157, 158, 185
Fusarium graminearum, 111
Fusarium oxysporum f. sp. *cucumerinum*, 130
Fusarium oxysporum f.sp. *cumini*, 86
Fusarium oxysporum f. sp. *lycopersici*, 159
Fusarium oxysporum, 3, 5, 8, 9, 64, 66, 86, 164, 165, 187
Fusarium sambucinum, 183
Fusarium solani, 23, 164, 165 , 183, 196, 197
Fusarium sp., 60, 61, 165, 170, 183

G

Gaeumannomyces graminis var. *tritici*, 60
Gaeumannomyces graminis, 158, 159
Gigaspora rosea, 186
Globodera pallida, 67
Globodera rostochiensis, 18
Glomus etunicatum, 185
Glomus fasciculatum, 183
Glomus intraradices, 68, 183, 186
Glomus mosseae, 68, 183, 186, 187
Glomus versiforme, 183
Glomus sp., 183
Glycine max, 56, 65, 66, 69, 83
Gnomonia comari, 7, 12
Gossypium spp., 56

H

Hebeloma crustuliniforme, 44
Helianthus annuus, 39, 57, 86
Helicobacter pylori, 49
Helicotylenchus dihystera, 64, 65
Helicotylenchus sp.,24, 65
Helminthosporium longirostratum, 37
Helminthosporium oryzae, 202
Heteranthera limosa, 104
Heterodera glycines, 24, 64, 65
Heterodera schachtii, 22, 62
Hintonia latiflora, 47

Hippeastrum hybridum, 48
Hordeum leporinum, 92
Hymenoscyphus ericae, 44
Hyrdilla verticilata, 107, 108

I

Imperata cylindrica, 107
Indigofera hirsuta, 17, 70, 71
Indigofera nummularifolia, 71
Indigofera spicata, 71
Indigofera suffruticosa, 71
Indigofera tinctoria, 71
inoculum, 12, 63, 83, 111, 125, 145, 183, 187
Ipomoea batatas, 54
Ipomoea tricolor, 46
Ipomoea sp., 57
Iva axillaris, 38

J

Juglans nigra, 82
Juncus sp., 51

K

Kocuria kristinae, 63
Kocuria varians, 63

L

Laccaria bicolor, 184
Lactuca sativa, 37, 41, 83
Lactuca serriola, 89
Lantana camara, 47
Larrea tridentata, 41
Lemna leucocephala, 114
Lemna paucicostaha, 108
Lemna perpusilla, 108
Leptinotarsa decemlineata, 37
Leptosphaeria maculans, 87
Leptosphaerulina trifolii, 85
Leucaena leucocephala, 35, 52, 106, 114
Leucaena sp., 53, 106
Leucas aspera, 114
Linum usitatissimum, 57
Lolium multiflorum, 92, 107
Lolium perenne, 84, 87
Lolium sp., 64
Lycopersicon esculentum, 47, 53, 70, 82, 108
Lythium salicaria, 107

M

Macaranga monandra, 9
Macrophomina sp., 183
Magnaporthe grisea, 202, 203
Manihot esculenta, 107
Medicago truncatula, 185
Meloidogyne arenaria, 18, 20, 24, 65
Meloidogyne hapla, 18
Meloidogyne incognita, 18, 20, 21, 23, 24, 25, 61, 63, 65, 66, 68, 69, 70
Meloidogyne javanica, 18, 20, 21, 22, 23, 68, 70
Meloidogyne spp, 19, 18, 24, 64, 65, 71
Microbacterium esteraromaticum, 63
Mucuna deeringiana, 17, 20, 62, 64, 66, 70
mycoherbicide, 111
Mycosphaerella sp, 111
Myriophyllum aquaticum, 111
Myriophyllum spicatum, 110
Myrothecium roridum, 111
Myrothecium sp., 68

N

Najas graminea, 108
necrotrophic, 182
necrotrophs, 79
nematophagous, 67, 69
Neochetina bruchii, 118
Neotyphodium coenophialum, 91
Neotyphodium lolii, 90, 94
Nicotiana tabaccum, 57, 86
Nimbya alternantherae, 111

O

Orobanche aegyptica, 57
Orobanche cernua, 57, 40
Orobanche crenata, 40, 57
Orobanche cumana, 40
Orobanche ramosa, 40, 57
Orobanche sp., 40, 57
Oryza sativa, 57, 194, 199

P

Paecilomyces sp., 20
Paenibacillus sp, 130
Panax cactorum, 164
Panax ginseng, 162
Panax quinquefolius, 157, 161
Panax spp, 161
Panicum miliaceum, 56
Panonychus citri, 196
Paratrichodorus minor, 19, 65
Paratichodorus sp., 24
Parthenium hysterophorous, 38, 53, 108, 114
Paspalum sp., 17
Pellicularia sasaki, 178
Penicillium sp., 68
Pennisetum americanum, 52
Phaseolus aureus, 57
Phaseolus mungo, 57
Phaseolus vulgaris, 47, 86, 183
Phaseolus sp., 41
Phoenix dactylifera, 86
Phoma lingam, 159

Phoma typhae-domingensis, 111
Phomopsis obscurans, 3, 5, 9
Phomopsis viticola, 3, 5, 9
Phomopsis sp., 8
phyllosphere, 62
phytoalexins, 3, 56, 85, 86, 159, 182, 184, 185, 186, 193, 194, 201, 202, 203
phytophagus, 34
Phytophthora cactorum, 7, 66, 164,165,166, 169, 170
Phytophthora cinnamomi, 61
Phytophthora citrophthora, 196, 197
Phytophthora infestans, 86, 166
Phytophthora nicotianae, 6, 9
Phytophthora parasitica, 183
Phytophthora sp., 5, 9, 166, 170, 183
phytoprotection, 132
phytotoxic, 25, 46, 52, 57, 64, 87, 106, 124, 160, 199
phytotoxin, 35, 42, 44, 45, 56, 57, 58, 59, 105, 111, 147, 150
Picea abies, 44
Pinus laricio, 47
Pistia stratiotes, 108, 109
Pisum sativa, 86
Plantago lanceolata, 43
Pluchea lanceolata, 42
Plutella xylostella, 54
Portulaca oleracea, 57, 83
Pratylenchus brachyurus, 68
Pratylenchus neglectus, 21
Pratylenchus penetrans, 63
Pratylenchus vulnus, 22
Pratylenchus sp., 19, 24, 65
predators, 69, 181
propagule, 147, 183
Prunus dulcis, 23
Psacalium decompositum, 37
Pseudomonas aeruginosa, 130
Pseudomonas aureofaciens, 127
Pseudomonas chlororaphis, 63
Pseudomonas fluorescens, 56, 63, 147
Pseudomonas putida, 56, 148
Pseudomonas syringae, 147
Pseudomonas sp., 60, 61, 63, 66, 147
Pseudopeziza coronata, 94
Pseudopeziza medicagnis, 94
Puccinia chondrillina, 83
Puccinia coronata f.sp. *avenae,* 86
Puccinia coronata f.sp. *lolii,* 88
Puccinia coronata, 89, 92
Puccinia graminis, 86, 92
Puccinia hordei, 92
Puccinia lagenophorae, 83
Puccinia pulsatillae, 84
Puccinia recondita, 92
Pusatilla pratensis, 84
Pyrenochaeta terrestris, 183
Pyricularia oryzae, 178, 199, 200, 202
Pythium aphanidermatum, 196, 197
Pythium irregulare, 164, 165, 166,168,169, 170
Pythium ultimum, 66, 164, 169
Pythium sp., 50, 159, 166, 168, 170

R

Radopholus similis, 65
Ralstonia solanacearum, 126
Ralastonia sp., 127
Ratibida mexicana, 46
Renibacterium salmoninarum, 61
Reynoutria sachalinensis, 3
Rhizobium etli, 67
Rhizobium japonicum, 111
Rhizobium leguminosarum biovar *phaseoli,* 41
Rhizobium leguminosarum, 183
Rhizobium sp., 41, 51, 67, 170
Rhizoctonia solani, 60, 64, 66, 92, 159, 164, 183, 185, 187, 203,195,199
rhizoplane, 108, 124, 131, 146, 147, 160, 169, 180, 184
rhizosphere, 17, 20, 23, 35, 36, 42, 44, 57, 58, 62, 67, 68, 108, 131, 133, 134, 144, 146, 147, 148, 151, 157, 169, 170, 183, 184, 185
Ricinus communis, 17, 70
rotation, 16, 17, 18, 19, 20, 22, 35, 41, 64, 66, 71, 117, 134, 150, 152
Rotylenchulus reniformis, 22, 24, 65, 70, 71
Rumex crispus, 43
Ruta graveolens, 8

S

Saccharum officinarum, 86
Salvinia auriculata, 111
Salvinia molesta, 108,109, 111
Schistocerca gregaria, 177
Sclerotinia sclerotiorum, 87, 92, 195
Sclerotinia sp., 3
Sclerotium cepivorum, 183
Scutellonema sp., 65
Sebastiania adenophora, 47
Secale cereale, 57, 64
Senecio vulgaris, 83
Septoria lycopersici, 159
Serratia plymuthica, 130
Sesamum bicolor, 17
Sesamum indicum, 17, 19, 57
Sesbania aculeata, 117
Sesbania exaltata, 107
Sesbania grandiflora, 114

Setaria faberi, 151
Setaria virdis, 148
Sicyos deppei, 47
Sida spinosa, 57
Simsia amplexicaulis, 41
Sinapis arvensis, 54
Solanum nigrum, 43
Solanum tuberosum, 62, 86
Solanum, 86
*Sorghum bicolor,*17, 19, 56, 57, 86
Sorghum sudanense, 56
Sorghum vulgare var, *sudanense,*17
Sorghum vulgare, 17 , 52
spermosphere, 144
Sphenoclea zeylanica, 106
Spirodella polyrhiza, 108
Stellaria media, 149
Stenotrophomonas sp., 130
Streptococcus pneumoniae, 61
Streptomyces hygroscopicus, 56, 112
Streptomyces viridichromogens, 56, 111
Streptomyces sp.,112, 130
Striga asiatica, 40, 56
Striga gesnerioides, 56
Striga hermonthica, 32, 56, 152
Striga sp., 40, 132
suppression, 17, 18, 20, 21, 34, 55, 60, 61, 70, 72, 84, 88, 89, 90, 91, 104, 105, 106, 107, 118, 128, 132, 133, 149, 150, 151, 152
susceptible, 4, 8, 13, 55, 68, 86, 88, 90, 93, 108, 112, 160
synergistic, 103, 118
systemic acquired resistance, 13

T

Tagetes erecta, 18, 53, 62
Tagetes patula, 18, 62
Tagetes signata, 18
Tagetes spp., 17, 63
Tephrosia adunca, 71
Thlaspi arvense, 149
Thymus vulgaris, 24
Tradescantia crassifolia, 41
Trianthema portulacastum, 114
Trichoderma hamatum, 165, 166, 169
Trichoderma harzianum, 165
Trichoderma viride, 165
Trichoderma sp., 68, 165, 169
Trifolium incarnatum, 57, 66
Trifolium pratense, 50, 54, 85
Trifolium repens, 84, 87
Trifolium subterraneum, 57, 83, 92
Trifolium sp., 64
Triticum aestivum, 42, 54, 64, 66, 70, 86
Tsukamurella paurometabolum, 63
Tylenchorhynchus claytoni, 64, 65
Tylenchulus semipenetrans, 21
Tylenchorhynchus sp., 24
Typha domingensis, 111

U

Uredo eichhorniae, 111
Uromyces troflii-repentis, 92
Urtica urens, 132

V

Vaccinium myrtillus, 44
Variovorax paradoxus, 127
Verticillium albo-atrum, 159
Verticillium biguttatum, 64
Vicia villosa, 64
Vicia sp., 17
Vigna catjang, 56
Vigna unguiculata, 64, 66
virulence, 93, 170
Vitex negundo, 35
Vitis vinifera, 3

X

Xanthium strumarium, 57
Xanthomonas campestris, 126
Xiphinema sp., 24

Z

Zea mays, 47, 50, 52, 56, 57, 83, 107

List of Contributors

- **Ana Luisa Anaya**
  Laboratorio de Alelopatía, Instituto de Ecología, Universidad Nacional Autónoma de México. Circuito Exterior, Ciudad Universitaria. 04510 México, D.F. Email: alanaya@miranda.ecologia.unam.mx

- **Mark A. Bernards**
  Department of Biology, University of Western Ontario, London, ON, Canada, N6A 5B7, Email: bernards@uwo.ca

- **Ramanathan Kathiresan**
  Professor, Department of Agronomy, Annamalai University, Annamalainagar, Tamilnadu 608 002, India. E-mail : rm.kathiresan@sify.com

- **Cilfford H. Koger**
  Weed Ecologist, Southern Weed Science Research Unit, USDA-ARS, P. O. Box 350, Stoneville, Mississippi 38776, USA.

- **Nancy Kokalis-Burelle**
  Research Ecologist, USDA, Agricultural Research Service, U.S. Horticultural Research Lab, Fort Pierce, FL, 34945, USA
  Email:NBurelle@ushrl.ars.usda.gov

- **Chuihua Kong**
  Institute of Applied Ecology, Chinese Academy of Sciences, Shenyang 110016, China, and South China Agricultural University, Guangzhou 510642, China.
  E-mail: kongch@mail.edu.cnw

- **Robert J. Kremer**
  U.S.D.A., Agricultural Research Service, Cropping Systems & Water Quality Unit, University of Missouri, Columbia, Missouri, U.S.A.
  Email: KremerR@missouri.edu

- **Scott W. Mattner**
  Department of Primary Industries, Knoxfield Centre, Victoria, Private Bag 15, Ferntree Gully Delivery Centre, 3156, VIC, Australia, scott.mattner@dpi.vic.gov.au

- **Robert W. Nicol**
  NovoBiotic Pharmaceuticals, Cambridge MA, 02138, USA
  Email: bernards@uwo.ca

- **Hiroyuki Nishimura**
  Department of Biosciences and Technology, School of Engineering, Hokkaido Tokai University, Sapporo 005-8601, Japan
  Email:nishimura@db.htokai.ac.jp

- **Krishna N. Reddy**
  Plant Physiologist, Southern Weed Science Research Unit, USDA-ARS, P. O. Box 350, Stoneville, Mississippi 38776, USA;
  Email: kreddy@ars.usda.gov

- **Rodrigo Rodríguez-Kábana**
  Distinguished University Professor, Auburn University, Auburn, AL, 36849, USA. Email : NBurelle@ushrl.ars.usda.gov

- **Atsushi Satoh**
  Department of Biosciences and Technology, School of Engineering, Hokkaido Tokai University, Sapporo 005-8601, Japan Email:nishimura@db.htokai.ac.jp

- **Barbara J. Smith**
  Small Fruit Research Station, 306 S. High St., Poplarville, MS 39470.
  Email : dwedge@olemiss.edu

- **Antony V. Sturz**
  Prince Edward Island Department of Agriculture, Fisheries, Aquaculture and Forestry, P.O. Box 1600, Charlottetown, P.E.I., C1A 7N3, Canada.
  Tel: 902-368-5664, FAX: 902-368-5661, E-mail:avsturz@gov.pe.ca

- **David E. Wedge**
  United States Department of Agriculture, Agricultural Research Service, Natural Products Utilization Research Unit, The Thad Cochran National Center for Natural Products Research, University of Mississippi, University, MS 38677, USA Email : dwedge@olemiss.edu

- **Lina F. Yousef**
  Department of Biology, University of Western Ontario, London, ON, Canada, N6A 5B7

- **Ren-Sen Zeng**
  Chemical Ecology Lab., Institute of Tropical & Subtropical Ecology, Agricultural College, South China Agricultural University Wushan, Tianhe District, Guangzhou, 510642 P.R. China
  Email: rszeng@scau.edu.cn, zengrs8@hotmail.com